de Gruyter Studies in Mathematics 26

Editors: Carlos Kenig · Andrew Ranicki · Michael Röckner

de Gruyter Studies in Mathematics

Heinz Bauer

Measure and Integration Theory

Translated from the German by Robert B. Burckel

Walter de Gruyter
Berlin · New York 2001

Author
Heinz Bauer
Mathematisches Institut
der Universität Erlangen-Nürnberg
Bismarckstraße 1 1/2
91054 Erlangen
Germany

Translator
Robert B. Burckel
Department of Mathematics
Kansas State University
137 Cardwell Hall
Manhattan, Kansas 66506-2602
USA

Series Editors

Carlos E. Kenig
Department of Mathematics
University of Chicago
5734 University Ave
Chicago, IL 60637
USA

Andrew Ranicki
Department of Mathematics
University of Edinburgh
Mayfield Road
Edinburgh EH9 3JZ
Scotland

Michael Röckner
Fakultät für Mathematik
Universität Bielefeld
Universitätsstraße 25
33615 Bielefeld
Germany

Mathematics Subject Classification 2000: 28-01; 28-02

Keywords: Product measures, measures on topological spaces, topological measure theory, introduction to measures and integration theory

ⓒ Printed on acid-free paper which falls within the guidelines of the ANSI to ensure permanence and durability.

Library of Congress − Cataloging-in-Publication Data

Bauer, Heinz, 1928−
    [Mass- und Integrationstheorie. English]
    Measure and integration theory / Heinz Bauer ; translated from the German by Robert B. Burckel.
      p.   cm. − (De Gruyter studies in mathematics ; 26)
    Includes bibliographical references and indexes.
    ISBN 3110167190 (acid-free paper)
    1. Measure theory. 2. Integrals, Generalized. I. Title. II. Series.
  QC20.7.M43 B4813   2001
  530.8′01−dc21
                                      2001028235

Die Deutsche Bibliothek − Cataloging-in-Publication Data

Bauer, Heinz:
Measure and integration theory / Heinz Bauer. Transl. from the German Robert B. Burckel. − Berlin ; New York : de Gruyter, 2001
    (De Gruyter studies in mathematics ; 26)
    Einheitssacht.: Mass- und Integrationstheorie ⟨engl.⟩
    ISBN 3-11-016719-0

Typesetting: Oldřich Ulrych, Prague, Czech Republic.
Printing and binding: Hubert & Co. GmbH & Co. KG, Göttingen.
Cover design: Rudolf Hübler, Berlin.

In memoriam
OTTO HAUPT
(5.3.1887 – 10.11.1988)
former Professor of Mathematics
at the University of Erlangen

Preface

More than thirty years ago my textbook *Wahrscheinlichkeitstheorie und Grundzüge der Maßtheorie* was published for the first time. It contained three introductory chapters on measure and integration as well as a chapter on measure in topological spaces, which was embedded in the probabilistic developments. Over the years these parts of the book were made the basis for lectures on measure and integration at various universities. Generations of students used the measure theory part for self-study and for examination preparations, even if their interests often did not extend as far as the probability theory.

When the decision was made to rewrite and extend the parts devoted to probability theory, it was also decided to publish the part on measure and integration theory as a separate volume. This volume had to serve two purposes. As before it had to provide the measure-theoretic background for my book on probability theory. Secondly, it should be a self-contained introduction into the field. The German edition of this book was published in 1990 (with a second edition in 1992), followed in 1992 by the rewritten book on probability theory. The latter was translated into English and the translation was published in 1995 as *Probability Theory* (Volume 23) in this series.

When offering now a translation of the book *Maß- und Integrationstheorie* we have two aims: To provide the reader of my book on probability theory with the necessary auxiliary results and, secondly, to serve as a secure entry into a theory which to an ever-increasing extent is significant not only for many areas within mathematics, but also for applications in physics, economics and computer science.

However, once again this book is much more than a pure translation of the German original and the following quotation of the preface of my book *Probability Theory*, applies a further time: "It is in fact a revised and improved version of that book. A translator, in the sense of the word, could never do this job. This explains why I have to express my deep gratitude to my very special translator, to my American colleague Professor Robert B. Burckel from Kansas State University. He had gotten to know my book by reading its very first German edition. I owe our friendship to his early interest in it. He expended great energy, especially on this new book, using his extensive acquaintance with the literature to make many knowledgeable suggestions, pressing for greater clarity and giving intensive support in bringing this enterprise to a good conclusion."

In addition I want to thank Dr. Oldřich Ulrych from Prague for his skill and patience in preparing the book manuscript in TEX for final processing. Many thanks are due to my family and Professor Niels Jacob, University of Swansea, for reasons they will know. Finally, I thank my publisher Walter de Gruyter & Co., and, above all, Dr. Manfred Karbe for publishing the translation of my book.

Erlangen, March 2001   *Heinz Bauer*

Introduction

Measure theory and integration are closely interwoven theories, both content-wise and in their historical developments. They form a unit. The development of analysis in the 19th century – here one is thinking especially about the theory of Fourier series and classical function theory – compelled the creation of a sufficiently general concept of the integral that discontinuous functions could also be integrated. The jump function of P. G. LEJEUNE DIRICHLET should be seen in this light. At that time only an integration theory due to CAUCHY, a precursor of Riemann's, was known. And it was not until B. RIEMANN'S Habilitation in 1854 (text published posthumously in 1867) that Cauchy's ideas were made sufficiently precise to integrate (certain) discontinuous functions. For the first time the need was felt for integrability criteria. Parallel to this a "theory of content" was evolving – primarily at the hands of G. PEANO and C. JORDAN – to measure the areas of plane and the volumes of spatial "figures".

But the decisive breakthrough occurred at the turn of the century, thanks to the French mathematicians ÉMILE BOREL and HENRI LEBESGUE. In 1898 Borel – coming from the direction of function theory – described the "σ-algebra" of sets that today bear his name, the Borel sets, and showed how to construct a "measure" on this σ-algebra that satisfactorily resolved the problems of measuring content. In particular, he recognized the significance of the "σ-additivity" of the measure. In his thesis (1902) LEBESGUE presented the integral concept, subsequently named after him, that proved decisive for the development of a general theory. At the same time he furnished the tools needed to make Borel's ideas more precise. From then on Lebesgue–Borel measure on the σ-algebra of Borel sets and Lebesgue measure on a somewhat larger σ-algebra – consisting of the sets which are "measurable" in Lebesgue's sense – became standard methods of analysis.

What was new about Lebesgue's integral concept was not just the way it was defined, but also – and this was the real reason for its fame – its great versatility as manifested in the way it behaved with respect to limit operations. Consequently the convergence theorems are at the center of the integration theory developed by Lebesgue and his intellectual progeny.

Subsequent developments are characterized by increasing recognition of the versatility of Lebesgue's concepts in dealing with new demands from mathematics and its applications. In the course of time (up to 1930) the general (abstract) measure concept crystallized, and a theory of integration built on it – after Lebesgue's model.

It is this theory that will be developed here in an introductory fashion, but far enough that from the platform so erected the reader can easily press ahead to deeper questions and the manifold applications. Areas in which measure and integration play a key role are, for example, ergodic theory, spectral theory, harmonic analysis on locally compact groups, and mathematical economics. But the foremost example is probability theory, which uses measure and integration as an indispensable tool and whose own specific kinds of questions and methods have in turn helped to shape the former. Even today the development of measure and integration theory is far from finished.

The book is comprised of four chapters. The first is devoted to the measure concept and in particular to the Lebesgue–Borel measure and its interplay with geometry. In the second chapter the integral determined by a measure, and in particular the Lebesgue integral, the one determined by Lebesgue–Borel measure, will be introduced and investigated. The short third chapter deals with the product of measures and the associated integration. An application of this which is very important in Fourier analysis is the convolution of measures. In the fourth and last chapter the abstract concept of measure is made more concrete in the form of Radon measures. As in the original example of Lebesgue–Borel measure, here the relation of the measure to a topology on the underlying set moves into the foreground. Essentially two kinds of spaces are allowed: Polish spaces and locally compact spaces. The topological tools needed for this will mostly be developed in the text, with the reader occasionally being given only a reference (very specific) to the standard textbook literature.

The examples accompanying the exposition of a theme have an important function. They are supposed to illuminate the concepts and illustrate the limitations of the theory. The reader should therefore work through them with care. Exercises also accompany the exposition. They are not essential to understanding later developments and, in particular, proofs are not superficially shortened by consigning parts to the exercises. But the exercises do serve to deepen the reader's understanding of the material treated in the text, and working them is strongly recommended.

Notations

Here we assemble some of the notation and phraseology which will be used in the text without further comment and which – with but a few exceptions – are in general use.

By $\mathbb{N}$, $\mathbb{Z}$, $\mathbb{Q}$, $\mathbb{R}$ we designate the sets of *natural numbers* 1,2,... (excluding 0), of *whole numbers*, of *rational numbers* and of *real numbers*, respectively. We always think of the field $\mathbb{R}$ as equipped with its usual (euclidean) metric and the topology that it determines. Thus $|x - y|$ is the euclidean distance between two numbers $x, y \in \mathbb{R}$. We also speak of the *number line* $\mathbb{R}$.

Via the adjunction of $(+)\infty$ and $-\infty$ to $\mathbb{R}$, the *extended* or *compactified number line* $\overline{\mathbb{R}}$ is produced. Addition with the improper numbers $+\infty$ and $-\infty$ is performed in the usual way: $a + (\pm\infty) = (\pm\infty) + a = \pm\infty$ for $a \in \mathbb{R}$, and as well $(+\infty) + (+\infty) = +\infty$ and $(-\infty) + (-\infty) = -\infty$. On the other hand $+\infty + (-\infty)$ and $-\infty + (+\infty)$ are not defined.

As usual too we set $a \cdot (\pm\infty) = \pm\infty$ for all real $a > 0$, including $a = +\infty$, and $a \cdot (\pm\infty) = \mp\infty$ for all real $a < 0$, including $a = -\infty$. Not so general but typical in measure theory are the additional conventions

$$0 \cdot (\pm\infty) = (\pm\infty) \cdot 0 = 0\,,$$

which mean that the product $a \cdot b$ is defined for all $a, b \in \overline{\mathbb{R}}$.

The notation $A := B$ or $B =: A$ means that this equation is the *definition* of A in terms of B.

The $<$ (resp., $\leq$) relation in $\mathbb{R}$ is extended to $\overline{\mathbb{R}}$ via the decree $-\infty < a < +\infty$ for all $a \in \mathbb{R}$. A plus sign affixed to $\mathbb{Z}$, $\mathbb{Q}$, $\mathbb{R}$ or $\overline{\mathbb{R}}$ as a subscript means the sets $\mathbb{Z}_+$, $\mathbb{Q}_+$, $\mathbb{R}_+$, $\overline{\mathbb{R}}_+$ of all *non-negative* whole, rational, real numbers, or – in the last case – all $a \in \overline{\mathbb{R}}$ with $0 \leq a \leq +\infty$.

Intervals in $\overline{\mathbb{R}}$ are designated as usual by $[a, b]$, $]a, b[$, $]a, b]$ and $[a, b[$. However, (a, b) will never be used for an open interval, but only for the ordered pair with first element a, second element b.

For every pair of elements $a, b \in \overline{\mathbb{R}}$

$$a \vee b := \max\{a, b\}\,, \qquad a \wedge b := \min\{a, b\}$$

designate their respective *maximum* and *minimum*. Obviously the equations

$$|a| = a \vee (-a) = a^+ + a^- \quad \text{and} \quad a = a^+ - a^-$$

hold without any restrictions on a if we set, as usual,

$$a^+ := a \vee 0 \quad \text{and} \quad a^- := (-a)^+ = -(a \wedge 0)\,.$$

Of course, $a^+ \geq 0$ and $a^- \geq 0$ for all a. For finitely many $a_1, \ldots, a_n \in \overline{\mathbb{R}}$ the corresponding expressions $a_1 \vee \ldots \vee a_n$ and $a_1 \wedge \ldots \wedge a_n$ stand for $\max\{a_1, \ldots, a_n\}$ and $\min\{a_1, \ldots, a_n\}$, respectively.

For the *set-theoretic operations* we use the usual symbols: $\cup$ or $\bigcup$ for *union*, $\cap$ or $\bigcap$ for *intersection*, and the prefix $\complement$ to signify *complementation*. The set-theoretic relation of *inclusion* is written $A \subset B$, and equality of the sets is not thereby excluded. For the *difference set* $A \cap \complement B$, the set of all $x \in A$ such that $x \notin B$, we also write $A \setminus B$. Sets A and B which have an empty intersection, that is, for which $A \cap B = \emptyset$, are said to be *disjoint*.

The *power set* $\mathscr{P}(\Omega)$ of a set Ω is the set of all subsets of Ω, including the *empty set* $\emptyset$. A set A will be called *countable* if it is either *finite* or *denumerably infinite*. In other words, we will be using "countable" in lieu of the equally popular expression "at most countable". Obviously the empty set is to be understood as a finite set. A set will be called *non-denumerable* or *uncountable* if it is neither finite nor denumerable.

Mappings of a set A into a set B will be denoted by $f : A \to B$ or by the mapping prescription $x \mapsto f(x)$ (with $x \in A$). In case $B = \mathbb{R}$ we speak of a *real function* or a *real-valued function* on A. Not universal, but useful for our purposes, is the designation *numerical function* on A for mappings $f : A \to \overline{\mathbb{R}}$ into the extended number line. The *restriction* of a mapping $f : A \to B$ to a subset A' of A will be denoted by $f \mid A'$. The *composition* of f with a mapping $g : B \to C$ will be denoted $g \circ f$ and the *pre-image* or *inverse-image* of a set $B' \subset B$ under the mapping f will be denoted $f^{-1}(B')$.

A *sequence* in a set A is a mapping $f : \mathbb{N} \to A$ of the set $\mathbb{N}$ of natural numbers into A. Designating the image element $f(n)$ by a_n, we also write $(a_n)_{n \in \mathbb{N}}$, $(a_n)_{n=1,2,\ldots}$ or simply (a_n) for the mapping f. If other index sets, e.g., $\mathbb{Z}_+ = \{0, 1, \ldots\}$, come up, this notation is appropriately modified to, e.g., $(a_n)_{n \in \mathbb{Z}_+}$ or $(a_n)_{n=0,1,\ldots}$. In the same way finite sets are often exhibited as $(a_i)_{i=1,\ldots,n}$, with $n \in \mathbb{N}$. Even more generally, we write mappings $f : I \to A$ of a set I into the set A as $(a_i)_{i \in I}$, understanding by a_i the element $f(i)$ of A. And we then speak of a *family* in A (with index set I).

If the terms of a sequence $(a_n)_{n \in \mathbb{N}}$ in a set A from some index $n_0 \in \mathbb{N}$ onwards possess a certain property, that is, if there are but finitely many exceptional indices, we say that *ultimately all* terms of the sequence have the property. The popular phrasing "almost all terms of the sequence possess the property" has to be avoided in measure theory because there the concept "almost all" is employed in another sense.

If f and g are real functions on a set X, then $f + g$, fg, etc., designate the real functions $x \mapsto f(x) + g(x)$, $x \mapsto f(x)g(x)$, etc., on X. Numerical functions are combined analogously, as long as $f(x) + g(x)$ is defined for every $x \in X$, there being no problem with $f(x)g(x)$ in this regard, thanks to the preceding conventions. If (f_n) is a sequence of real or numerical functions on X such that the series $\sum\limits_{n=1}^{\infty} f(x)$

converges in $\overline{\mathbb{R}}$ for every $x \in X$, then $\sum\limits_{n=1}^{\infty} f_n$, or simply $\sum f_n$, designates the function $x \mapsto \sum\limits_{n=1}^{\infty} f_n(x)$. Also, functions like $\sup\limits_{n\in\mathbb{N}} f_n$, $\inf\limits_{n\in\mathbb{N}} f_n$, $\limsup\limits_{n\to\infty} f_n$, $\liminf\limits_{n\to\infty} f_n$, $\lim\limits_{n\to\infty} f_n$ are defined "pointwise" via $x \mapsto \sup\limits_{n\in\mathbb{N}} f_n(x)$, $x \mapsto \inf\limits_{n\in\mathbb{N}} f(x)$, etc.; whereby, of course, use of $\lim f_n$ presupposes the convergence in $\overline{\mathbb{R}}$ of the sequence $(f_n(x))$ for each $x \in X$.

For numerical functions $f_1, \ldots, f_n$ on a set X

$$f_1 \vee \ldots \vee f_n \qquad \text{and} \qquad f_1 \wedge \ldots \wedge f_n$$

designate the functions

$$x \mapsto f_1(x) \vee \ldots \vee f_n(x) \qquad \text{and} \qquad x \mapsto f_1(x) \wedge \ldots \wedge f_n(x).$$

At each point $x \in X$ they assume, respectively, the largest and smallest of the function values $f_1(x), \ldots, f_n(x)$. These two functions are called, respectively, the *upper* and the *lower envelopes* of $f_1, \ldots, f_n$. Correspondingly, $\sup\limits_{n\in\mathbb{N}} f_n$ and $\inf\limits_{n\in\mathbb{N}} f_n$ are called the upper and lower envelopes of the sequence (f_n) of numerical functions on X.

A numerical function defined on a subset of $\overline{\mathbb{R}}$ is called *isotone*, resp., *antitone*, if it is weakly increasing, resp., decreasing. We use this terminology also for numerical functions $f : A \to \overline{\mathbb{R}}$ when A is a (partially) *ordered set*. That is, if from $x, y \in A$ and $x \leq y$ always follows $f(x) \leq f(y)$, resp., $f(x) \geq f(y)$, then f is called isotone, resp., antitone. If from $x < y$ always follows $f(x) < f(y)$, resp., $f(x) > f(y)$, then f is called *strictly isotone*, resp., *strictly antitone*.

For sequences (a_n) in $\overline{\mathbb{R}}$ the symbolisms

$$a_n \uparrow a , \qquad a_n \downarrow a$$

express that the sequence is isotone, resp., antitone, and that $a \in \overline{\mathbb{R}}$ is its supremum, resp., its infimum.

The end of a proof is signaled by the symbol $\square$.

References of the form "RADON [1913]" are to the bibliography at the end of the book.

Section 18, labelled with *, can be skipped over in a first reading.

Table of Contents

Chapter I

Measure Theory

To geometrically simple subsets of the line, the plane, and 3-dimensional space, elementary geometry assigns "numerical measures" called length, area and volume. At first all that is intuitively clear is how the length of a segment, the area of a rectangle and the volume of a box should be defined. Proceeding from these we can determine by elementary geometric methods the lengths, areas, and volumes of more complicated sets if we accept certain calculational rules for dealing with such numerical measures.

If one thinks for example about the elementary determination of the area of a (topologically) open triangle, one begins by decomposing it via one of its altitudes into two open right triangles and the altitude itself. One further recalls that every right triangle arises from insertion of a diagonal into an appropriate rectangle. Every line segment is assigned numerical measure 0 when considered as a surface. The following two rules of calculation therefore lead to the determination of the areas of triangles:

(A) If the set A has numerical measure α, and B is congruent to A, then B also has numerical measure α.

(B) If A and B are disjoint sets with numerical measures α and β, resp., then $A \cup B$ has numerical measure $\alpha + \beta$.

The limits of such elementary geometric considerations are already reached in defining the area of an open disk K, to which end one proceeds thus: A sequence of open $3 \cdot 2^{n-1}$-gons E_n ($n \in \mathbb{N}$) is inscribed in K, with E_1 being an open equilateral triangle, and the vertices of E_{n+1} being those of E_n together with the intersections of the circle with the radii perpendicular to the sides of E_n. Thus E_{n+1} consists of E_n together with its $3 \cdot 2^{n-1}$ edges and the open isosceles triangles which have these edges as hypotenuses and vertices on the circle. Since K is the union of all the E_n, it looks like a "mosaic of triangles", that is, like a union of disjoint open triangles and segments (namely, common sides of various triangles). The following broader formulation of (B) therefore leads to a definition of the area of the disk K:

(C) If (A_n) is a sequence of pairwise disjoint sets, and A_n has numerical measure α_n ($n \in \mathbb{N}$), then $\bigcup_{n=1}^{\infty} A_n$ has numerical measure $\sum_{n=1}^{\infty} \alpha_n$.

If we replace K and every E_n by its topological closure, this method would not lead to a plausible definition of the area of a closed disk $\overline{K}$, because $\overline{K}$ is not the union of the closures $\overline{E}_n$ of the above constructed polygons E_n. A peculiarity and disadvantage of the elementary geometric procedure is precisely the necessity of choosing a special mode of decomposition tailored to the set K being considered in order to arrive at a numerical measure.

The question of a general method by means of which as many subsets of $\mathbb{R}^d$ (for arbitrary $d \in \mathbb{N}$) "as possible" could in a natural way be assigned a d-dimensional volume as numerical measure is what finally led to the mathematical discipline called measure theory. The primary content of this chapter is an exposition of the answer which measure theory furnished to this question. It will be seen that the key to the answer lies in rule (C), and that this rule is obeyed by much more general "numerical measures" which arise in situations quite remote from the original intuitive geometric one. It is just the latter reason that explains the variety of opportunities for applying measure theory in analysis, geometry and stochastics.

§1. σ-algebras and their generators

Let Ω be an arbitrary set, $\mathscr{P}(\Omega)$ its *power set*, that is, the set of all subsets of Ω. Then along with every family $(A_i)_{i \in I}$ of sets from $\mathscr{P}(\Omega)$, its union $\bigcup_{i \in I} A_i$ and its intersection $\bigcap_{i \in I} A_i$ are also in $\mathscr{P}(\Omega)$. Furthermore, $\mathscr{P}(\Omega)$ contains the complement $\complement A$ of every set A which it contains. In what follows we will be interested in subsystems $\mathscr{A} \subset \mathscr{P}(\Omega)$ which have the corresponding properties, at least for countable index sets I. According to the conventions set out in the introduction, such index sets are those that are either finite or denumerably infinite.

1.1 Definition. A system $\mathscr{A}$ of subsets of a set Ω is called a *σ-algebra* (in Ω) if it has the following properties:

(1.1) $$\Omega \in \mathscr{A} ;$$

(1.2) $$A \in \mathscr{A} \quad \Rightarrow \quad \complement A \in \mathscr{A} ;$$

(1.3) $$(A_n)_{n \in \mathbb{N}} \subset \mathscr{A} \quad \Rightarrow \quad \bigcup_{n \in \mathbb{N}} A_n \in \mathscr{A} .$$

Examples. 1. $\mathscr{P}(\Omega)$ is always a σ-algebra.

2. For any set Ω the system of all its subsets which are either countable or co-countable, that is, the $A \subset \Omega$ such either A of $\complement A$ is countable, constitute a σ-algebra. Property (1.3) is confirmed as follows: If each A_n is countable, then so is the union $\bigcup_{n \in \mathbb{N}} A_n$. If some A_m is not countable, then its complement is, and $\complement \bigcup_{n \in \mathbb{N}} A_n = \bigcap_{n \in \mathbb{N}} \complement A_n \subset \complement A_m$ is likewise countable.

3. If $\mathscr{A}$ is a σ-algebra in a set Ω and Ω' is a subset of Ω, then

(1.4) $$\Omega' \cap \mathscr{A} := \{\Omega' \cap A : A \in \mathscr{A}\}$$

is a σ-algebra in Ω', called the *trace* of $\mathscr{A}$ in Ω'. In case $\Omega' \in \mathscr{A}$, $\Omega' \cap \mathscr{A}$ consists simply of all the subsets of Ω' which are elements of $\mathscr{A}$.

4. Let Ω, Ω' be sets, $\mathscr{A}'$ a σ-algebra in Ω', and $T : \Omega \to \Omega'$ a mapping. Then the system of sets

$$(1.5) \qquad\qquad T^{-1}(\mathscr{A}) := \{T^{-1}(A') : A' \in \mathscr{A}'\}$$

is a σ-algebra in Ω, as follows from the known behavior of the set-theoretic operations under inverse mappings (like T^{-1} here).

Every σ-algebra $\mathscr{A}$ has properties "dual" to (1.1) and (1.3), namely:

$$(1.6) \qquad\qquad \emptyset \in \mathscr{A},$$

$$(1.7) \qquad\qquad (a_n)_{n\in\mathbb{N}} \subset \mathscr{A} \quad\Rightarrow\quad \bigcap_{n\in\mathbb{N}} A_n \in \mathscr{A}.$$

These follow from (1.1)–(1.3) and the identities $\emptyset = \complement\Omega$ and $\bigcap A_n = \complement(\bigcup \complement A_n)$. Moreover,

$$A_1 \cup \ldots \cup A_n = A_1 \cup \ldots \cup A_n \cup \emptyset \cup \emptyset \cup \ldots$$

and

$$A_1 \cap \ldots \cap A_n = A_1 \cap \ldots \cap A_n \cap \Omega \cap \Omega \cap \ldots$$

Therefore, along with any finite number of sets which $\mathscr{A}$ contains, it also contains their union and their intersection. From this observation and (1.2) follows as well:

$$(1.8) \qquad\qquad A, B \in \mathscr{A} \quad\Rightarrow\quad A \setminus B = A \cap \complement B \in \mathscr{A}.$$

For constructing σ-algebras the following theorem is important:

1.2 Theorem. *The intersection $\bigcap_{i\in I} \mathscr{A}_i$ of any family $(\mathscr{A}_i)_{i\in I}$ of σ-algebras in a common set Ω is itself a σ-algebra in Ω.*

Its proof is just a routine check of properties (1.1)–(1.3). It follows that for every system $\mathscr{E}$ of subsets of Ω there is a smallest σ-algebra $\boldsymbol{\sigma}(\mathscr{E})$ which contains $\mathscr{E}$; that is, $\boldsymbol{\sigma}(\mathscr{E})$ is a σ-algebra in Ω with the defining properties

(i) $\mathscr{E} \subset \boldsymbol{\sigma}(\mathscr{E})$,
(ii) for every σ-algebra $\mathscr{A}$ in Ω with $\mathscr{E} \subset \mathscr{A}$, $\boldsymbol{\sigma}(\mathscr{E}) \subset \mathscr{A}$.

For a proof, consider the system Σ of all σ-algebras $\mathscr{A}$ in Ω with $\mathscr{E} \subset \mathscr{A}$; for example, $\mathscr{P}(\Omega)$ is an element of Σ. Then $\boldsymbol{\sigma}(\mathscr{E})$ is the intersection of all the $\mathscr{A} \in \Sigma$, which according to 1.2 possesses all the desired properties.

$\boldsymbol{\sigma}(\mathscr{E})$ is called the *σ-algebra generated by* $\mathscr{E}$ (in Ω) and $\mathscr{E}$ is called a *generator* of $\boldsymbol{\sigma}(\mathscr{E})$.

Examples. 5. If $\mathscr{E}$ itself is a σ-algebra in Ω, then $\mathscr{E} = \boldsymbol{\sigma}(\mathscr{E})$.

6. If $\mathscr{E}$ consists of a single set $A \subset \Omega$, then $\boldsymbol{\sigma}(\mathscr{E}) = \{\emptyset, A, \complement A, \Omega\}$.

7.   The σ-algebra in Example 2 is generated by the system of all finite subsets of Ω.

Several systems of sets possessing some of the properties of σ-algebras frequently occur as generators. Of special interest are rings of sets.

1.3 Definition. A system $\mathscr{R}$ of subsets of a set Ω is called a *ring* (in Ω) if it has the following properties:

(1.9) $$\emptyset \in \mathscr{R};$$

(1.10) $$A, B \in \mathscr{R} \quad \Rightarrow \quad A \setminus B \in \mathscr{R};$$

(1.11) $$A, B \in \mathscr{R} \quad \Rightarrow \quad A \cup B \in \mathscr{R}.$$

If in addition

(1.12) $$\Omega \in \mathscr{R}$$

then $\mathscr{R}$ is called an *algebra* (in Ω).

A ring contains with each two of its sets (and so, with each finite collection of its sets) not only their union, but also their intersection. This is because $A \cap B = A \setminus (A \setminus B)$.

1.4 Theorem. *A system $\mathscr{R}$ of subsets of a set Ω is an algebra if and only if it has properties (1.1), (1.2) and (1.11).*

Proof. By definition an algebra has properties (1.1) and (1.11) and (1.10), and from the latter follows (1.2). The converse follows from the fact that $\emptyset = \complement\Omega$, together with the set-theoretic identity

$$A \setminus B = A \cap \complement B = \complement(B \cup \complement A). \quad \square$$

Examples. 8. Every σ-algebra is an algebra.

9.   For any set Ω the system of all sets $A \subset \Omega$ which are either finite or co-finite (i.e., have finite complement in Ω) is an algebra, but is a σ-algebra only if Ω is finite.

10.   The system of all finite subsets of a set Ω is a ring, but is an algebra only if Ω itself is finite.

11.   The smallest ring of subsets of a set Ω is the empty set $\emptyset$.

Exercises.

1. For every system $\mathscr{E}$ of subsets of a set Ω there exists a smallest ring $\rho(\mathscr{E})$ in Ω which contains $\mathscr{E}$. It is called the *ring generated* by $\mathscr{E}$. Prove this existence assertion. Determine $\rho(\mathscr{E})$ and $\sigma(\mathscr{E})$ in the case where $\mathscr{E}$ consists of two subsets A, B of Ω. When does $\rho(\mathscr{E}) = \sigma(\mathscr{E})$ hold in this latter case; when does it hold for general $\mathscr{E}$?

2. For sets A and B
$$A \triangle B := (A \setminus B) \cup (B \setminus A)$$

is called their *symmetric difference*. Prove that it obeys the following rules of calculation (in which A, B, C are arbitrary sets):

(a) $\qquad\qquad\qquad\qquad A \triangle B = B \triangle A;$

(b) $\qquad\qquad\qquad (A \triangle B) \triangle C = A \triangle (B \triangle C);$

(c) $\qquad\qquad\qquad A \triangle A = \emptyset; \qquad A \triangle \emptyset = A;$

(d) $\qquad\qquad\qquad\quad \complement A \triangle \complement B = A \triangle B;$

(e) $\qquad\qquad\quad (A \triangle B) \cap C = (A \cap C) \triangle (B \cap C);$

(f) $\qquad\qquad \left(\bigcup_{n \in \mathbb{N}} A_n\right) \triangle \left(\bigcup_{n \in \mathbb{N}} B_n\right) \subset \bigcup_{n \in \mathbb{N}} (A_n \triangle B_n)$

(for arbitrary sequences (A_n) and (B_n) of sets).

3. Deduce from exercise 2 that $\mathscr{R} \subset \mathscr{P}(\Omega)$ is a ring in a set Ω if and only if with respect to the operation $\triangle$ (as addition) and $\cap$ (as multiplication) $\mathscr{R}$ constitutes a commutative ring in the sense that the algebraists use that term.

4. A subset $\mathscr{N}$ of a ring $\mathscr{R}$ in a set Ω is called an *ideal* if it satisfies

(a) $\qquad\qquad\qquad\qquad\quad \emptyset \in \mathscr{N};$

(b) $\qquad\qquad N \in \mathscr{N},\, M \in \mathscr{R},\, M \subset N \quad \Rightarrow \quad M \in \mathscr{N};$

(c) $\qquad\qquad M, N \in \mathscr{N} \quad \Rightarrow \quad M \cup N \in \mathscr{N}.$

Continuing with exercise 3, show that $\mathscr{N} \subset \mathscr{R}$ is an ideal in $\mathscr{R}$ if and only if it is an ideal in the algebraists' sense in the commutative ring $\mathscr{R}$. Every ideal in $\mathscr{R}$ is itself a ring in Ω.

5. Let $\Omega := \mathbb{N}$ and for each $n \in \mathbb{N}$, $\mathscr{A}_n$ denote the σ-algebra in Ω generated by the system $\mathscr{E}_n$ comprised of the singletons $\{1\}$, $\{2\}, \ldots, \{n\}$. Show that $\mathscr{A}_n$ consists of all subsets of Ω which are either contained in $\{1, 2, \ldots, n\}$ or contain the complement of this set. Obviously $\mathscr{A}_n \subset \mathscr{A}_{n+1}$ for every $n \in \mathbb{N}$. Why is $\bigcup_{n \in \mathbb{N}} \mathscr{A}_n$ nevertheless not a σ-algebra in $\Omega = \mathbb{N}$?

[*Hint*: It is generally true of any isotone sequence $(\mathscr{R}_n)_{n \in \mathbb{N}}$ of rings in a set Ω that the union of all of them constitutes a σ-algebra if and only if they are equal from some index onward. Cf. OVERDIJK, SIMONS and THIEMANN [1979] and, for the special case of σ-algebras, BROUGHTON and HUFF [1977].]

§2. Dynkin systems

It is often difficult to directly determine whether a given system of sets is a σ-algebra. The following concept, which goes back to DYNKIN [1961] but in inchoate form even to SIERPINSKI [1928], helps to get around some of these difficulties.

2.1 Definition. A system $\mathscr{D}$ of subsets of a set Ω is called a *Dynkin system* (in Ω) if it has the following properties:

(2.1) $$\Omega \in \mathscr{D};$$

(2.2) $$D \in \mathscr{D} \quad \Rightarrow \quad \complement D \in \mathscr{D};$$

(2.3) $$D_n \text{ pairwise disjoint } \in \mathscr{D} \ (n \in \mathbb{N}) \quad \Rightarrow \quad \bigcup_{n \in \mathbb{N}} D_n \in \mathscr{D}.$$

Every Dynkin system $\mathscr{D}$ thus contains the empty set $\emptyset = \complement \Omega$, and then (2.3) also insures that $\mathscr{D}$ contains the union of every finite, pairwise disjoint collection of its sets.

Examples. 1. Every σ-algebra is obviously a Dynkin system.

2. Let Ω be a finite set with an even number $2n$ of elements ($n \in \mathbb{N}$). Then the system $\mathscr{D}$ of all $D \subset \Omega$ which contain an even number of elements is a Dynkin system. In case $n > 1$, $\mathscr{D}$ is not an algebra, hence certainly not a σ-algebra.

The precise connection between the concepts of σ-algebra and Dynkin system is elucidated in the following considerations:

2.2 Lemma. *Every Dynkin system $\mathscr{D}$ is closed with respect to the formation of proper complements, meaning that*

(2.2′) $$D, E \in \mathscr{D}, \ D \subset E \quad \Rightarrow \quad E \setminus D \in \mathscr{D}.$$

Proof. According to what was noted right after definition 2.1, the set $D \cup \complement E$, being the union of the disjoint sets D and $\complement E$ from $\mathscr{D}$, lies in $\mathscr{D}$. But then the complement of this set with respect to Ω, that is, $E \cap \complement D = E \setminus D$, lies in $\mathscr{D}$. $\square$

Consequently, Dynkin systems can also be defined via properties (2.1), (2.2′) and (2.3).

2.3 Theorem. *A Dynkin system is a σ-algebra just if it contains the intersection of any two of its sets.*

Proof. What needs to be shown is that every Dynkin system $\mathscr{D}$ which is closed under finite intersections is a σ-algebra. Of the defining properties of a σ-algebra, only (1.3) needs to be confirmed and we do that thus: According to (2.2′) and the closure hypothesis, $A \setminus B = A \setminus (A \cap B)$ lies in $\mathscr{D}$ whenever $A, B \in \mathscr{D}$. Since $(A \setminus B) \cap B = \emptyset$ and $A \cup B = (A \setminus B) \cup B$, $\mathscr{D}$ contains the union of any two, hence the union of any finitely many, of its elements. For any sequence $(D_n)_{n \in \mathbb{N}} \subset \mathscr{D}$, we have

$$\bigcup_{n=1}^{\infty} D_n = \bigcup_{n=0}^{\infty} (D'_{n+1} \setminus D'_n)$$

in which $D_0' := \emptyset$ and $D_n' := D_1 \cup \ldots \cup D_n$ for each $n \in \mathbb{N}$. The sets $D_{n+1}' \setminus D_n'$ are pairwise disjoint and, thanks to $(2.2')$ and what has already been proved, they lie in $\mathscr{D}$. According to (2.3) then the union of the sets D_n lies in $\mathscr{D}$. □

Just as for σ-algebras, algebras and rings, every system $\mathscr{E} \subset \mathscr{P}(\Omega)$ lies in a smallest Dynkin system. It is, of course, called the *Dynkin system generated by* $\mathscr{E}$, and is denoted $\boldsymbol{\delta}(\mathscr{E})$.

The significance of Dynkin systems lies primarily in the following fact:

2.4 Theorem. *Every* $\mathscr{E} \subset \mathscr{P}(\Omega)$ *which is closed with respect to finite intersection satisfies*

$$(2.4) \qquad\qquad \boldsymbol{\delta}(\mathscr{E}) = \boldsymbol{\sigma}(\mathscr{E}) .$$

Proof. Since every σ-algebra is a Dynkin system, $\boldsymbol{\sigma}(\mathscr{E})$ is a Dynkin system containing $\mathscr{E}$ and consequently $\boldsymbol{\delta}(\mathscr{E}) \subset \boldsymbol{\sigma}(\mathscr{E})$. If conversely, $\boldsymbol{\delta}(\mathscr{E})$ were known to be a σ-algebra, the dual relation $\boldsymbol{\sigma}(\mathscr{E}) \subset \boldsymbol{\delta}(\mathscr{E})$ would also follow. In view of 2.3 therefore it suffices to show that $\boldsymbol{\delta}(\mathscr{E})$ is closed under intersection. To prove this, we introduce for every $D \in \boldsymbol{\delta}(\mathscr{E})$ the system

$$\mathscr{D}_D := \{ Q \in \mathscr{P}(\Omega) : Q \cap D \in \boldsymbol{\delta}(\mathscr{E}) \} .$$

A routine check confirms that $\mathscr{D}_D$ is a Dynkin system. For every $E \in \mathscr{E}$ the hypothesis on $\mathscr{E}$ insures that $\mathscr{E} \subset \mathscr{D}_E$ and therewith that $\boldsymbol{\delta}(\mathscr{E}) \subset \mathscr{D}_E$. Thus for every $D \in \boldsymbol{\delta}(\mathscr{E})$ and every $E \in \mathscr{E}$ we have $E \cap D \in \boldsymbol{\delta}(\mathscr{E})$; that is, $\mathscr{E} \subset \mathscr{D}_D$, and consequently $\boldsymbol{\delta}(\mathscr{E}) \subset \mathscr{D}_D$, holding for every $D \in \boldsymbol{\delta}(\mathscr{E})$. But this is just the property of $\boldsymbol{\delta}(\mathscr{E})$ that had to be confirmed. □

Systems of subsets which are closed under intersections (respectively, unions) of two, hence of any finite number, of their sets will from now on be described as ∩-*stable* (respectively, ∪-*stable*).

Exercise.

Determine the Dynkin system generated by the system consisting of just two subsets A, B of Ω. Show that $\boldsymbol{\delta}(\mathscr{E})$ and $\boldsymbol{\sigma}(\mathscr{E})$ coincide just in case one of the sets $A \cap B$, $A \cap \complement B$, $B \cap \complement A$ of $\complement A \cap \complement B$ is empty.

§3. Contents, premeasures, measures

Combining the concepts of ring and σ-algebra with the properties (B) and (C) of lengths, areas and volumes that we encountered in the introduction leads to the basic concepts of measure theory.

3.1 Definition. Let $\mathscr{R}$ be a ring in Ω and μ a function on $\mathscr{R}$ with values in $[0, +\infty]$. μ is called a *premeasure* on $\mathscr{R}$ if

$$(3.1) \qquad\qquad \mu(\emptyset) = 0$$

and for every sequence (A_n) of pairwise disjoint sets from $\mathscr{R}$ whose union lies in $\mathscr{R}$

$$(3.2) \qquad\qquad \mu\Big(\bigcup_{n=1}^{\infty} A_n\Big) = \sum_{n=1}^{\infty} \mu(A_n) \qquad\qquad (\sigma\text{-additivity})$$

holds. μ is called a *content* if instead of (3.2) it only satisfies

$$(3.3) \qquad\qquad \mu\Big(\bigcup_{i=1}^{n} A_i\Big) = \sum_{i=1}^{n} \mu(A_i) \qquad\qquad (\text{finite additivity})$$

(for every two and therewith) for every finitely many pairwise disjoint sets $A_1, \ldots, A_n \in \mathscr{R}$.

Due to (3.1) every premeasure is evidently a content. To see this, you have only to take $A_{n+1} = A_{n+2} = \ldots = \emptyset$ in (3.2).

Examples. 1. For every ring $\mathscr{R}$ in Ω and every point $\omega \in \Omega$ the function ε_ω defined on $\mathscr{R}$ by

$$\varepsilon_\omega(A) := \begin{cases} 1 & \text{if } \omega \in A \\ 0 & \text{if } \omega \notin A \end{cases}$$

is a premeasure. It is called the premeasure defined by *unit mass* at ω.

2.   Let $\mathscr{A}$ be the σ-algebra defined in Example 2 of §1, for an uncountable set Ω, say for $\Omega = \mathbb{R}$. Set $\mu(A) := 0$ or 1 according as A of $\complement A$ is countable. Since of two disjoint subsets of Ω at most one can have a countable complement, property (3.2) is easily confirmed; thus μ is a premeasure on $\mathscr{A}$.

3.   Let $\mathscr{A}$ be the algebra defined in Example 9 of §1, for a countably infinite set Ω. Set $\mu(A) := 0$ or 1 according as A or $\complement A$ is finite. Then μ is a content but not a premeasure. The first assertion has a proof analogous to that in the preceding example, the second follows from the fact that Ω is the disjoint union of countably many 1-element sets.

4.   Let $\mu_1, \mu_2, \ldots$ be a sequence of contents (premeasures) on a ring $\mathscr{R}$, and let $\alpha_1, \alpha_2, \ldots$ be a sequence of non-negative real numbers, Then

$$\mu := \sum_{n=1}^{\infty} \alpha_n \mu_n$$

is also a content (premeasure) on $\mathscr{R}$.

Every content μ on a ring $\mathscr{R}$ enjoys the following further properties (in which $A, B, A_1, B_1, \ldots \in \mathscr{R}$):

$$(3.5) \qquad \mu(A \cup B) + \mu(A \cap B) = \mu(A) + \mu(B)\,;$$

$$(3.6) \qquad A \subset B \quad \Rightarrow \quad \mu(A) \leq \mu(B) \qquad \text{(isotoneity)};$$

$$(3.7) \qquad A \subset B, \mu(A) < +\infty \quad \Rightarrow \quad \mu(B \setminus A) = \mu(B) - \mu(A) \quad \text{(subtractivity)};$$

$$(3.8) \qquad \mu\left(\bigcup_{i=1}^{n} A_i\right) \leq \sum_{i=1}^{n} \mu(A_i) \qquad \text{(subadditivity)};$$

for every sequence (A_n) of pairwise disjoint sets from $\mathscr{R}$ whose union lies in $\mathscr{R}$

$$(3.9) \qquad \sum_{n=1}^{\infty} \mu(A_n) \leq \mu\left(\bigcup_{n=1}^{\infty} A_n\right).$$

Proof. For arbitrary $A, B \in \mathscr{R}$

$$A \cup B = A \cup (B \setminus A) \qquad \text{and} \qquad B = (A \cap B) \cup (B \setminus A)\,.$$

Because of finite additivity, it follows from these that

$$\mu(A \cup B) = \mu(A) + \mu(B \setminus A) \qquad \text{and} \qquad \mu(B) = \mu(A \cap B) + \mu(B \setminus A)\,,$$

and from addition of the last two equations

$$\mu(A \cup B) + \mu(A \cap B) + \mu(B \setminus A) = \mu(A) + \mu(B) + \mu(B \setminus A)\,.$$

In case $\mu(B \setminus A)$ is finite, (3.5) follows from this. In case $\mu(B \setminus A) = +\infty$, the formulas for $\mu(A \cup B)$, $\mu(B)$ show that each of them must also equal $+\infty$, and (3.5) consequently holds in this case too. If $A \subset B$, the preceding formula for $\mu(B)$ reads

$$\mu(B) = \mu(A) + \mu(B \setminus A)\,,$$

which, thanks to $\mu \geq 0$, delivers both (3.6) and (3.7). If we set $B_1 := A_1$, $B_2 := A_2 \setminus A_1, \ldots, B_n := A_n \setminus (A_1 \cup \ldots \cup A_{n-1})$, then $B_1, \ldots, B_n$ are pairwise disjoint sets from $\mathscr{R}$, which entails that

$$\mu\left(\bigcup_{i=1}^{n} B_i\right) = \sum_{i=1}^{n} \mu(B_i)\,.$$

From the facts that $B_i \subset A_i$ $(i = 1, \ldots, n)$, μ is isotone, and $\bigcup_{i=1}^{n} B_i = \bigcup_{i=1}^{n} A_i$ now follows (3.8). To prove (3.9) we only have to observe that for every sequence $(A_n)_{n \in \mathbb{N}}$ of pairwise disjoint sets from $\mathscr{R}$ with $A := \bigcup_{n \in \mathbb{N}} A_n \in \mathscr{R}$

$$\mu(A_1) + \ldots + \mu(A_m) = \mu(A_1 \cup \ldots \cup A_m) \leq \mu(A) \qquad (m \in \mathbb{N})$$

and let $m \to \infty$. $\square$

Finally, if μ is a premeasure on $\mathcal{R}$, then for any sets $A_0, A_1, \ldots \in \mathcal{R}$

$$(3.10) \qquad A_0 \subset \bigcup_{n=1}^{\infty} A_n \quad \Rightarrow \quad \mu(A_0) \le \sum_{n=1}^{\infty} \mu(A_n).$$

Because of $A_0 = \bigcup(A_0 \cap A_n)$ and (3.6), we can assume, in verifying (3.10), that $A_0 = \bigcup A_n$. Then set $B_1 := A_1$, $B_2 := A_2 \setminus A_1, \ldots, B_n := A_n \setminus (A_1 \cup \ldots \cup A_{n-1})$ and proceed as in the proof of (3.8).

In particular, we now have

$$(3.10') \qquad \mu\left(\bigcup_{n=1}^{\infty} A_n\right) \le \sum_{n=1}^{\infty} \mu(A_n)$$

whenever all the sets A_n as well as their union lie in $\mathcal{R}$.

The following theorem characterizes premeasures via other properties related to the σ-additivity. Its formulation is facilitated by the notations:

$$(3.11) \qquad E_n \uparrow E \qquad \text{and} \qquad E_n \downarrow E$$

which mean that the sets $E_1 \subset E_2 \subset \ldots$ satisfy $E = \bigcup E_n$, or that the sets $E_1 \supset E_2 \supset \ldots$ satisfy $E = \bigcap E_n$. In other words, the sequence (E_n) either increases isotonically to E or decreases antitonically to E.

3.2 Theorem. *For a content μ on a ring $\mathcal{R}$ consider the following statements:*
(a) *μ is a premeasure.*
(b) *$A_n, A \in \mathcal{R}$ with $A_n \uparrow A \Rightarrow \lim_{n\to\infty} \mu(A_n) = \mu(A)$* (continuity from below).
(c) *$A_n, A \in \mathcal{R}$ with $A_n \downarrow A$ and $\mu(A_n) < +\infty$ for all $n \Rightarrow$*

$$\lim_{n\to\infty} \mu(A_n) = \mu(A) \qquad \text{(continuity from above).}$$

(d) *$A_n \in \mathcal{R}$ with $A_n \downarrow \emptyset$ and $\mu(A_n) < +\infty$ for all $n \Rightarrow$*

$$\lim_{n\to\infty} \mu(A_n) = 0 \qquad \text{(continuity at $\emptyset$).}$$

Then the following implications hold:

$$\text{(a)} \Leftrightarrow \text{(b)} \quad \Rightarrow \quad \text{(c)} \Leftrightarrow \text{(d)}.$$

If μ is finite on $\mathcal{R}$, that is, $\mu(A) < +\infty$ for all $A \in \mathcal{R}$, then all four statements (a)–(d) are equivalent.

Proof. (a)$\Rightarrow$(b): Defining $A_0 := \emptyset$, the sets $B_n := A_n \setminus A_{n-1}$ $(n \in \mathbb{N})$ are pairwise disjoint, lie in $\mathcal{R}$ and satisfy

$$A = \bigcup_{n=1}^{\infty} B_n, \qquad A_n = B_1 \cup \ldots \cup B_n.$$

Therefore on account of the σ-additivity of μ

$$\mu(A) = \sum_{n=1}^{\infty} \mu(B_n) = \lim_{n \to \infty} \sum_{i=1}^{n} \mu(B_i) = \lim_{n \to \infty} \mu(A_n).$$

(b)$\Rightarrow$(a): Let (A_n) be a sequence of pairwise disjoint sets from $\mathscr{R}$ whose union $A := \bigcup A_n$ also is in $\mathscr{R}$. If we set $B_n := A_1 \cup \ldots \cup A_n$, then $B_n \in \mathscr{R}$ and $B_n \uparrow A$; therefore $\mu(A) = \lim \mu(B_n)$. As a result of the finite additivity of μ

$$\mu(B_n) = \mu(A_1) + \ldots + \mu(A_n)$$

and therefore $\mu(A) = \sum \mu(A_n)$. Thus μ is σ-additive, and consequently is a premeasure.

(b)$\Rightarrow$(c): According to (3.7), $\mu(A_1 \setminus A_n) = \mu(A_1) - \mu(A_n)$ for every $n \in \mathbb{N}$. From $A_n \downarrow A$ follows $A_1 \setminus A_n \uparrow A_1 \setminus A$, and all the sets appearing here are in $\mathscr{R}$. From (b) therefore

$$\mu(A_1 \setminus A) = \lim_{n \to \infty} \mu(A_1 \setminus A_n) = \mu(A_1) - \lim_{n \to \infty} \mu(A_n).$$

From this follows (c), because $A \subset A_n$ means that also $\mu(A) < +\infty$ and so $\mu(A_1 \setminus A) = \mu(A_1) - \mu(A)$.

(c)$\Rightarrow$(d): Here there is nothing to prove!

(d)$\Rightarrow$(c): From $A_n \downarrow A$ follows $A_n \setminus A \downarrow \emptyset$. Since $A_n \setminus A \subset A_n$, the isotoneity of μ means that along with $\mu(A_n)$, $\mu(A_n \setminus A)$ is finite too. Hence by (d), $\lim \mu(A_n \setminus A) = 0$. But then (c) follows because $\mu(A) \leq \mu(A_n) < +\infty$, causing $\mu(A_n \setminus A)$ to equal $\mu(A_n) - \mu(A)$.

To finish off, let us consider the case that μ is finite, and show that then

(d)$\Rightarrow$(b): If (A_n) is a sequence of sets from $\mathscr{R}$ and $A_n \uparrow A \in \mathscr{R}$, then $A \setminus A_n \downarrow \emptyset$. Taking account of the finiteness of μ, it therefore follows that $0 = \lim \mu(A \setminus A_n) = \lim[\mu(A) - \mu(A_n)]$ and therewith (b). $\square$

Remark. If one modifies Example 3 of this section by making $\mu(A) := 0$ for all finite sets and $\mu(A) := +\infty$ for all cofinite sets, then he gets a content that is continuous at $\emptyset$ but is not a premeasure. Thus without the finiteness hypothesis in the preceding theorem, statements (a)–(d) are not generally equivalent. On the other hand, in (c) and (d) it is enough to explicitly hypothesize $\mu(A_n) < +\infty$ for some $n \in \mathbb{N}$, as then $\mu(A_m) < +\infty$ for all $m \geq n$ (isotoneity).

The concepts of content and premeasure are preliminary to the central concept of this book, that of a measure.

3.3 Definition. A premeasure defined on a σ-algebra $\mathscr{A}$ of subsets of a set Ω is called a *measure* (on $\mathscr{A}$). The function value $\mu(A)$ of μ at an $A \in \mathscr{A}$ is called the $(\mu\text{-})measure$ or the $(\mu\text{-})mass$ of A. If $\mu(\Omega) < +\infty$ (and consequently $\mu(A) < +\infty$ for every $A \in \mathscr{A}$), the measure μ is called *finite*.

Thus a measure is a non-negative, numerical function μ defined on a σ-algebra $\mathscr{A}$ and enjoying properties (3.1) and (3.2). The constant function $\mu = 0$ is a measure on every σ-algebra, the so-called *zero-measure*. The examples that follow are still of a rather formal nature. But as early as §6 and then quite a bit later we will become acquainted with an abundance of important examples.

Examples. 5. If for the ring $\mathcal{R}$ in Example 1 one takes a σ-algebra $\mathcal{A}$ in Ω, then ε_ω is a measure on $\mathcal{A}$, called the measure defined by a unit point mass at ω, or more briefly the *unit mass at ω*, and also the *Dirac measure* at ω. These designations derive from interpreting a measure μ on a σ-algebra in Ω as a mass distribution over Ω. Accordingly for $A \in \mathcal{A}$, $\mu(A)$ is viewed as the mass that has been "smeared" over A. The Dirac measure at ω has, in so far as the one-element set $\{\omega\}$ lies in $\mathcal{A}$, all of its (unit) mass concentrated at the point ω: $\varepsilon_\omega(\{\omega\}) = 1$, $\varepsilon_\omega(\complement\{\omega\}) = 0$.

6.   Let Ω be an arbitrary set. For every $A \in \mathscr{P}(\Omega)$ let $|A|$ denote the number of elements in A in case A is finite, and otherwise $+\infty$. Then $\zeta(A) := |A|$ defines a measure on $\mathscr{P}(\Omega)$, called the *counting measure* on Ω (or on $\mathscr{P}(\Omega)$). Its restriction to a σ-algebra $\mathcal{A}$ in Ω is called *counting measure on $\mathcal{A}$*.

7.   The premeasure defined in Example 2 is a measure.

Next we derive a not-so-obvious consequence of the σ-additivity of measures.

3.4 Lemma. *Let μ be a measure on a σ-algebra $\mathcal{A}$ and $(A_n)_{n\in\mathbb{N}}$ a sequence of sets from $\mathcal{A}$. Suppose there is a $k \in \mathbb{N}$ such that the sets A_m and A_n are disjoint whenever their indices satisfy $|m - n| \geq k$. Then*

$$(3.12) \qquad \sum_{n=1}^{\infty} \mu(A_n) \leq k\mu\left(\bigcup_{n=1}^{\infty} A_n\right).$$

When $k = 1$ this is, in view of (3.10′), just the σ-additivity requirement of a measure.

Proof. Designate the union of all the A_n as C. For each $r = 1, \ldots, k$ the sets $(A_{r+mk})_{m\in\mathbb{N}_0}$ are pairwise disjoint. So if we set

$$F_r := \bigcup_{m=0}^{\infty} A_{r+mk},$$

then

$$\sum_{m=0}^{\infty} \mu(A_{r+mk}) = \mu(F_r) \leq \mu(C)$$

because $F_r \subset C$. Since the sum of a series of non-negative terms in independent of the ordering of the terms, it follows that

$$\sum_{n=1}^{\infty} \mu(A_n) = \sum_{r=1}^{k} \mu(F_r).$$

From this equality and the preceding inequality the asserted inequality can be read off. □

Exercises.

1. Let Ω be a finite, non-empty set. Show that the counting measure ζ on $\mathscr{P}(\Omega)$ coincides with $\sum_{w \in \Omega} \varepsilon_w$. Show further that every measure μ on $\mathscr{P}(\Omega)$ has the form $\mu = \sum_{w \in \Omega} \alpha_w \varepsilon_w$, with each $\alpha_w := \mu(\{w\})$.

2. For a finite content μ on a ring $\mathscr{R}$ establish the following *input-output formula* generalizing equality (3.5): For all $n \in \mathbb{N}$, $A_1, \ldots, A_n \in \mathscr{R}$

$$\mu\left(\bigcup_{i=1}^n A_i\right) = \sum_{i=1}^n \mu(A_i) - \sum_{1 \le i < j \le n} \mu(A_i \cap A_j) + \sum_{1 \le i < j < k \le n} \mu(A_i \cap A_j \cap A_k)$$
$$- + \ldots + (-1)^{n-1} \mu(A_1 \cap \ldots \cap A_n).$$

3. For a premeasure μ on a ring $\mathscr{R}$ in Ω define

$$\tilde{\mathscr{R}} := \{A \in \mathscr{P}(\Omega) : A \cap R \in \mathscr{R} \text{ for every } R \in \mathscr{R}\}$$
$$\tilde{\mu}(A) := \sup\{\mu(R) : R \subset A, R \in \mathscr{R}\}, \quad \text{for } A \in \tilde{\mathscr{R}}.$$

Show that $\tilde{\mathscr{R}}$ is an algebra in Ω which contains $\mathscr{R}$, and that $\tilde{\mu}$ is a premeasure on $\tilde{\mathscr{R}}$ which extends μ.

4. Suppose that $(\mu_n)_{n \in \mathbb{N}}$ is a sequence of premeasures on a common ring $\mathscr{R}$ which is isotone, that is, satisfies $\mu_n(A) \le \mu_{n+1}(A)$ for all $A \in \mathscr{R}$, $n \in \mathbb{N}$. Show that via $\mu(A) := \sup_{n \in \mathbb{N}} \mu_n(A)$ a premeasure μ is defined on $\mathscr{R}$.

5. Let μ be a measure on a σ-algebra $\mathscr{A}$ in Ω, and denote by $\mathscr{N}_\mu$ the set of all μ-*null* (or μ-*negligible*) sets, that is, the $N \in \mathscr{A}$ for which $\mu(N) = 0$. Check that $\mathscr{N}_\mu$ has the following properties:

(a) $$\emptyset \in \mathscr{N}_\mu;$$

(b) $$N \in \mathscr{N}_\mu, M \in \mathscr{A}, M \subset N \quad \Rightarrow \quad M \in \mathscr{N}_\mu;$$

(c) $$(N_n)_{n \in \mathbb{N}} \subset \mathscr{N}_\mu \quad \Rightarrow \quad \bigcup_{n \in \mathbb{N}} N_n \in \mathscr{N}_\mu.$$

Subsets of $\mathscr{A}$ with these properties are called σ-*ideals* in $\mathscr{A}$. Thus $\mathscr{N}_\mu$ is always a σ-ideal. (Cf. Exercise 4 of §1.)

6. Every σ-ideal $\mathscr{N}$ in a σ-algebra $\mathscr{A}$ is the σ-ideal $\mathscr{N}_\mu$ of μ-null sets of an appropriate measure μ on $\mathscr{A}$. To get such a μ, define

$$\mu(A) := \begin{cases} 0 & \text{if } A \in \mathscr{A} \\ +\infty & \text{if } A \in \mathscr{A} \setminus \mathscr{N}. \end{cases}$$

As a special case, on the power set $\mathscr{P}(\Omega)$ of any set Ω there is a measure μ such that $\mu(A) = 0$ precisely if A is a countable subset of Ω.

7. Let μ be a finite content on a ring $\mathscr{R}$. Show that

$$d_\mu(A, B) := \mu(A \triangle B) \qquad (A, B \in \mathscr{R})$$

defines a pseudometric on $\mathscr{R}$, that is, d_μ has all the properties of a metric on $\mathscr{R}$ with one possible exception: $d_\mu(A, B) = 0$ can happen without $A = B$. (Cf. Exercise 3 of §15.)

§4. Lebesgue premeasure

Now we specialize Ω to be the d-dimensional number-line $\mathbb{R}^d$ ($d \in \mathbb{N}$). For every two points $a = (\alpha_1, \ldots, \alpha_d)$ and $b = (\beta_1, \ldots, \beta_d) \in \mathbb{R}^d$ we write $a \leq b$ (resp., $a \lhd b$) if $\alpha_i \leq \beta_i$ for all $i = 1, \ldots, d$ (resp., $\alpha_i < \beta_i$ for all $i = 1, \ldots, d$). Every set of the form

$$(4.1) \qquad [a, b[:= \{x \in \mathbb{R}^d : a \leq x \lhd b\},$$

where $a, b \in \mathbb{R}^d$ and $a \leq b$, is called a *right half-open interval* in $\mathbb{R}^d$. Geometrically described, these are parallelepiped "open on the right" and having sides parallel to the coordinate axes. Clearly $[a, b[$ is nonempty if and only if $a \lhd b$, and in this case the interval $[a, b[$ uniquely determines the points a, b.

For every such interval $[a, b[$ the real number

$$(4.2) \qquad (\beta_1 - \alpha_1) \cdot \ldots \cdot (\beta_d - \alpha_d)$$

is called its *d-dimensional elementary content*. It equals 0 just when $[a, b[= \emptyset$, that is, when $a \lhd b$ fails (although $a \leq b$ holds, a prerequisite to employing interval notation).

From now on, $\mathscr{I}^d$ shall designate the set of all right half-open intervals in $\mathbb{R}^d$, and $\mathscr{F}^d$ the system of all finite unions of such intervals, so $\mathscr{I}^d \subset \mathscr{F}^d$. The elements of $\mathscr{F}^d$ are called *d-dimensional figures*.

4.1 Lemma. *For all $I, J \in \mathscr{I}^d$*

$$I \cap J \in \mathscr{I}^d \qquad and \qquad J \setminus I \in \mathscr{F}^d.$$

Every figure is a union of finitely many pairwise disjoint intervals from $\mathscr{I}^d$.

Proof. Let $I = [a, b[$, $J = [a', b'[$ with $a \leq b$, $a' \leq b'$, and let the corresponding coordinates of these points be α_i, β_i, α_i', β_i'. If we let e and f denote those points in $\mathbb{R}^d$ whose coordinates are $\max\{\alpha_i, \alpha_i'\}$ and $\min\{\beta_i, \beta_i'\}$ ($i = 1, \ldots, d$), respectively, then $I \cap J = [e, f[$ in case $e \leq f$ and otherwise $I \cap J = \emptyset$. Consequently, $I \cap J$ is already in $\mathscr{I}^d$. Because $J \setminus I = J \setminus (I \cap J)$ and we now have $I \cap J \in \mathscr{I}^d$, in proving the second claim we may assume that $I \neq \emptyset$ and $I \subset J$. Then I and J determine the points a, b, a', b' uniquely and they satisfy $a' \leq a \lhd b \leq b'$.

Create new points from $a = (\alpha_1, \ldots, \alpha_d)$ and $b = (\beta_1, \ldots, \beta_d)$ by replacing α_i by α_i' and β_i by α_i, or by replacing α_i by β_i and β_i by β_i', and do this in all possible ways. More precisely, make such replacements for the i coming from each non-empty subset of $\{1, \ldots, d\}$. The points so created give rise to at most $3^d - 1$ pairwise disjoint intervals from $\mathscr{I}^d$ whose union is $J \setminus I$. Thus $J \setminus I$ is a figure and is representable as a finite union of pairwise disjoint sets from $\mathscr{I}^d$. That this obtains as well for every figure $F = I_1 \cup \ldots \cup I_n \in \mathscr{F}^d$ with $I_1, \ldots, I_n \in \mathscr{I}^d$ can now be seen as follows:

$$F = I_1 \cup (I_2 \setminus I_1) \cup (I_3 \setminus I_1 \cup I_2) \cup \ldots \cup (I_n \setminus I_1 \cup \ldots \cup I_{n-1})$$

exhibits F as a union of n pairwise disjoint sets, each of the form $I \setminus J_1 \cup \ldots \cup J_m$ with $I, J_1, \ldots, J_m$ intervals from $\mathscr{I}^d$. Thus it suffices to show that every set of this form is the union of finitely many pairwise disjoint intervals from $\mathscr{I}^d$. But this follows from

$$I \setminus J_1 \cup \ldots \cup J_m = \bigcap_{i=1}^{m} (I \setminus J_i)$$

when, using what has already been proved, we write each $I \setminus J_i$ as a union of finitely many pairwise disjoint intervals from $\mathscr{I}^d$ and distribute the intersection through these unions. □

4.2 Theorem. $\mathscr{F}^d$ is a ring in $\mathbb{R}^d$.

Proof. The only thing that is not obvious is property (1.10) of a ring, according to which along with any sets $F, G \in \mathscr{F}^d$ their difference $F \setminus G$ must also be in $\mathscr{F}^d$. By definition there exist intervals $I_1', \ldots, I_m', I_1'', \ldots, I_n'' \in \mathscr{I}^d$ such that

$$F = \bigcup_{i=1}^{m} I_i' \quad \text{and} \quad G = \bigcup_{j=1}^{n} I_j''.$$

But then

$$F \setminus G = \bigcup_{i=1}^{m} \left(\bigcap_{j=1}^{n} (I_i' \setminus I_j'') \right)$$

and so it only has to be shown that each set $\bigcap_{j=1}^{m} (I_i' \setminus I_j'')$ is a figure. According to 4.1 $I_i' \setminus I_j''$ is always a figure. So it further suffices to demonstrate that the intersection of two (whence, of any finite number of) figures is itself a figure. If however F and G are two figures represented as above, then thanks to distributivity $F \cap G$ is just the union of the sets $I_i' \cap I_j''$ $(i = 1, \ldots, m; \ j = 1, \ldots, n)$, which by another appeal to 4.1 is a figure. □

By definition every figure is a union of finitely many intervals from $\mathscr{I}^d$. Consequently, $\mathscr{F}^d \subset \mathscr{R}$ for every ring $\mathscr{R}$ in $\mathbb{R}^d$ such that $\mathscr{I}^d \subset \mathscr{R}$. So theorem 4.2 really says that $\mathscr{F}^d$ is the ring generated by $\mathscr{I}^d$.

Our geometric intuition now suggests the validity of the following theorem:

4.3 Theorem. *There exists exactly one content λ on $\mathscr{F}^d$ with the property that $\lambda(I)$ coincides with the d-dimensional elementary content of I, for each $I \in \mathscr{I}^d$. This content is real-valued.*

Proof. According to 4.1, every figure $F \in \mathscr{F}^d$ has a representation $F = I_1 \cup \ldots \cup I_n$ as a union of finitely many pairwise disjoint intervals from $\mathscr{I}^d$. Every content λ on $\mathscr{F}^d$ therefore satisfies

$$\lambda(F) = \lambda(I_1) + \ldots + \lambda(I_n),$$

which shows that λ is determined throughout $\mathscr{F}^d$ just by its values on $\mathscr{I}^d$ and is necessarily real-valued. Thus all we have to do is settle the existence question. To this end we first define λ only on $\mathscr{I}^d$ as it must be defined, namely $\lambda(I)$ shall be the d-dimensional elementary content of I for each $I \in \mathscr{I}^d$. Then we have:

(a) Let $I = [a, b[\in \mathscr{I}^d$, $a = (\alpha_1, \ldots, \alpha_d)$ and $b = (\beta_1, \ldots, \beta_d)$ and γ a real number satisfying $\alpha_i \leq \gamma \leq \beta_i$ for a fixed $i \in \{1, \ldots, d\}$. The hyperplane with equation $\xi_i = \gamma$ divides I into two disjoint intervals $I_1 := [a', b[$ and $I_2 := [a, b'[$, a' being a with its ith coordinate replaced by γ, and b' being b with its ith coordinate replaced by γ. From (4.2) then follows that $\lambda(I) = \lambda(I_1) + \lambda(I_2)$. Induction therefore yields

(b) If $I \in \mathscr{I}^d$ is decomposed by finitely many hyperplanes in the manner described in (a) into pairwise disjoint intervals $I_1, \ldots, I_n \in \mathscr{I}^d$, then $\lambda(I) = \lambda(I_1) + \ldots + \lambda(I_n)$. More generally:

(c) For any finitely many, pairwise disjoint $I_1, \ldots, I_n \in \mathscr{I}^d$ with $I_0 := I_1 \cup \ldots \cup I_n \in \mathscr{I}^d$, $\lambda(I_0) = \lambda(I_1) + \ldots + \lambda(I_n)$. In proving this we can obviously assume that each I_j is not empty. Then there are points $a_j = (\alpha_{j1}, \ldots, \alpha_{jd})$ and $b_j = (\beta_{j1}, \ldots, \beta_{jd})$ from $\mathbb{R}^d$ with $a_j \lhd b_j$ and $I_j = [a_j, b_j[$, $j = 0, 1, \ldots, n$. The hyperplanes whose respective equations are $\xi_i = \alpha_{ji}$ or $\xi_i = \beta_{ji}$ for $i \in \{1, \ldots, d\}$, $j \in \{1, \ldots, n\}$ decompose I_0 into pairwise disjoint intervals $I'_1, \ldots, I'_m \in \mathscr{I}^d$. Each of $I_1, \ldots, I_n$ also decomposes into certain of these $I'_1, \ldots, I'_m$. The claimed equality therefore follows from $(n + 1)$ citations of (b).

(d) If now

$$F = I_1 \cup \ldots \cup I_n = J_1 \cup \ldots \cup J_m$$

are two representations of the figure $F \in \mathscr{F}^d$, each a union of pairwise disjoint intervals, then

$$\lambda(I_1) + \ldots + \lambda(I_n) = \lambda(J_1) + \ldots + \lambda(J_m).$$

Indeed, $I_j = I_j \cap F = \bigcup_{i=1}^{m} (I_j \cap J_i)$ is a representation of I_j as a union of the pairwise disjoint intervals $I_j \cap J_1, \ldots, I_j \cap J_m$ and thanks to (c)

$$\lambda(I_j) = \sum_{i=1}^{m} \lambda(I_j \cap J_i) \qquad (j = 1, \ldots, n).$$

Upon interchanging the roles of i and j, one gets analogously

$$\lambda(J_i) = \sum_{j=1}^{n} \lambda(I_j \cap J_i) \qquad\qquad (i = 1, \ldots, m).$$

Together these last two equations entail the equality $\sum \lambda(I_j) = \sum \lambda(J_i)$.

(e) Thus for every $F \in \mathscr{F}^d$ the number $\sum \lambda(I_j)$ is independent of the special representation

$$F = I_1 \cup \ldots \cup I_n$$

of F as a union of finitely many pairwise disjoint $I_1, \ldots, I_n \in \mathscr{I}^d$. Therefore the decree

$$\lambda(F) := \lambda(I_1) + \ldots + \lambda(I_n)$$

well defines an extension, to be denoted still by λ, of the original function on $\mathscr{I}^d$ to one on $\mathscr{F}^d$. This function is real-valued, non-negative, and according to (d) finitely-additive. Since $\emptyset \in \mathscr{I}^d$ and $\lambda(\emptyset) = 0$, a content with the sought-for properties is at hand. $\square$

4.4 Theorem. *The content λ on $\mathscr{F}^d$ is a premeasure.*

Proof. Because λ is finite, 3.2 says that we only need to prove the continuity of λ at $\emptyset$. To this end, let (F_n) be an antitone sequence of figures from $\mathscr{F}^d$. We will show that from the assumption that

$$\delta := \lim_{n \to \infty} \lambda(F_n) = \inf_{n \in \mathbb{N}} \lambda(F_n) > 0$$

follows

$$\bigcap_{n \in \mathbb{N}} F_n \neq \emptyset .$$

Each F_n being a union of finitely many pairwise disjoint intervals from $\mathscr{I}^d$, it should be clear that by a slight leftward shift of the right endpoints of each of these intervals a new figure $G_n \in \mathscr{F}^n$ is created, whose topological closure $\overline{G}_n$ is still a subset of F_n, and

$$\lambda(F_n) - \lambda(G_n) \leq 2^{-n}\delta .$$

If we set $H_n := G_1 \cap \ldots \cap G_n$, then (H_n) is a sequence of sets from $\mathscr{F}^d$ satisfying $H_n \supset H_{n+1}$, $\overline{H}_n \subset \overline{G}_n \subset F_n$ for all n. Because F_n is bounded its closed subset H_n is compact. As soon as we succeed in showing that each H_n is not empty, it will follow from the finite-intersection property of compacta (WILLARD [1970], p. 118, KELLEY [1955], p. 136) that $\bigcap_{n \in \mathbb{N}} \overline{H}_n \neq \emptyset$ and so a fortiori $\bigcap_{n \in \mathbb{N}} F_n \neq \emptyset$. So let us prove that no H_n is empty. For every $n \in \mathbb{N}$

$$(*) \qquad\qquad \lambda(H_n) \geq \lambda(F_n) - (1 - 2^{-n})\delta ,$$

as we will confirm by induction. The inequality holds for $n = 1$ because $H_1 = G_1$, and by choice of G_1, $\lambda(F_1) - \lambda(G_1) \leq 2^{-1}\delta$. Suppose the inequality valid for some n. Since $H_{n+1} = G_{n+1} \cap H_n$ and everything is finite, (3.5) gives

$$\lambda(H_{n+1}) = \lambda(G_{n+1}) + \lambda(H_n) - \lambda(G_{n+1} \cup H_n).$$

From the induction hypothesis $\lambda(H_n) \geq \lambda(F_n) - (1 - 2^{-n})\delta$; from the choice of G_{n+1}, $\lambda(G_{n+1}) \geq \lambda(F_{n+1}) - 2^{-n-1}\delta$ and $G_{n+1} \cup H_n \subset F_{n+1} \cup F_n = F_n$, so that $\lambda(G_{n+1} \cup H_n) \leq \lambda(F_n)$. Combining these observations completes the inductive step in the confirmation of (*):

$$\lambda(H_{n+1}) \geq \lambda(F_{n+1}) - 2^{-n-1}\delta - (1 - 2^{-n})\delta = \lambda(F_{n+1}) - (1 - 2^{-n-1})\delta.$$

Recalling that $\lambda(F_n) \geq \delta$ by definition of δ, we infer from (*) the inequality $\lambda(H_n) \geq 2^{-n}\delta > 0$ and therewith the fact that $H_n \neq \emptyset$, the last link that had to be accounted for in the logical chain. $\square$

4.5 Definition. The premeasure λ on the ring $\mathscr{F}^d$ of d-dimensional figures in $\mathbb{R}^d$ is called *Lebesgue premeasure* in $\mathbb{R}^d$ or d-dimensional Lebesgue premeasure. From now on it will be denoted by λ^d.

Here we encounter for the first time the name of the French mathematician H. LEBESGUE (1875–1941), the inventor of the measure and integration concepts that today are named after him. The development of the theory of measure and integration was spurred above all by his investigations and those of his countryman É. BOREL (1871–1956). For the history of Lebesgue integration see DIEUDONNÉ [1978] and HAWKINS [1970].

Exercises.

1. Show that on $\mathscr{F}^1$ there is exactly one content μ that assigns to the right half-open interval $[\alpha, \beta[$, $\alpha, \beta \in \mathbb{R}$, the following values

$$\mu([\alpha, \beta[) = \begin{cases} 1 & \text{if } \alpha < 0 \leq \beta \\ 0 & \text{in all other cases.} \end{cases}$$

Is μ σ-additive?

2. Two intervals $I_0, J \in \mathscr{I}^d$ with $I_0 \subset J$ are given. Prove the existence of $k \leq 2^d$ intervals $I_1, \ldots, I_k \in \mathscr{I}^d$ with the following two properties: (i) $I_0 \cup \ldots \cup I_j \in \mathscr{I}^d$ for each $j \in \{0, \ldots, k\}$; (ii) $J = I_0 \cup \ldots \cup I_k$. [*Hint*: Proceed by induction on the dimension d.]

§5. Extension of a premeasure to a measure

Lebesgue premeasure is not a measure because its domain of definition, the ring $\mathscr{F}^d$ of d-dimensional figures, is not a σ-algebra. For example, the whole space $\mathbb{R}^d$ is not in $\mathscr{F}^d$, every d-dimensional figure being a bounded subset of $\mathbb{R}^d$.

The elementary geometric considerations sketched at the beginning of this chapter however suggest that the domain of the premeasure λ^d be so enlarged that a "numerical measure" gets assigned also to more complicated subsets of $\mathbb{R}^d$. The most satisfactory such result would say that λ^d can be extended in exactly one way to a measure on an appropriate σ-algebra $\mathscr{A}$ in $\mathbb{R}^d$ with $\mathscr{F}^d \subset \mathscr{A}$.

Here we encounter the following general problem: A ring $\mathscr{R}$ in a set Ω and a content μ on $\mathscr{R}$ are given. Under what conditions does there exist a σ-algebra $\mathscr{A}$ in Ω and a measure $\tilde{\mu}$ on $\mathscr{A}$ such that μ is the restriction of $\tilde{\mu}$ to $\mathscr{R}$? An obvious necessary condition for this is that μ be a premeasure on $\mathscr{R}$. The designation "premeasure" will turn out to be justified if we can show the converse: For every premeasure μ on a ring $\mathscr{R}$ there exists a σ-algebra $\mathscr{A}$ in Ω with $\mathscr{R} \subset \mathscr{A}$, and a measure $\tilde{\mu}$ on $\mathscr{A}$ satisfying $\tilde{\mu} \mid \mathscr{R} = \mu$. It suffices to take for $\mathscr{A}$ the σ-algebra $\sigma(\mathscr{R})$ generated in Ω by $\mathscr{R}$.

5.1 Theorem (Extension theorem). *Every premeasure μ on a ring $\mathscr{R}$ in Ω can be extended in at least one way to a measure $\tilde{\mu}$ on the σ-algebra $\sigma(\mathscr{R})$ generated by $\mathscr{R}$ in Ω.*

Proof. For each subset $Q \subset \Omega$ designate by $\mathscr{U}(Q)$ the set of all sequences $(A_n)_{n \in \mathbb{N}}$ of sets from $\mathscr{R}$ which cover Q, that is, which satisfy

$$Q \subset \bigcup_{n \in \mathbb{N}} A_n \, .$$

Then the numerical function μ^* may be defined on $\mathscr{P}(\Omega)$ via

$$(5.1) \qquad \mu^*(Q) := \begin{cases} \inf\{\sum_{n=1}^{\infty} \mu(A_n) : (A_n) \in \mathscr{U}(Q)\}, & \text{in case } \mathscr{U}(Q) \neq \emptyset \\ +\infty, & \text{in case } \mathscr{U}(Q) = \emptyset. \end{cases}$$

It has the following properties:

$$(5.2) \qquad\qquad\qquad \mu^*(\emptyset) = 0 \, ;$$

$$(5.3) \qquad\qquad Q_1 \subset Q_2 \quad \Rightarrow \quad \mu^*(Q_1) \leq \mu^*(Q_2) \, ;$$

$$(5.4) \qquad (Q_n)_{n \in \mathbb{N}} \subset \mathscr{P}(\Omega) \quad \Rightarrow \quad \mu^*\left(\bigcup_{n=1}^{\infty} Q_n\right) \leq \sum_{n=1}^{\infty} \mu^*(Q_n) \, .$$

Equality (5.2) follows from the observation that the constant sequence $\emptyset, \emptyset, \ldots$ is in $\mathscr{U}(\emptyset)$. The observation that $\mathscr{U}(Q_2) \subset \mathscr{U}(Q_1)$ follows from $Q_1 \subset Q_2$, serves to confirm (5.3). For the proof of (5.4) it can evidently be assumed that $\mu^*(Q_n)$ is finite and so in particular $\mathscr{U}(Q_n) \neq \emptyset$, for every $n \in \mathbb{N}$. For an arbitrary $\varepsilon > 0$ then, each $\mathscr{U}(Q_n)$ contains a sequence $(A_{nm})_{m \in \mathbb{N}}$ such that

$$\sum_{m=1}^{\infty} \mu(A_{nm}) \leq \mu^*(Q_n) + 2^{-n}\varepsilon \, .$$

The double sequence $(A_{nm})_{n,m\in\mathbb{N}}$ lies in $\mathscr{U}\left(\bigcup_{n=1}^{\infty} Q_n\right)$ and as a consequence the definition of μ^* gives

$$\mu^*\left(\bigcup_{n=1}^{\infty} Q_n\right) \leq \sum_{n,m\in\mathbb{N}} \mu(A_{nm}) \leq \sum_{n=1}^{\infty} \mu^*(Q_n) + \varepsilon$$

and (5.4) follows from this and the arbitrariness of $\varepsilon > 0$. It is immediate from the definition that

(5.5) $$\mu^* \geq 0.$$

Decisive for what follows is the fact that every $A \in \mathscr{R}$ satisfies

(5.6) $$\mu^*(Q) \geq \mu^*(Q\cap A) + \mu^*(A\cap\complement A) \quad\text{for every } Q \in \mathscr{P}(\Omega),$$

as well as

(5.7) $$\mu^*(A) = \mu(A).$$

In proving (5.6) we can again assume $\mu^*(Q) < +\infty$, so that $\mathscr{U}(Q) \neq \emptyset$. First of all we have

$$\sum_{n=1}^{\infty} \mu(A_n) = \sum_{n=1}^{\infty} \mu(A_n \cap A) + \sum_{n=1}^{\infty} \mu(A_n \setminus A)$$

for every sequence (A_n) from $\mathscr{U}(A)$, due to the finite additivity of μ. Moreover, the sequence $(A_n \cap A)$ lies in $\mathscr{U}(Q\cap A)$ and the sequence $(A_n \setminus A)$ lies in $\mathscr{U}(Q\setminus A)$. Consequently,

$$\sum_{n=1}^{\infty} \mu(A_n) \geq \mu^*(Q\cap A) + \mu^*(Q \setminus A)$$

for every such sequence (A_n), and from this fact (5.6) is immediate. Equality (5.7) follows on the one hand from (3.10), according to which $\mu(A) \leq \mu^*(A)$, and on the other hand from consideration of the sequence $A, \emptyset, \emptyset, \ldots$ which lies in $\mathscr{U}(A)$.

The significance of what has been proven lies in the fact, which we will establish, that the system $\mathscr{A}^*$ of all sets $A \in \mathscr{P}(\Omega)$ satisfying (5.6) is a σ-algebra in Ω and the restriction of μ^* to $\mathscr{A}^*$ is a measure. Now (5.6) as just proved says that $\mathscr{R} \subset \mathscr{A}^*$, and so we shall have $\sigma(\mathscr{R}) \subset \mathscr{A}^*$. Then according to (5.7) $\tilde{\mu} := \mu^* \mid \sigma(\mathscr{R})$ is an extension of μ to a measure on $\sigma(\mathscr{R})$. The definition and theorem which follow will therefore complete the present proof. $\square$

5.2 Definition. A numerical function μ^* on the power set $\mathscr{P}(\Omega)$ having properties (5.2)–(5.4) is called an *outer measure* on the set Ω. A subset A of Ω is called μ^*-*measurable* if it satisfies (5.6).

Notice that $\mu^* \geq 0$ always prevails, an immediate consequence of (5.2) and (5.3) together.

The idea in the proof of the measure-extension theorem, which goes back to C. CARATHÉODORY (1873–1950), consists in associating via definition (5.1) an outer measure to the premeasure μ on $\mathscr{R}$ and then invoking the following theorem.

5.3 Theorem (Carathéodory). *Let μ^* be an outer measure on a set Ω. Then the system $\mathscr{A}^*$ of all μ^*-measurable sets $A \subset \Omega$ is a σ-algebra in Ω. Moreover, the restriction of μ^* to $\mathscr{A}^*$ is a measure.*

Proof. First let us note that the requirement (5.6) for a subset A of Ω to lie in $\mathscr{A}^*$ is equivalent to

$$(5.6') \qquad\qquad \mu^*(Q) = \mu^*(Q \cap A) + \mu^*(Q \setminus A) \qquad \text{for all } Q \in \mathscr{P}(\Omega),$$

because from (5.4) applied to the sequence $Q \cap A, Q \setminus A, \emptyset, \emptyset, \ldots$ follows the reverse of inequality (5.6), for every $Q \in \mathscr{P}(\Omega)$. From either (5.6) or (5.6') it is immediate that $\Omega \in \mathscr{A}^*$, and because of their symmetry in A and $\complement A$, whenever A lies in $\mathscr{A}^*$, so does $\complement A$. The following considerations will show that with each two of its sets A and B, $\mathscr{A}^*$ also contains their union $A \cup B$, and so $\mathscr{A}^*$ is an algebra. $B \in \mathscr{A}^*$ entails that

$$\mu^*(Q) = \mu^*(Q \cap B) + \mu^*(Q \setminus B)$$

for every $Q \in \mathscr{P}(\Omega)$. Replacing Q here first by $Q \cap A$, then by $Q \setminus A = Q \cap \complement A$, we get two new equalities (valid for all $Q \in \mathscr{P}(\Omega)$) which, when inserted into (5.6'), lead to

$$\mu^*(Q) = \mu^*(Q \cap A \cap B) + \mu^*(Q \cap A \cap \complement B) + \mu^*(Q \cap \complement A \cap B) + \mu^*(Q \cap \complement A \cap \complement B).$$

Replacing Q here by $Q \cap (A \cup B)$ gives

$$(5.8) \quad \mu^*(Q \cap (A \cup B)) = \mu^*(Q \cap A \cap B) + \mu^*(Q \cap A \cap \complement B) + \mu^*(Q \cap \complement A \cap B),$$

which in conjunction with the preceding equality yields

$$\mu^*(Q) = \mu^*(Q \cap (A \cup B)) + \mu^*(Q \cap \complement A \cap \complement B) = \mu^*(Q \cap (A \cup B)) + \mu^*(Q \setminus (A \cup B)).$$

This being valid for all $Q \in \mathscr{P}(\Omega)$ affirms that $A \cup B \in \mathscr{A}^*$.

Now let (A_n) be a sequence of pairwise disjoint sets from $\mathscr{A}^*$ and A be their union. The choice of $A := A_1$, $B := A_2$ in (5.8) produces

$$\mu^*(Q \cap (A_1 \cup A_2)) = \mu^*(Q \cap A_1) + \mu^*(Q \cap A_2).$$

An induction argument generalizes this to

$$\mu^*\left(Q \cap \bigcup_{i=1}^{n} A_i\right) = \sum_{i=1}^{n} (Q \cap A_i)$$

for all $Q \in \mathscr{P}(\Omega)$, all $n \in \mathbb{N}$. Recalling that $B_n := \bigcup_{i=1}^{n} A_i$ has already been proven to be in $\mathscr{A}^*$, and that $Q \setminus B_n \supset Q \setminus A$, so that $\mu^*(Q \setminus B_n) \geq \mu^*(Q \setminus A)$, we obtain

$$\mu^*(Q) = \mu^*(Q \cap B_n) + \mu^*(Q \setminus B_n) \geq \sum_{i=1}^{n} \mu^*(Q \cap A_i) + \mu^*(Q \setminus A)$$

for all $n \in \mathbb{N}$. From this and an application of (5.4) follows

$$\mu^*(Q) \geq \sum_{n=1}^{\infty} \mu^*(Q \cap A_n) + \mu^*(Q \setminus A) \geq \mu^*(Q \cap A) + \mu^*(Q \setminus A)$$

and consequently, as noted at the beginning of the proof, we actually have equality throughout:

$$\mu^*(Q) = \sum_{n=1}^{\infty} \mu^*(Q \cap A_n) + \mu^*(Q \setminus A) = \mu^*(Q \cap A) + \mu^*(Q \setminus A),$$

holding for all $Q \in \mathscr{P}(\Omega)$. Thus A lies in $\mathscr{A}^*$. After all this we recognize that the algebra $\mathscr{A}^*$ is an $\cap$-stable Dynkin system and therefore by Theorem 2.3 a σ-algebra. If in the last pair of equalities we take $Q := A$, we get

$$\mu^*(A) = \sum_{n=1}^{\infty} \mu^*(A_n),$$

proving that the restriction of μ^* to $\mathscr{A}^*$ is a measure. $\square$

It can be further shown that in many important cases the measure $\tilde{\mu}$ from Theorem 5.1 is uniquely determined. As a preliminary we give a proof that is a typical application of the technique of Dynkin systems. (Cf. also Exercise 9.)

5.4 Theorem (Uniqueness theorem). *Let $\mathscr{E}$ be an $\cap$-stable generator of a σ-algebra $\mathscr{A}$ in Ω and suppose that (E_n) is a sequence in $\mathscr{E}$ with $\bigcup_{n \in \mathbb{N}} E_n = \Omega$. Then measures μ_1 and μ_2 on $\mathscr{A}$ which satisfy*

(i) $$\mu_1(E) = \mu_2(E) \qquad\qquad \textit{for all } E \in \mathscr{E}$$

and

(ii) $$\mu_1(E_n) = \mu_2(E_n) < +\infty \qquad\qquad \textit{for all } n \in \mathbb{N}$$

must in fact be identical.

Proof. Denote by $\mathscr{E}_f$ the system of all sets $E \in \mathscr{E}$ satisfying $\mu_1(E) = \mu_2(E) < +\infty$. For a given $E \in \mathscr{E}_f$ consider the system

$$\mathscr{D}_E := \{D \in \mathscr{A} : \mu_1(E \cap D) = \mu_2(E \cap D)\}.$$

We will show that it is a Dynkin system. Obviously $\Omega \in \mathscr{D}_E$. If $D \in \mathscr{D}_E$, then $\mu_1(E \cap D) = \mu_2(E \cap D) < +\infty$ (since $E \in \mathscr{E}_f$), and so (3.7) shows that

$$\mu_1(E \cap \complement D) = \mu_1(E \setminus E \cap D) = \mu_1(E) - \mu_1(E \cap D) = \mu_2(E) - \mu_2(E \cap D) = \mu_2(E \cap \complement D),$$

which says that $\complement D \in \mathscr{D}_E$. The remaining property of Dynkin systems (2.3) follows at once from the σ-additivity of the measures μ_1, μ_2. Because $\mathscr{E}$ is $\cap$-stable, $\mathscr{E} \subset \mathscr{D}_E$ follows from (i) and the definition of $\mathscr{D}_E$. But then $\delta(\mathscr{E}) \subset \mathscr{D}_E$ because $\delta(\mathscr{E})$ is the smallest Dynkin system which contains $\mathscr{E}$. From Theorem 2.4 however,

$\delta(\mathscr{E}) = \sigma(\mathscr{E}) = \mathscr{A}$. Therefore $\delta(\mathscr{E}) \subset \mathscr{D}_E \subset \mathscr{A}$ entails $\mathscr{D}_E = \mathscr{A}$. Thus

(5.9) $$\mu_1(E \cap A) = \mu_2(E \cap A)$$

holds for all $E \in \mathscr{E}_f$ and $A \in \mathscr{A}$. On account of (ii) then in particular

(5.9') $$\mu_1(E_n \cap A) = \mu_2(E_n \cap A) \qquad (n \in \mathbb{N}, A \in \mathscr{A}).$$

In analogy with the proofs of (3.8) and (3.10) we set

$$F_1 := E_1, \ F_2 := E_2 \setminus E_1, \ldots, \ F_n := E_n \setminus (E_1 \cup \ldots \cup E_{n-1}), \ldots,$$

and get a sequence (F_n) of pairwise disjoint sets from $\mathscr{A}$ satisfying $F_n \subset E_n$ for all $n \in \mathbb{N}$ and $\bigcup_{n \in \mathbb{N}} F_n = \bigcup_{n \in \mathbb{N}} E_n = \Omega$. Since $F_n \cap A \in \mathscr{A}$ it follows from (5.9') that

$$\mu_1(F_n \cap A) = \mu_1(E_n \cap F_n \cap A) = \mu_2(E_n \cap F_n \cap A) = \mu_2(F_n \cap A)$$

for all $A \in \mathscr{A}$ and all $n \in \mathbb{N}$. But then the fact that

$$A = \bigcup_{n \in \mathbb{N}} (F_n \cap A)$$

combines with the σ-additivity of μ_1 and μ_2 to deliver

$$\mu_1(A) = \sum_{n=1}^{\infty} \mu_1(F_n \cap A) = \sum_{n=1}^{\infty} \mu_2(F_n \cap A) = \mu_2(A)$$

for every $A \in \mathscr{A}$, which says that the measures μ_1, μ_2 are identical.   □

For *finite* measures some other natural stability properties of the generator $\mathscr{E}$ (e.g., its closure under set-differences) also insure uniqueness. See, for example, Robertson [1967].

In order to be able to formulate a useful sufficient condition for the uniqueness of the measure $\tilde{\mu}$ from Theorem 5.1, we make the

5.5 Definition. A content μ on a ring $\mathscr{R}$ in Ω is called *σ-finite* when a sequence $(A_n)_{n \in \mathbb{N}}$ of sets from $\mathscr{R}$ exists such that $\bigcup_{n \in \mathbb{N}} A_n = \Omega$ and $\mu(A_n) < +\infty$ for every $n \in \mathbb{N}$.

Examples. 1. Suppose that the content μ on the ring $\mathscr{R}$ in Ω is finite, that is, $\mu(A) < +\infty$ for every $A \in \mathscr{R}$. The σ-finiteness of μ is the equivalent to the existence of a sequential covering (A_n) of Ω by sets $A_n \in \mathscr{R}$. But the latter condition does not automatically hold, as the trivial example $\Omega \neq \emptyset$, $\mathscr{R} := \{\emptyset\}$ illustrates.

In general, the σ-finiteness of a content μ on a ring $\mathscr{R}$ is equivalent to the existence of a sequence (A'_n) of sets in $\mathscr{R}$ with $\mu(A'_n) < +\infty$ for all n and $A'_n \uparrow \Omega$. In fact, if (A_n) is merely a covering of Ω by sets in $\mathscr{R}$ having finite μ-measure, then the sets $A'_n := A_1 \cup \ldots \cup A_n$, $n \in \mathbb{N}$, furnish a sequence of the desired kind.

2. Lebesgue premeasure in $\mathbb{R}^d$ is σ-finite (as well as finite). For if we denote by $\mathbf{n}$ the point in $\mathbb{R}^d$ whose coordinates are all equal to n, then $I_n := [-\mathbf{n}, \mathbf{n}[$ is an interval from $\mathscr{I}^d$, $\lambda^d(I_n) < +\infty$ $(n \in \mathbb{N})$, and $I_n \uparrow \mathbb{R}^d$.

3.   The counting measure on a set Ω, defined in Example 6 of §3 is σ-finite (resp., finite) just when Ω is countable (resp., finite).

In summary we have

5.6 Theorem. *Every σ-finite premeasure μ on a ring $\mathscr{R}$ in a set Ω can be extended in exactly one way to a measure $\tilde{\mu}$ on $\boldsymbol{\sigma}(\mathscr{R})$.*

Proof. Only the uniqueness of $\tilde{\mu}$ has to be proved. But this follows immediately from 5.4: thanks to the σ-finiteness of μ, the ring $\mathscr{R}$ has all the properties required of the generator $\mathscr{E}$ in the hypothesis of 5.4.   □

Remark. The hypothesis of σ-finiteness of μ on 5.6 can not be dispensed with. It suffices to look, as in Example 1, at a non-empty set Ω and to take for $\mathscr{R}$ the ring consisting just of the empty set. On $\boldsymbol{\sigma}(\mathscr{R}) = \{\emptyset, \Omega\}$ two different measures having the same restriction to $\mathscr{R}$ are defined by $\mu(\emptyset) = \nu(\emptyset) := 0$ and $\mu(\Omega) := 0 =: 1 - \nu(\Omega)$.

The uniqueness of the measure $\tilde{\mu}$ which extends the σ-finite premeasure μ in 5.6 is expressed more dramatically by the following approximation property. For simplicity we formulate it only for finite measures on an algebra.

5.7 Theorem (Approximation property). *Let μ be a finite measure on a σ-algebra $\mathscr{A}$ in Ω which is generated by an algebra $\mathscr{A}_0$ in Ω. Then for each $A \in \mathscr{A}$ there is a sequence $(C_n)_{n\in\mathbb{N}}$ in $\mathscr{A}_0$ satisfying*

$$(5.10) \qquad \lim_{n\to\infty} \mu(A \triangle C_n) = 0\,.$$

Here $\triangle$ designates the symmetric difference defined in Exercise 2 of §1. Exercise 7 of §3 is the real justification for the terminology "approximation property".

Proof. Let $A \in \mathscr{A}$, $\varepsilon > 0$ be given. At issue is the existence of a $C \in \mathscr{A}_0$ with $\mu(A \triangle C) < \varepsilon$. According to 5.1 and 5.6, especially the equation (5.1) which extends $\mu|\mathscr{A}_0$ to $\mathscr{A}$, there exists a sequence $(A_n)_{n\in\mathbb{N}}$ in $\mathscr{A}_0$ which covers A and satisfies

$$(5.11) \qquad 0 \le \sum_{n=1}^{\infty} \mu(A_n) - \mu(A) < \frac{\varepsilon}{2}\,.$$

If we set $C_n := \bigcup_{i=1}^{n} A_i$, $n \in \mathbb{N}$, then $A' := \bigcup_{n\in\mathbb{N}} A_n$ satisfies

$$C_n \uparrow A' \qquad \text{and} \qquad A' \setminus C_n \downarrow \emptyset\,.$$

Since μ is finite, and consequently continuous at $\emptyset$, an $n_0 \in \mathbb{N}$ exists for which

$$(5.12) \qquad \mu(A' \setminus C_{n_0}) < \frac{\varepsilon}{2}\,.$$

Let us show that the set $C := C_{n_0} \in \mathscr{A}_0$ does what is wanted:

$$A \triangle C = (A \setminus C) \cup (C \setminus A) \subset (A' \setminus C) \cup (A' \setminus A)\,,$$

and so the subadditivity of μ yields

$$\mu(A \triangle C) \leq \mu(A' \setminus C) + \mu(A' \setminus A) = \mu(A' \setminus C) + \mu(A') - \mu(A)$$

$$\leq \mu(A' \setminus C) + \sum_{n=1}^{\infty} \mu(A_n) - \mu(A)$$

$$\leq \varepsilon/2 + \varepsilon/2, \qquad\qquad\qquad \text{by (5.11) and (5.12),}$$

which establishes the claim $\mu(A \triangle C) < \varepsilon$. □

It should also be noted that

$$(5.13) \qquad\qquad \lim_{n \to \infty} \mu(C_n) = \mu(A)$$

follows immediately from (5.10). The inequalities

$$(5.14) \qquad\qquad |\mu(C_n) - \mu(A)| \leq \mu(A \triangle C_n) \qquad\qquad (n \in \mathbb{N})$$

make this obvious: For $C, D \in \mathscr{A}$, $C \subset D \cup (C \setminus D)$, so that $\mu(C) - \mu(D) \leq \mu(C \setminus D) \leq \mu(C \triangle D)$. As C and D may be interchanged here, (5.14) is confirmed. □

Exercises.

1. Let $\mu = \varepsilon_\omega$ be the premeasure on a ring $\mathscr{R}$ in Ω defined by putting unit mass at the point $\omega \in \Omega$. Under the hypothesis that $\{\omega\}$ can be realized as the intersection of a sequence from $\mathscr{R}$ and Ω as the union of such a sequence, prove that: (a) The outer measure μ^* defined from μ via (5.1) assigns to every set $A \in \mathscr{P}(\Omega)$ the value 1 or 0, according as $\omega \in A$ or $\omega \in \complement A$. (b) Every subset of Ω is μ^*-measurable. (c) μ^* is the measure ε_ω on $\mathscr{P}(\Omega)$.

2. Consider the measure μ in Examples 2 and 7 of §3, say for $\Omega := \mathbb{R}$, and prove that: (a) The outer measure μ^* defined from μ via (5.1) assigns to every set $A \in \mathscr{P}(\Omega)$ the value 0 or 1, according as A is countable or not. (b) μ is not a measure on $\mathscr{P}(\Omega)$, not even a content. (c) The only μ^*-measurable sets are those in the σ-algebra $\mathscr{A}$ on which μ is defined.

3. Let $\mathscr{A}$ be the σ-algebra generated by an algebra $\mathscr{A}_0$ on the set Ω, μ and ν measures on $\mathscr{A}$. Show that the validity of $\mu(A) \leq \nu(A)$ for all $A \in \mathscr{A}_0$ need not imply its validity for all $A \in \mathscr{A}$. [*Hint:* $\mathscr{A}_0 := \mathscr{F}^1$, ν counting measure, $\mu := 2\nu$.] Find supplemental hypotheses that will render such an implication true.

4. Show that the sequence required in Definition 5.5 of the σ-finiteness of the content μ on the ring $\mathscr{R}$ in Ω, can always be chosen to be a sequence of pairwise disjoint sets from $\mathscr{R}$ which cover Ω and each have finite measure.

5. Let μ be a σ-finite measure on a σ-algebra $\mathscr{A}$ in Ω, and μ^* the outer measure defined by (5.1). Then to every set $Q \in \mathscr{P}(\Omega)$ corresponds an $A \in \mathscr{A}$, called a *measurable hull* of Q, with the properties that $Q \subset A$, $\mu^*(Q) = \mu(A)$, and $\mu(B) = 0$ for all $B \in \mathscr{A}$ such that $B \subset A \setminus Q$. [*Hint:* In case $\mu^*(Q) < +\infty$, show that there exists a sequence (A_n) in $\mathscr{A}$ with $Q \subset A_n$ and $\mu(A_n) < \mu^*(A) + n^{-1}$ for every $n \in \mathbb{N}$. Then $A := \bigcap_{n \in \mathbb{N}} A_n$ has the desired properties.]

6. A measure μ on a σ-algebra $\mathscr{A}$ in Ω is called *complete* if every subset of a μ-null set (cf. Exercise 5, §3) belongs to $\mathscr{A}$, and consequently is itself a μ-null set. Show that:

(a) The measure $\mu^*|\mathscr{A}^*$ from Theorem 5.3 is complete.

(b) The measure in Examples 2 and 7 of §3 is complete.

(c) If $\mathscr{A}$ is a σ-algebra in a set Ω, $\omega \in \Omega$ and $\{\omega\} \in \mathscr{A}$, then the Dirac measure ε_ω on $\mathscr{A}$ is complete just when $\mathscr{A} = \mathscr{P}(\Omega)$.

7. (a) Show that every measure μ on a σ-algebra $\mathscr{A}$ in a set Ω can be completed. That is, μ can be extended to a complete measure μ_0 on a σ-algebra $\mathscr{A}_0$ in Ω, $\mathscr{A} \subset \mathscr{A}_0$, in such a way that every complete measure μ' on a σ-algebra $\mathscr{A}'$ in Ω, $\mathscr{A} \subset \mathscr{A}'$, which extends μ is also an extension of μ_0. The (obviously unique) σ-algebra $\mathscr{A}_0$ is called the μ-*completion* of $\mathscr{A}$; the triple $(\Omega, \mathscr{A}_0, \mu_0)$ is called the *completion* of $(\Omega, \mathscr{A}, \mu)$. [For such triples the term *measure space* will be introduced in §7.]

(b) Determine the completion of $(\Omega, \mathscr{A}, \varepsilon_\omega)$ from Exercise 6(c).

(c) Show that the μ-completion $\mathscr{A}_0$ of a σ-algebra $\mathscr{A}$ in Ω consists of all sets $A \cup N$ with $A \in \mathscr{A}$ and N a subset of a μ-null set. For every such set, $\mu_0(A \cup N) = \mu(A)$.

(d) Characterize the sets in $\mathscr{A}_0$ as follows: A set $A_0 \subset \Omega$ lies in $\mathscr{A}_0$ just if sets $A_1, A_2 \in \mathscr{A}$ exist such that $A_1 \subset A_0 \subset A_2$ and $\mu(A_2 \setminus A_1) = 0$.

8. Let μ be a σ-finite measure on a σ-algebra $\mathscr{A}$ in Ω, μ^* the outer measure it determines via (5.1), and $\mathscr{A}^*$ the σ-algebra of all μ^*-measurable subsets of Ω. With the help of Exercises 5 and 7, show that $(\Omega, \mathscr{A}^*, \mu^*|\mathscr{A}^*)$ is the completion of $(\Omega, \mathscr{A}, \mu)$.

9. The proof of Theorem 5.4 only uses condition (i) for sets $A \in \mathscr{E}$ which satisfy $\mu_1(E) = \mu_2(E) < +\infty$. Clarify this observation by showing that under the hypotheses of Theorem 5.4 the system $\mathscr{E}_f$ of all sets $E \in \mathscr{E}$ satisfying $\mu_1(E) = \mu_2(E) < +\infty$ is likewise an $\cap$-stable generator of $\mathscr{A}$.

§6. Lebesgue–Borel measure and measures on the number line

We are going to pursue further the investigations in §4. So as before $\mathscr{I}^d$ will be the set of all right half-open intervals in $\mathbb{R}^d$, $\mathscr{F}^d$ the ring of all d-dimensional figures, and λ^d the Lebesgue premeasure on $\mathscr{F}^d$. We have already noted that λ^d is σ-finite. According to 5.6, λ^d can be extended in exactly one way to a measure on $\sigma(\mathscr{F}^d)$, which measure will also be denoted by λ^d from now on. Since every figure is a union of finitely many intervals $I \in \mathscr{I}^d$, we have

(6.1) $$\sigma(\mathscr{F}^d) = \sigma(\mathscr{I}^d).$$

6.1 Definition. The elements of the σ-algebra generated in $\mathbb{R}^d$ by the system $\mathscr{I}^d$ of half-open intervals are called the *Borel* subsets of the space $\mathbb{R}^d$. Correspondingly $\sigma(\mathscr{I}^d)$ is called the *σ-algebra of Borel subsets* of $\mathbb{R}^d$; it will henceforth be denoted $\mathscr{B}^d$.

The results reviewed in the introduction can, following 4.3, be expressed thus:

6.2 Theorem. *There is exactly one measure λ^d on $\mathscr{B}^d$ which assigns to every right half-open interval in $\mathbb{R}^d$ its d-dimensional elementary content.*

6.3 Definition. The measure λ^d in Theorem 6.2 is called the *Lebesgue–Borel measure* (L-B *measure*, for short) on $\mathbb{R}^d$. For every Borel set $B \in \mathscr{B}^d$, $\lambda^d(B)$ will also be called the *d-dimensional Lebesgue measure of B.*

It is expedient to expand this definition: For every set $C \in \mathscr{B}^d$ the trace σ-algebra $C \cap \mathscr{B}^d$ consists of all Borel subsets of C (cf. (1.4)). The restriction λ_C^d of λ^d to $C \cap \mathscr{B}^d$ is a measure. It will also be called the L-B *measure on C.*

Like the Lebesgue premeasure of which it is an extension, the L-B measure λ^d is σ-finite (cf. Example 2 of §5). More generally

$$(6.2) \qquad \qquad \lambda^d(B) < +\infty$$

for every bounded set $B \in \mathscr{B}^d$, since such a B lies in an interval in $\mathscr{I}^d$; e.g., excepting finitely many n, B lies in each interval I_n from Example 2, §5, with the result that $\lambda^d(B) \leq \lambda^d(I_n) < +\infty$.

Let us recall the question formulated in the introduction to Chapter I of finding a unified method for assigning a numerical measure of d-dimensional volume to as many subsets of $\mathbb{R}^d$ as possible. Step by step we will come to recognize that Theorem 6.2 answers this question in a most satisfactory way: for every Borel set B in $\mathbb{R}^d$ its d-dimensional measure in the number we were seeking.

First of all it seems desirable to get a deeper insight into the σ-algebra $\mathscr{B}^d$ of Borel sets. In particular, the question naturally comes up whether topologically interesting sets, like the open, closed, or compact ones are Borel. The characterization of $\mathscr{B}^d$ via such sets in the next theorem is often taken as the definition of the σ-algebra $\mathscr{B}^d$.

6.4 Theorem. *Let $\mathscr{O}^d$, $\mathscr{C}^d$, $\mathscr{K}^d$ denote the system of all open, closed, compact subsets of $\mathbb{R}^d$, respectively. Then*

$$(6.3) \qquad \qquad \mathscr{B}^d = \sigma(\mathscr{O}^d) = \sigma(\mathscr{C}^d) = \sigma(\mathscr{K}^d).$$

Proof. $\mathscr{K}^d \subset \mathscr{C}^d \subset \sigma(\mathscr{C}^d)$, so $\sigma(\mathscr{K}^d) \subset \sigma(\mathscr{C}^d)$. Every set $C \in \mathscr{C}^d$ is the union of a sequence of sets $C_n \in \mathscr{K}^d$; for example, if K_n are the compact balls with a fixed center and radii $n \in \mathbb{N}$, then the sets $C_n := C \cap K_n$ furnish such a sequence. Thus by (1.3), $\mathscr{C}^d \subset \sigma(\mathscr{K}^d)$, whence $\sigma(\mathscr{C}^d) \subset \sigma(\mathscr{K}^d)$ and so finally the equality of these two σ-algebras. Since the open sets are the complements of the closed ones, the equality $\sigma(\mathscr{O}^d) = \sigma(\mathscr{C}^d)$ is obvious; therewith the last two equalities in (6.3) are confirmed.

We finish up by showing that $\sigma(\mathcal{O}^d) = \mathcal{B}^d$. We will, as usual, use the term bounded open interval in $\mathbb{R}^d$ for every set of the form

$$(6.4) \qquad]a, b[:= \{x \in \mathbb{R}^d : a \lhd x \lhd b\},$$

where $a, b \in \mathbb{R}^d$ satisfy $a \leq b$. Every right half-open interval $[a, b[\in \mathscr{I}^d$ is the intersection of a sequence of bounded open intervals, namely, for

$$a := (\alpha_1, \ldots, \alpha_d) \qquad \text{and} \qquad a_n := (\alpha_1 - n^{-1}, \ldots, \alpha_d - n^{-1}) \quad (n \in \mathbb{N})$$

we have

$$]a_n, b[\downarrow [a, b[.$$

Therefore $\mathscr{I}^d \subset \sigma(\mathcal{O}^d)$ by (1.7) and consequently $\mathcal{B}^d = \sigma(\mathscr{I}^d) \subset \sigma(\mathcal{O}^d)$. Every open set in $\mathbb{R}^d$ can be exhibited as the union of countably many bounded open intervals (e.g., all those which it contains whose endpoints have only rational coordinates). Moreover, every bounded open interval $]a, b[$ is the union of a sequence of intervals from $\mathscr{I}^d$, namely

$$(6.5) \qquad [a_n, b[\uparrow]a, b[$$

if we set

$$(6.5') \qquad a_n := (\min\{\alpha_1 + n^{-1}, \beta_1\}, \ldots, \min\{\alpha_d + n^{-1}, \beta_d\}) \qquad (n \in \mathbb{N}),$$

$\alpha_1, \ldots, \alpha_d$ and $\beta_1, \ldots, \beta_d$ being the coordinates of a and b, respectively. Every open set is therefore the union of a sequence of intervals from $\mathscr{I}^d$, and so $\mathcal{O}^d \subset \sigma(\mathscr{I}^d) = \mathcal{B}^d$. It thus follows that $\sigma(\mathcal{O}^d) \subset \mathcal{B}^d$ and, as the reverse inequality has already been established, equality $\mathcal{B}^d = \sigma(\mathcal{O}^d)$ is confirmed. $\square$

We will become acquainted with some deeper properties of L-B measure in §8. In particular, there the existence of non-Borel sets, that is, the assertion

$$\mathcal{B}^d \neq \mathscr{P}(\mathbb{R}^d)$$

will be proved. For the moment we content ourselves with computing the Lebesgue measure $\lambda^d(B)$ of some geometrically simple Borel sets B.

Examples. 1. *Every hyperplane H orthogonal to one of the coordinate axes in $\mathbb{R}^d$ is an L-B-null set*, i.e., a Borel set with $\lambda^d(H) = 0$. Let, say H be orthogonal to the ith coordinate axis, $i \in \{1, \ldots, d\}$, that is, be of the form

$$(6.6) \qquad H := \{x = (\xi_1, \ldots, \xi_d) \in \mathbb{R}^d : \xi_i = \alpha\}$$

for an appropriate $\alpha \in \mathbb{R}$. H is a closed set, and so is Borel. For each $n \in \mathbb{N}$, let x_n, y_n be those points in $\mathbb{R}^d$ whose coordinates are $-n$ or n, respectively, at every index except i and whose i^{th} coordinates are α or $\alpha + 2^{-n}(2n)^{1-d}\varepsilon$, respectively, where $\varepsilon > 0$. Evidently

$$H \subset \bigcup_{n \in \mathbb{N}} [x_n, y_n[$$

and

$$\lambda^d([x_n, y_n[) = 2^{-n}\varepsilon, \qquad\qquad n \in \mathbb{N}.$$

From (3.10) we therefore get

$$\lambda^d(H) \leq \sum_{n=1}^{\infty} \lambda^d([x_n, y_n[) = \varepsilon.$$

Since this is true for every $\varepsilon > 0$, $\lambda^d(H) = 0$ follows.

Due to the isotoneity of measures, we consequently also have $\lambda^d(B) = 0$ for every Borel subset B of such a hyperplane H.

2. *Every countable subset of $\mathbb{R}^d$ is an* L-B-*null set.* Because of σ-additivity of measures, it suffices to treat the case of one-point sets $\{x\} \subset \mathbb{R}^d$. Being a closed set, it is Borel; moreover for an appropriate hyperplane H of the form (6.6) we have $\{x\} \subset H$.

3. For points $a, b \in \mathbb{R}^d$ with $a \leq b$ consider besides the intervals $[a, b[$ and $]a, b[$ already defined, the compact interval

$$[a, b] := \{x \in \mathbb{R}^d : a \leq x \leq b\}$$

and, in contrast to $[a, b[$, the left half-open interval

$$]a, b] := \{x \in \mathbb{R}^d : a \lhd x \leq b\}.$$

Then

(6.7) $$\lambda^d([a, b[) = \lambda^d(]a, b[) = \lambda^d([a, b]) = \lambda^d(]a, b]).$$

First of all the intervals $[a, b[$, $]a, b[$ and $[a, b]$ are Borel sets by Theorem 6.4. As in its proof, we can show that

(6.8) $$]a, b_n[\downarrow]a, b] \quad \text{and} \quad [a, b_n[\downarrow [a, b]$$

for appropriate sequences (b_n) in $\mathbb{R}^d$ converging to b. Again from 6.4 we then get that $]a, b]$ is Borel. From (6.5) follows

$$\lambda^d(]a, b[) = \lim_{n \to \infty} \lambda^d([a_n, b[) = \lambda^d([a, b[)$$

the first equality using the continuity from below of a measure, and the second using $\lim_{n \to \infty} a_n = a$ (from (6.5′)) and the continuous dependence on c and d of the elementary content of the interval $[c, d[$. Analogously, with the help of (6.8), we conclude that

$$\lambda^d(]a, b]) = \lambda^d([a, b]),$$

this time citing the continuity of measures from above. Thus finally from the inclusions $]a, b[\subset]a, b] \subset [a, b]$ the remaining equality in (6.7) follows.

The choice of right half-open intervals for the construction of λ^d is now seen to have been due solely to the fact that the ring $\mathscr{F}^d$ they generate is so simple to describe.

In a second step a large class of measures on the σ-algebra $\mathscr{B}^1$ of Borel subsets of the line will now be presented. These are the Borel measures. In general for $d \in \mathbb{N}$, a measure μ defined on $\mathscr{B}^d$ is called a *Borel measure on* $\mathbb{R}^d$ if

$$\mu(K) < +\infty \qquad \text{for every compact } K \subset \mathbb{R}^d$$

or, equivalently, if $\mu(B) < +\infty$ for every *bounded* set $B \in \mathscr{B}^d$. L-B measure λ^d is such a measure, according to (6.2).

The point of departure for defining λ^1 is the determination of $\lambda^1([a, b[)$ for intervals $[a, b[\in \mathscr{I}^1$, namely as $b - a$. It suggests itself that this opening move might be generalized as follows: One has a function $F : \mathbb{R} \to \mathbb{R}$ and asks for conditions on it which guarantee the existence of a measure μ on $\mathscr{B}^1$ with the property

$$(6.9) \qquad \mu([a, b[) = F(b) - F(a) \qquad \text{for all } a, b \in \mathbb{R} \text{ with } a \le b.$$

Thanks to the uniqueness theorem 5.4 such a measure is already thereby, i.e., by its values on $\mathscr{I}^1$, *uniquely* specified. Since $\mu([a, b[) \ge 0$, (6.9) entails that the function F must be isotone. Moreover, F has to be left-continuous. This is because for every $x \in \mathbb{R}$ and every sequence (x_n) in $\mathbb{R}$ with $x_n \uparrow x$, the corresponding interval behavior is $[x_1, x_n[\uparrow [x_1, x[$, and since μ must be continuous from below, it follows that

$$\lim_{n \to \infty} F(x_n) - F(x_1) = \lim_{n \to \infty} \mu([x_1, x_n[) = \mu([x_1, x[) = F(x) - F(x_1)$$

that is, $\lim_{n \to \infty} F(x_n) = F(x)$, F is left-continuous at x.

Functions $F : \mathbb{R} \to \mathbb{R}$ which are *isotone and left-continuous* will be called *measure-generating* (or *measure-defining*) *functions* (on $\mathbb{R}$). Of course, whenever F is such a function, so is $aF + b$ for any $a \in \mathbb{R}_+$, $b \in \mathbb{R}$. The designation "measure-generating" is justified by the next theorem, which answers completely the earlier question of what are the appropriate conditions on F.

6.5 Theorem. *To every measure-generating function F on $\mathbb{R}$ there corresponds exactly one measure μ_F on $\mathscr{B}^1$ having property (6.9), that is, satisfying*

$$(6.9') \qquad \mu_F([a, b[) = F(b) - F(a) \qquad \text{for all } [a, b[\in \mathscr{I}^1.$$

The measure μ_G determined by the measure-generating function G satisfies $\mu_G = \mu_F$ if and only if $G = F + c$ for some constant $c \in \mathbb{R}$. Every μ_F is a Borel measure on $\mathbb{R}$, and every Borel measure on $\mathbb{R}$ is a μ_F for an appropriate F.

Proof. The techniques employed in the proof of Theorem 4.3 can be repeated to show that corresponding to F there is a unique content μ on the ring $\mathscr{F}^1$ of 1-dimensional figures which has property (6.9). That part of the proof used only the isotoneity of F. From the left-continuity of F it follows that for every $I = [a, b[\in \mathscr{I}^1$ and every $\varepsilon > 0$ there is a $J = [a, c[\in \mathscr{I}^1$ with $\overline{J} \subset I$ and

$$\mu(I) - \mu(J) = \mu([c, b[) = F(b) - F(c) \le \varepsilon.$$

But then the technique employed in the proof of Theorem 4.4 shows that μ is a σ-finite (as well as finite) premeasure on $\mathscr{F}^1$.

According to 5.6 it can be extended in exactly one way to a measure on $\mathscr{B}^1$. This measure does what is wanted, is a μ_F. Its uniqueness with respect to its prescription on $\mathscr{I}^1$ via F was settled in the deliberations preceding the present theorem. From $\mu_F = \mu_G$ we get $G(b) - G(a) = F(b) - F(a)$ whenever $a \leq b$. Upon applying this with $a = 0 \leq b$ as well as with $a < 0 = b$, we learn that $G = F + c$, with $c := G(0) - F(0)$. Every μ_F is a Borel measure, because every bounded $B \in \mathscr{B}^1$ is contained in $[-n, n[$ for some $n \in \mathbb{N}$ and so $\mu_F(B) \leq \mu_F([-n, n[) = F(n) - F(-n) < +\infty$.

If conversely, μ is an arbitrary Borel measure on $\mathbb{R}$, we can define

$$F(x) := \begin{cases} \mu([0, x[) & \text{if } x \geq 0 \\ -\mu([x, 0[) & \text{if } x < 0 \end{cases}$$

and get a function on $\mathbb{R}$ having property (6.9) and therewith, in light of the discussion preceding this proof, measure-generating. In fact, for real numbers $0 \leq a \leq b$ the subtractivity (3.7) of measures entails that

$$\mu([a, b[) = \mu([0, b[\setminus [0, a[) = F(b) - F(a),$$

and (6.9) is confirmed analogously when $a \leq b \leq 0$. In the remaining case $a < 0 < b$ we get (6.9) from $[a, b[= [a, 0[\cup [0, b[$ and the additivity of μ. The uniqueness already proved leads finally to the equality of μ with the measure μ_F derived from F. $\square$

Notice that L-B measure λ^1 has the form μ_F, with F the identity map $x \mapsto x$ on $\mathbb{R}$.

Of special importance are the finite measures on $\mathscr{B}^1$. Every one is a Borel measure on $\mathbb{R}$. Because $0 \leq \mu(B) \leq \mu(\mathbb{R}) < +\infty$ for all $B \in \mathscr{B}^1$, a *finite Borel measure* μ on $\mathbb{R}$ is either the zero measure $\mu = 0$, or $0 < \mu(\mathbb{R}) < +\infty$ and $\nu := \dfrac{1}{\mu(\mathbb{R})}\mu$ is a measure on $\mathscr{B}^1$ with $\nu(\mathbb{R}) = 1$. Measures normalized this way play a fundamental role in probability theory. This explains the following vocabulary:

A measure μ on a σ-algebra $\mathscr{A}$ in a set Ω is called a *probability measure* (abbreviated to *p-measure*) if $\mu(\Omega) = 1$. Because of the isotoneity property every p-measure satisfies

(6.10) $$0 \leq \mu(A) \leq 1 = \mu(\Omega) \qquad \text{for all } A \in \mathscr{A}.$$

Consider now a p-measure μ on $\mathscr{B}^1$. The open interval $[-\infty, x[$ lies in $\mathscr{B}^1$ for each $x \in \mathbb{R}$, so a real function F_μ with values in $[0, 1]$ is defined by

(6.11) $$F_\mu(x) := \mu(] - \infty, x[) \qquad (x \in \mathbb{R}).$$

It is called the *distribution function* of μ. For example, the distribution of the Dirac measure ε_0 equals 0 throughout $] - \infty, 0]$ and 1 throughout $]0, +\infty[$.

Since $] - \infty, b[\setminus] - \infty, a[= [a, b[$ whenever $a \leq b$,

$$\mu([a, b[) = F_\mu(b) - F_\mu(a) \qquad \text{for all } [a, b[\in \mathscr{I}^1.$$

Therefore (6.11) uniquely defines a measure-generating function, which obviously satisfies

(6.12)
$$\mu_{F_\mu} = \mu$$

in the notation introduced in Theorem 6.5. Among the infinitely many measure-generating functions F that satisfy $\mu_F = \mu$ for a given p-measure μ the distribution function F_μ is characterized as follows:

6.6 Theorem. *A real function F on $\mathbb{R}$ is the distribution function of a – necessarily uniquely determined – p-measure μ on $\mathscr{B}^1$ if and only if it is measure-generating (that is, isotone and left-continuous) and satisfies*

(6.13)
$$\lim_{x \to -\infty} F(x) = 0 \quad and \quad \lim_{x \to +\infty} F(x) = 1\,.$$

Proof. The distribution function F_μ of a p-measure μ on $\mathscr{B}^1$ is always measure-generating, as (6.12) shows. Properties (6.13) follow from the continuity at $\emptyset$ and the continuity from below of every finite measure, respectively, since for sequences (x_n) in $\mathbb{R}$ with $x_n \downarrow -\infty$, resp., $x_n \uparrow +\infty$ we have $]-\infty, x_n[\downarrow \emptyset$, resp., $]-\infty, x_n[\uparrow \mathbb{R}$.

If conversely F is a measure-generating function satisfying (6.13), then according to 6.5 μ_F is the only Borel measure on $\mathbb{R}$ with property (6.9), in particular, with $\mu_F([-n, n[) = F(n) - F(-n)$ for all $n \in \mathbb{N}$. When $n \to +\infty$ here, the normalization condition $\mu(\mathbb{R}) = 1$ follows from (6.13). Thus μ_F is a probability measure. F is then the distribution function of μ_F, because for $x \in \mathbb{R}$ and all $n \in \mathbb{N} \cap [-x, +\infty[$

$$\mu_F([-n, x[) = F(x) - F(-n) \quad \text{and} \quad [-n, x[\uparrow]-\infty, x[$$

so that

$$F(x) = \lim_{n \to \infty} \mu_F([-n, x[) + \lim_{n \to \infty} F(-n) = \mu(]-\infty, x[) = F_{\mu_F}(x)\,. \quad \square$$

Via $\mu \mapsto F_\mu$ the set of p-measures on $\mathscr{B}^1$ is thus bijectively mapped onto the set of measure-generating functions F on $\mathbb{R}$ having property (6.13). This is the significance of the preceding theorem.

Remarks. 1. Measure-generating functions are also called "*Stieltjes measure functions*". This is because, even before the invention of the measure concept, T.J. STIELTJES (1856–1894) had used such functions to extend the ideas behind the Riemann integral (cf. Remark 2 in §12).

2. Measure-generating functions (and distribution functions) also make sense in $\mathbb{R}^d$. But they are difficult to deal with and that is not the least reason why they are of less significance. A function $F : \mathbb{R}^d \to \mathbb{R}$ is called *measure-generating* if in each of its d variables $\xi_1, \ldots, \xi_d$, when the others are held fixed, it is left-continuous and satisfies the additional condition

$$\Delta_{\alpha_d}^{\beta_d} \ldots \Delta_{\alpha_1}^{\beta_1} F \geq 0 \qquad \text{for all } a, b \in \mathbb{R}^d \text{ with } a \leq b.$$

Here α_k, β_k $(k = 1, \ldots, d)$ are the coordinates of a, b, resp., and $\Delta_{\alpha_1}^{\beta_1} F$ is the function defined on $\mathbb{R}^{d-1}$ via $(\xi_1, \ldots, \xi_d) \mapsto F_2(\xi_2, \ldots, \xi_d) := F(\beta_1, \xi_2, \ldots, \xi_d) - F(\alpha_1, \xi_2, \ldots, \xi_d)$. Then $\Delta_{\alpha_2}^{\beta_2} F_2 = \Delta_{\alpha_2}^{\beta_2} \Delta_{\alpha_1}^{\beta_1} F$ is defined and the further "difference operators" $\Delta_{\alpha_k}^{\beta_k}$ are inductively brought into play. There is a theorem analogous to 6.5: To every measure-generating function F on $\mathbb{R}^d$ corresponds a unique Borel measure μ_F on $\mathbb{R}^d$ which satisfies the iterated difference condition

$$(6.14) \qquad \mu_F([a, b[) = \Delta_{\alpha_d}^{\beta_d} \ldots \Delta_{\alpha_1}^{\beta_1} F \qquad \text{for all } [a, b[\in \mathscr{I}^d.$$

For $d = 1$ this reduces simply to $(6.9')$. As an example, for the function $F_0(\xi_1, \ldots, \xi_d) := \xi_1 \cdot \ldots \cdot \xi_d$

$$\Delta_{\alpha_d}^{\beta_d} \ldots \Delta_{\alpha_1}^{\beta_1} F_0 = (\beta_1 - \alpha_1) \cdot \ldots \cdot (\beta_d - \alpha_d) \qquad \text{for } a, b \in \mathbb{R}^d \text{ with } a \leq b.$$

This function is consequently measure-generating, and generates the L-B measure λ^d in the sense that $\mu_{F_0} = \lambda^d$. Details can be found in RICHTER [1966], TUCKER [1967] and GNEDENKO [1988].

Exercises.

1. Prove that a Borel set $B \in \mathscr{B}^d$ is an L-B-null set if and only if one of the two following conditions (which are hence equivalent) is satisfied: (a) For every $\varepsilon > 0$ there is a covering of B by countably many open intervals $I_n \subset \mathbb{R}^d$ such that $\sum_{n=1}^{\infty} \lambda^d(I_n) < \varepsilon$. (b) There is a covering of B by countably many open intervals I_n such that $\sum_{n=1}^{\infty} \lambda^d(I_n) < +\infty$ and every point of B lies in I_n for infinitely many n. Both characterizations remain valid if the I_n are allowed to be half-open or compact, instead of open. [*Hint* for (a): Utilize (5.1).]

2. Write $\mathbb{R}^d$ in the form $\mathbb{R}^d = \mathbb{R}^p \times \mathbb{R}^q$ with $p, q \in \mathbb{N}$, $p + q = d$, by grouping the first p coordinates of a point $x \in \mathbb{R}^d$ into a point in $\mathbb{R}^p$ and the last q coordinates into a point in $\mathbb{R}^q$. Denoting by $\mathbf{0}$ the zero of the vector space $\mathbb{R}^q$, show that for a set $A \subset \mathbb{R}^p$, $A \times \{\mathbf{0}\} \in \mathscr{B}^d$ precisely when $A \in \mathscr{B}^p$.

3. Let μ be a p-measure on $\mathscr{B}^1$ and F_μ its distribution function. Show that F_μ is continuous at the point $x \in \mathbb{R}$ just if $\mu(\{x\}) = 0$.

4. Determine the p-measure on $\mathscr{B}^1$ which has $x \mapsto 0 \vee (x \wedge 1)$ as distribution function, and answer anew the question in Exercise 1 of §4.

5. Show that every σ-finite measure μ on $\mathscr{B}^d$ can be represented in the form $\mu = \sum_{n=1}^{\infty} \alpha_n \mu_n$, where for each $n \in \mathbb{N}$, $\alpha_n \in \mathbb{R}_+$ and μ_n is a p-measure on $\mathscr{B}^d$. The supplemental condition that for every bounded set $B \in \mathscr{B}^d$, $\mu_n(B) \neq 0$ for only finitely many $n \in \mathbb{N}$ can be imposed if and only if μ is a Borel measure.

§7. Measurable mappings and image measures

The following considerations can be more simply formulated if we introduce some shorthand terminology. If Ω is a set and $\mathscr{A}$ a σ-algebra in Ω, the pair $(\Omega, \mathscr{A})$ will be called a *measurable space* and the sets in $\mathscr{A}$ *measurable sets*. If in addition a measure μ is defined on the σ-algebra $\mathscr{A}$, then the triple $(\Omega, \mathscr{A}, \mu)$ arising from the measurable space $(\Omega, \mathscr{A})$ is called a *measure space* (cf. Exercise 7 of §5). If μ is a p-measure, the measure space $(\Omega, \mathscr{A}, \mu)$ is called a *probability space* (*p-space* for short). Correspondingly, one speaks of a σ-*finite* measure space $(\Omega, \mathscr{A}, \mu)$ if the measure μ is σ-finite.

The measurable space $(\mathbb{R}^d, \mathscr{B}^d)$ will henceforth be called the d-dimensional *Borel measurable space*. The measure space $(\mathbb{R}^d, \mathscr{B}^d, \lambda^d)$ will correspondingly be called the d-dimensional *Lebesgue–Borel measure space* abbreviated to *L-B measure space*).

The concept measurable space exhibits a formal analogy to that of topological space. For a topological space is also a pair, consisting of a set and a system of its subsets, namely, the open ones. In the sense of this analogy the next concept, that of a measurable mapping, corresponds to the concept of continuity in topology.

7.1 Definition. Let $(\Omega, \mathscr{A})$ and $(\Omega', \mathscr{A}')$ be measurable spaces, and $T : \Omega \to \Omega'$ a mapping of Ω into Ω'. T is called $\mathscr{A}$-$\mathscr{A}'$-*measurable* if

$$(7.1) \qquad\qquad T^{-1}(A') \in \mathscr{A} \qquad\qquad \text{for every } A' \in \mathscr{A}'.$$

We express the $\mathscr{A}$-$\mathscr{A}'$-measurability of T symbolically by

$$T : (\Omega, \mathscr{A}) \to (\Omega', \mathscr{A}')$$

and speak of a measurable mapping of the first measurable space into the second. Using the notation introduced in (1.5), (7.1) can be written as

$$(7.1') \qquad\qquad T^{-1}(\mathscr{A}') \subset \mathscr{A}.$$

Examples. 1. Every *constant* mapping $T : \Omega \to \Omega'$ is $\mathscr{A}$-$\mathscr{A}'$-measurable.

2. Every *continuous* mapping $T : \mathbb{R}^d \to \mathbb{R}^{d'}$ $(d, d' \in \mathbb{N})$ is $\mathscr{B}^d$-$\mathscr{B}^{d'}$-measurable, briefly put, *Borel measurable*. According to 6.4 the system $\mathscr{O}^{d'}$ of all open subsets of $\mathbb{R}^{d'}$ is a generator of $\mathscr{B}^{d'}$. Because of the continuity of T, $T^{-1}(O) \in \mathscr{O}^d \subset \mathscr{B}^d$ for every $O \in \mathscr{O}^{d'}$. The asserted measurability of T therefore follows from the next theorem.

7.2 Theorem. Let $(\Omega, \mathscr{A})$ and $(\Omega', \mathscr{A}')$ be measurable spaces; further, let $\mathscr{E}'$ be a generator of $\mathscr{A}'$. A mapping $T : \Omega \to \Omega'$ is measurable just if

$$(7.2) \qquad\qquad T^{-1}(E') \in \mathscr{A} \qquad\qquad \text{for every } E' \in \mathscr{E}'.$$

Proof. The system $\mathcal{Q}'$ of all sets $Q \in \mathscr{P}(\Omega')$ for which $T^{-1}(Q') \in \mathscr{A}$ is a σ-algebra in Ω'. Consequently, $\mathscr{A}' \subset \mathcal{Q}'$ holds just if $\mathscr{E}' \subset \mathcal{Q}'$ does. $\mathscr{A}' \subset \mathcal{Q}'$ is equivalent to the measurability of T, while $\mathscr{E}' \subset \mathcal{Q}'$ is equivalent to (7.2). $\square$

Concerning the composition of measurable mappings, what the earlier analogy with topology suggests, prevails:

7.3 Theorem. *If $T_1 : (\Omega_1, \mathscr{A}_1) \to (\Omega_2, \mathscr{A}_2)$ and $T_2 : (\Omega_2, \mathscr{A}_2) \to (\Omega_3, \mathscr{A}_3)$ are measurable mappings, then the composite mapping $T_2 \circ T_1$ is $\mathscr{A}_1$-$\mathscr{A}_3$-measurable.*

Proof. The claim follows from the validity of the equation $(T_2 \circ T_1)^{-1}(A) = T_1^{-1}(T_2^{-1}(A))$ for all $A \in \mathscr{P}(\Omega_3)$, in particular, from its validity for all $A \in \mathscr{A}_3$. $\square$

Next consider a family of measurable spaces $((\Omega_i, \mathscr{A}_i))_{i \in I}$ and a family $(T_i)_{i \in I}$ of mappings $T_i : \Omega \to \Omega_i$ of some fixed set Ω into the individual sets Ω_i. Obviously the σ-algebra in Ω generated by $\bigcup_{i \in I} T_i^{-1}(\mathscr{A}_i)$ is the smallest σ-algebra $\mathscr{A}$ with respect to which every T_i is $\mathscr{A}$-$\mathscr{A}_i$-measurable. We designate this σ-algebra $\boldsymbol{\sigma}(T_i : i \in I)$, that is, we define

$$(7.3) \qquad \boldsymbol{\sigma}(T_i : i \in I) := \boldsymbol{\sigma}\left(\bigcup_{i \in I}(T_i^{-1}(\mathscr{A}_i))\right)$$

and call it the *σ-algebra generated by the mappings T_i* (and the measurable spaces $(\Omega_i, \mathscr{A}_i)$). In the case of the finite index set $I := \{1, \ldots, n\}$, we also use the notation $\boldsymbol{\sigma}(T_1, \ldots, T_n)$.

For $n = 1$ we clearly have $\boldsymbol{\sigma}(T_1) = T_1^{-1}(\mathscr{A}_1)$. If therefore a σ-algebra $\mathscr{A}$ in a set Ω is given, then a mapping $T_1 : \Omega \to \Omega_1$ being $\mathscr{A}$-$\mathscr{A}_1$-measurable *is equivalent to*

$$(7.4) \qquad\qquad \boldsymbol{\sigma}(T_1) \subset \mathscr{A}.$$

Cf. (7.1').

As a further application of 7.2 we will demonstrate:

7.4 Theorem. *Let $(T_i)_{i \in I}$ be a family of mappings $T_i : \Omega \to \Omega_i$ of a set Ω into measurable spaces $(\Omega_i, \mathscr{A}_i)$. Further, let $S : \Omega_0 \to \Omega$ be a mapping of a measurable space $(\Omega_0, \mathscr{A}_0)$ into Ω. The mapping S is then $\mathscr{A}_0$-$\boldsymbol{\sigma}(T_i : i \in I)$-measurable if and only if each mapping $T_i \circ S$ $(i \in I)$ is $\mathscr{A}_0$-$\mathscr{A}_i$-measurable.*

Proof. According to Theorem 7.3 the condition is necessary. The following considerations show that it is also sufficient. By (7.3) the system

$$\mathscr{E} := \bigcup_{i \in I} T_i^{-1}(\mathscr{A}_i)$$

is a generator of $\boldsymbol{\sigma}(T_i : i \in I)$. Each set $E \in \mathscr{E}$ has the form $E = T_i^{-1}(A_i)$ for some $i \in I$, $A_i \in \mathscr{A}_i$. Thus $S^{-1}(E) = (T_i \circ S)^{-1}(A_i) \in \mathscr{A}_0$ because of the hypothesized measurability of $T_i \circ S$. From 7.2 therefore, S is $\mathscr{A}_0$-$\boldsymbol{\sigma}(T_i : i \in I)$-measurable. $\square$

Finally, with the aid of measurable mappings, measures can be mapped:

7.5 Theorem. *Let $T : (\Omega, \mathscr{A}) \to (\Omega', \mathscr{A}')$ be a measurable mapping. Then for every measure μ on $\mathscr{A}$,*

(7.5) $$A' \mapsto \mu(T^{-1}(A'))$$

defines a measure μ' on $\mathscr{A}'$.

Proof. We only have to observe that for every sequence $(A'_n)_{n \in \mathbb{N}}$ of pairwise disjoint sets from $\mathscr{A}'$, $(T^{-1}(A'_n))_{n \in \mathbb{N}}$ is a sequence of pairwise disjoint sets from $\mathscr{A}$, and that

$$T^{-1}\left(\bigcup_{n \in \mathbb{N}} A'_n\right) = \bigcup_{n \in \mathbb{N}} T^{-1}(A'_n). \quad \square$$

7.6 Definition. In the situation described in 7.5, the measure μ' is called the *image of μ under the mapping T* and is denoted by $T(\mu)$.

Thus according to this definition

(7.5') $$T(\mu)(A') := \mu(T^{-1}(A')) \qquad \text{for all } A' \in \mathscr{A}'.$$

The formation of image measures is *transitive*, that is,

(7.6) $$(T_2 \circ T_1)(\mu) = T_2(T_1(\mu)),$$

whenever we are in the situation of 7.3 and μ is a measure on $\mathscr{A}_1$: For every $A \in \mathscr{A}_2$, $T := T_2 \circ T_1$ satisfies $T^{-1}(A) = T_1^{-1}(T_2^{-1}(A))$, and $T_2^{-1}(A) \in \mathscr{A}_2$. Therefore, setting $\mu' := T_1(\mu)$, $\mu'' := T_2(\mu')$ for short, it follows that

$$T(\mu)(A) = \mu(T_1^{-1}(T_2^{-1}(A))) = \mu'(T_2^{-1}(A)) = \mu''(A),$$

for all $A \in \mathscr{A}_3$, showing that $T(\mu) = \mu''$ and confirming (7.6).

Examples. 3. Let $(\Omega, \mathscr{A}) = (\Omega', \mathscr{A}') := (\mathbb{R}^d, \mathscr{B}^d)$ be the d-dimensional Borel measurable space and $\mu := \lambda^d$ the associated L-B measure. For every point $a \in \mathbb{R}^d$, the *translation mapping* $T_a : \mathbb{R}^d \to \mathbb{R}^d$ is defined by

$$T_a(x) := a + x \qquad\qquad x \in \mathbb{R}^d.$$

It is continuous and so (Example 2) measurable. We inquire into the image measure $\lambda' := T_a(\lambda^d)$.

The mapping T_a is bijective, and $T_a^{-1} = T_{-a}$. So for every interval $[b, c[\in \mathscr{I}^d$, $T_a^{-1}([b, c[) = [b - a, c - a[$, whence $\lambda'([b, c[) = \lambda^d([b - a, c - a[) = \lambda^d([b, c[)$. Both measures λ^d and λ' thus assign to every interval from $\mathscr{I}^d$ its d-dimensional elementary content. According to 6.2 therefore $\lambda^d = \lambda'$, that is,

(7.7) $$T_a(\lambda^d) = \lambda^d \qquad \text{for every } a \in \mathbb{R}^d.$$

This property of λ^d is called its *translation-invariance*. If we set, as is customary

(7.8) $$a + A = A + a := T_a(A) = \{a + x : x \in A\}$$

for sets $A \in \mathscr{P}(\mathbb{R}^d)$ and points $a \in \mathbb{R}^d$, then $T_a(\lambda^d)(A) = \lambda^d(-a+A)$ for arbitrary $A \in \mathscr{B}^d$. Property (7.7) can therefore also be expressed as

$$(7.7') \qquad\qquad \lambda^d(a + A) = \lambda^d(A) \qquad\qquad \text{for all } A \in \mathscr{B}^d, \, a \in \mathbb{R}^d.$$

4. In the context of Example 3, each non-zero real number α and each $i \in \{1, \ldots, d\}$ determine a continuous, hence Borel measurable, linear mapping $D_\alpha^{(i)}$ which assigns to the point $x = (x_1, \ldots, x_d) \in \mathbb{R}^d$ the image point $x' \in \mathbb{R}^d$ having coordinates $x_i' := \alpha x_i$, and $x_j' = x_j$ for all $j \neq i$, a *dilation* of x. It satisfies

$$(7.9) \qquad\qquad D_\alpha^{(i)}(\lambda^d) = |\alpha|^{-1} \lambda^d.$$

For, every open interval $]a, b[\subset \mathbb{R}^d$ has $D_\alpha^{(i)}$-pre-image equal to $]a', b'[$, where the coordinates of a', b' except the ith are those of a, b, the ith being α^{-1} times those of a, b if $\alpha > 0$, and α^{-1} times those of b, a if $\alpha < 0$. Hence

$$\lambda^d((D_\alpha^{(i)})^{-1}(]a, b[)) = |\alpha|^{-1} \lambda^d(]a, b[).$$

$D_\alpha^{(i)}(\lambda^d)$ and $|\alpha|^{-1} \lambda^d$ are therefore measures on $\mathscr{B}^d$ which coincide on all bounded open intervals. Thanks to 6.4 such intervals constitute a generator of $\mathscr{B}^d$, which obviously has with respect to each of these measures all the properties of the generator $\mathscr{E}$ in the uniqueness Theorem 5.4. From that theorem (7.9) therefore follows.

5. If we set $H_r := D_r^{(1)} \circ \ldots \circ D_r^{(d)}$ for real $r \neq 0$, we obtain the linear mapping $H_r(x) = rx$ $(x \in \mathbb{R}^d)$, called a *homothety*. Because of the transitivity of image measures, it follows from (7.9) that

$$(7.10) \qquad\qquad H_r(\lambda^d) = |r|^{-d} \lambda^d.$$

For $r = -1$ we get $H_{-1}(\lambda^d) = \lambda^d$. Because H_{-1} is reflection through the origin, this property is called the *reflection-invariance* of λ^d.

Exercises.

1. For $\Omega := \mathbb{R}$, let $(\Omega, \mathscr{A}, \mu)$ be the measure space of Example 2, §3. For $\Omega' := \{0, 1\}$ and $\mathscr{A}' := \mathscr{P}(\Omega)$ define the mapping $T : \Omega \to \Omega'$ by $T(\omega) := 0$ if ω is rational, $T(\omega) := 1$ if ω is irrational. Show that T is $\mathscr{A}$-$\mathscr{A}'$-measurable and determine the image measure $T(\mu)$.

2. Show that for any sets Ω, Ω', any mapping $T : \Omega \to \Omega'$, and any system of sets $\mathscr{E}' \subset \mathscr{P}(\Omega')$, $T^{-1}(\sigma(\mathscr{E}')) = \sigma(T^{-1}(\mathscr{E}'))$.

3. Let K be a compact subset of $\mathbb{R}^d$ with the property that the intersection $H_r(K) \cap H_{r'}(K)$ of every two homothetic images of K with $0 < r < r' < 1$ is an L-B-null set. (This property is enjoyed by every sphere $S_\alpha(\mathbf{0})$ of radius $\alpha > 0$ and center $\mathbf{0} := (0, \ldots, 0)$, that is, the set of $x \in \mathbb{R}^d$ having euclidean distance α from $\mathbf{0}$.) Show that $\lambda^d(K) = 0$. [*Hint:* For all $r \in]0, 1]$, $H_r(K) \subset \tilde{K} := \{tx : 0 \leq t \leq 1, x \in K\}$, which is a compact set. Hence $\lambda^d(\tilde{K}) < +\infty$.]

4. Let $\mathbb{T} := \{(x, y) \in \mathbb{R}^2 : x^2 + y^2 = 1\}$ denote the unit circle, that is, the sphere $S_1(\mathbf{0})$ in $\mathbb{R}^2$. Prove the existence of a finite non-zero measure ν on the σ-algebra $\mathscr{B}(\mathbb{T}) := \mathbb{T} \cap \mathscr{B}^2$ which is invariant under all rotations of $\mathbb{T}$. [*Hint:* Take for ν an image of λ_C^1 for an appropriate interval $C \subset \mathbb{R}$.]

§8. Mapping properties of the Lebesgue–Borel measure

L-B measure λ^d on $\mathscr{B}^d$ is, as was shown in Example 3 of the preceding section, translation-invariant. Of the greatest significance is the fact that λ^d is uniquely determined by this invariance property, together with a simple normalization. For the d-dimensional *unit cube*, defined by

$$(8.1) \qquad\qquad W := [\mathbf{0}, \mathbf{1}[\,,$$

where $\mathbf{0} = (0, \ldots, 0) \in \mathbb{R}^d$ and $\mathbf{1} := (1, \ldots, 1) \in \mathbb{R}^d$, 6.2 insures that

$$(8.2) \qquad\qquad \lambda^d(W) = 1 \,.$$

Along with λ^d each non-negative multiple $\alpha\lambda^d$ ($\alpha \in \mathbb{R}_+$) of it is a translation-invariant measure μ on $\mathscr{B}^d$, which satisfies $\mu(W) = \alpha < +\infty$. The following converse of this also holds, and contains the aforementioned characterization of λ^d as a special case.

8.1 Theorem. *Every measure μ on $\mathscr{B}^d$ which is translation-invariant, i.e., satisfies $T_a(\mu) = \mu$ for every translation $x \mapsto T_a(x) := a + x$ of $\mathbb{R}^d$, and which assigns finite measure*

$$(8.3) \qquad\qquad \alpha := \mu(W) < +\infty$$

to the unit cube W, has the form

$$(8.4) \qquad\qquad \mu = \alpha\lambda^d \,.$$

Proof. Let $a_n := n^{-1}\mathbf{1}$, the point in $\mathbb{R}^d$ all of whose coordinates are $1/n$. Then $W_n := [\mathbf{0}, a_n[\, \in \mathscr{I}^d$ is a cube with

$$\mu(W_n) = \alpha/n^d \,.$$

In fact: The interval $[0, 1[\, \in \mathscr{I}^1$ is the union of the pairwise disjoint intervals $\left[\dfrac{\nu}{n}, \dfrac{\nu+1}{n}\right[$ with $\nu = 0, 1, \ldots, n-1$. If therefore G_n denotes the set of points $(\varrho_1, \ldots, \varrho_d) \in \mathbb{R}^d$ whose coordinates all come from the set $\{\nu/n : \nu = 0, \ldots, n-1\}$, then

$$W = \bigcup_{r \in G_n} [r, r + a_n[\,,$$

a union of n^d pairwise disjoint intervals. Because $[r, r+a_n[= T_r([\mathbf{0}, a_n[) = T_r(W_n)$ and because of the translation-invariance of μ, it follows from this representation of W that $\alpha = n^d \mu(W_n)$.

A repetition of these considerations will show that

$$\mu([a, b[) = \alpha \lambda^d([a, b[)$$

holds for every interval $[a, b[\in \mathscr{I}^d$ in which the points a, b have only rational coordinates. Obviously in proving this we can assume that $a \lhd b$, and due to the translation-invariance of both measures we can further assume that $a = \mathbf{0}$. Then $b = (m_1/n, \ldots, m_d/n)$ for appropriate $m_1, \ldots, m_d, n \in \mathbb{N}$, and therefore $[\mathbf{0}, b[$ is the union of the $m_1 \cdot \ldots \cdot m_d$ pairwise disjoint intervals $[r, r + a_n[$ with $r = (\varrho_1/n, \ldots, \varrho_d/n)$ and $\varrho_i \in \{0, \ldots, m_i - 1\}$ for each i. As before, this yields $m_1 \cdot \ldots \cdot m_d \mu(W_n) = \mu([\mathbf{0}, b[)$, hence

$$\mu([\mathbf{0}, b[) = \alpha \cdot \frac{m_1}{n} \cdot \ldots \cdot \frac{m_d}{n} = \alpha \lambda^d([\mathbf{0}, b[) \,.$$

Now the set $\mathscr{I}^d_{\mathrm{rat}}$ of all intervals $[a, b[\in \mathscr{I}^d$ for which a, b have only rational coordinates is an $\cap$-stable system. The technique used in the proof of Theorem 6.4 shows that $\mathscr{I}^d_{\mathrm{rat}}$ is, just like $\mathscr{I}^d$, a generator of $\mathscr{B}^d$. Because the measures μ and $\alpha \lambda^d$ coincide on $\mathscr{I}^d_{\mathrm{rat}}$ and for $\mathbf{n} := (n, \ldots, n)$ with $n \in \mathbb{N}$ the intervals $[-\mathbf{n}, \mathbf{n}[$ lie in $\mathscr{I}^d_{\mathrm{rat}}$ and increase to $\mathbb{R}^d$, our claim (8.4) follows from the uniqueness theorem 5.4. $\square$

For $\alpha = 1$ we immediately get from 8.1.

8.2 Corollary. *Lebesgue–Borel measure λ^d is the only translation-invariant measure μ on $\mathscr{B}^d$ which satisfies*

(8.2′) $\mu(W) = 1 \,.$

This corollary says that λ^d is, in the theory of locally-compact groups, a Haar measure on the additive group $\mathbb{R}^d$. That theory provides an analogous non-zero invariant measure on every locally compact abelian group G; it is unique to within a positive scalar factor and is called *Haar measure* on G. The reader interested in its theory should consult NACHBIN [1965]. (Cf. also Exercise 4 of §7 and Exercise 8 of §17.)

The conclusion of the theorem and its corollary remain valid if in the normalization (8.2) and (8.2′) the unit cube W is replaced by its open interior $]\mathbf{0}, \mathbf{1}[$ or its compact closure $[\mathbf{0}, \mathbf{1}]$. This is immediate from (6.7). However, if $\mu(W) = +\infty$ is allowed, μ need not be a multiple of λ^d. See HENLE and WAGON [1983].

Example. 1. Besides the $D_\alpha^{(i)}$ of Example 4, §7 there is another basic class of linear mappings in $\mathbb{R}^d$, those that *skew* one coordinate by means of another. Specifically, for each $i, k \in \{1, \ldots, d\}$ with $i \neq k$ we define

$$S^{(i,k)}(x_1, \ldots, x_d) := (x_1, \ldots, x_{i-1}, x_i + x_k, x_{i+1}, \ldots, x_d) \,.$$

Evidently this mapping is continuous. It is also invertible, with inverse $D^{(k)}_{-1} \circ S^{(i,k)} \circ D^{(k)}_{-1}$, as is easily seen. In view of Example 2 in §7, the image measure $S^{(i,k)}(\lambda^d)$ may therefore be formed. We want to show that λ^d is also invariant under these mappings, that is, that

$$(8.5) \qquad S^{(i,k)}(\lambda^d) = \lambda^d \qquad \text{for all } i, k \in \{1, \ldots, d\} \text{ with } i \neq k.$$

Fix such a pair (i,k) and write simply S for $S^{(i,k)}$. Since this is a linear mapping, $S(\lambda^d)$ is a translation-invariant measure on $\mathscr{B}^d$, so (8.5) will follow from 8.2 if we succeed in showing that $S(\lambda^d)(W) = 1$, that is, $\lambda^d(S^{-1}(W)) = 1$. In view of (7.9) and the equality $S^{-1} = D^{(k)}_{-1} \circ S \circ D^{(k)}_{-1}$, it suffices to show instead that

$$(8.5') \qquad \lambda^d(S(W)) = 1.$$

Let a denote the vector in $\mathbb{R}^d$ whose only non-zero coordinate is the ith one, it being -1. Introduce

$$W' := \{(x_1, \ldots, x_d) : 0 \leq x_j < 1 \text{ for } j \neq i, 0 \leq x_i < x_k\} \quad \text{and}$$
$$W'' := \{(x_1, \ldots, x_d) : 0 \leq x_j < 1 \text{ for } j \neq i, 1 + x_k \leq x_i < 2\}.$$

Notice that

$$T_a(W'') = \{(x_1, \ldots, x_d) : 0 \leq x_j < 1 \text{ for } j \neq i, x_k \leq x_i < 1\}.$$

Clearly

$$(8.6) \qquad W = W' \cup T_a(W'') \quad \text{disjointly.}$$

W' is the intersection of W with the open set $\{(x_1, \ldots, x_d) \in \mathbb{R}^d : x_i < x_k\}$, so W' is a Borel set. Similarly $T_a(W'')$ is the intersection of W with the closed set $\{(x_1, \ldots, x_d) \in \mathbb{R}^d : x_k \leq x_i\}$, so it is a Borel set. Thus W'', its preimage under T_a, is also a Borel set. Since S is a homeomorphism, $S(W)$ is a Borel set. Next notice that

$$(8.7) \qquad W', W'' \text{ and } S(W) \quad \text{are pairwise disjoint.}$$

For the conditions on the ith coordinate that define each set in (8.7) are obviously incompatible with those that define the other two sets. Moreover,

$$(8.8) \qquad W' \cup W'' \cup S(W) = D^{(i)}_2(W).$$

Here the inclusion "$\subset$" is obvious from the coordinate inequalities defining the sets. A typical point x of $D^{(i)}_2(W)$ has j^{th} coordinate $x_j \in [0, 1[$ if $j \neq i$ and i^{th} coordinate $t \in [0, 2[= [0, x_k[\cup [x_k, 1 + x_k[\cup [1 + x_k, 2[$. If t lies in the first (third) interval, then $x \in W'$ ($x \in W''$). Otherwise, $x_i := t - x_k \in [0, 1[$, and

$$x = (x_1, \ldots, x_{i-1}, t, x_{i+1}, \ldots, x_d) = (x_1, \ldots, x_{i-1}, x_i + x_k, x_{i+1}, \ldots, x_d) \in S(W).$$

This confirms (8.8). Combining all that we have learned gives the desired (8.5′) as follows:

$$\begin{aligned}
2 = 2\lambda^d(W) &= \lambda^d(D_2^{(i)}(W)) \quad \text{by (7.9)} \\
&= \lambda^d(W') + \lambda^d(W'') + \lambda^d(S(W)) \quad \text{by (8.7) and (8.8)} \\
&= \lambda^d(W') + \lambda^d(T_a(W'')) + \lambda^d(S(W)) \quad \text{by (7.7)} \\
&= \lambda^d(W) + \lambda^d(S(W)) \quad \text{by (8.6)} \\
&= 1 + \lambda^d(S(W)) \quad \text{by (8.2).}
\end{aligned}$$

One usually thinks of the space $\mathbb{R}^d$ as equipped with the *euclidean scalar-product*

$$\langle x, y \rangle := \sum_{i=1}^{d} \xi_i \eta_i$$

and the *euclidean metric* derived from it by

$$\varrho(x, y) := \sqrt{\langle x - y, x - y \rangle}\,,$$

where $x := (\xi_1, \ldots, \xi_d)$, $y := (\eta_1, \ldots, \eta_d)$. Every mapping $T : \mathbb{R}^d \to \mathbb{R}^d$ which leaves this metric invariant, that is, satisfies

(8.9) $\varrho(T(x), T(y)) = \varrho(x, y)$ for all $x, y \in \mathbb{R}^d$,

is called a *motion* (or an *isometry*) in $\mathbb{R}^d$. It is obviously continuous, hence Borel measurable.

Suppose in addition that T fixes $\mathbf{0}$, that is,

$$T(\mathbf{0}) = \mathbf{0}\,.$$

Using the linearity of $\langle,\rangle$ in each of its positions, we get

$$\begin{aligned}
\varrho^2(T(x), T(y)) &= \langle T(x) - T(y), T(x) - T(y) \rangle \\
&= \langle T(x), T(x) \rangle - 2\langle T(x), T(y) \rangle + \langle T(y), T(y) \rangle \\
&= \varrho^2(T(x), T(\mathbf{0})) - 2\langle T(x), T(y) \rangle + \varrho^2(T(y), T(\mathbf{0}))\,,
\end{aligned}$$

so that in view of (8.9)

$$\varrho^2(x, y) = \varrho^2(x, \mathbf{0}) - 2\langle T(x), T(y) \rangle + \varrho^2(y, \mathbf{0})\,.$$

Replacing T with the identity mapping here shows that (8.9) may be supplemented with

(8.9′) $\langle T(x), T(y) \rangle = \langle x, y \rangle$ for all $x, y \in \mathbb{R}^d$ if $T(\mathbf{0}) = \mathbf{0}$.

Consider $\lambda \in \mathbb{R}$, $x, y \in \mathbb{R}^d$ and again suppose that $T(\mathbf{0}) = \mathbf{0}$. Using the linearity properties of $\langle,\rangle$ once more we expand

(∗) $\varrho^2(\lambda T(x) + T(y) - T(\lambda x + y), \lambda T(x) + T(y) - T(\lambda x + y))$

into a linear combination of expressions $\langle T(a), T(b) \rangle$ with $a, b \in \{x, y, \lambda x + y\}$. Equation (8.9′) allows T to be replaced by the identity mapping in every such expression. Upon doing so and re-assembling the terms, we get back a single expression like $(*)$ but with the identity mapping in place of T. That is, we get 0. In other words,

$$\lambda T(x) + T(y) - T(\lambda x + y) = \mathbf{0}\,,$$
$$T(\lambda x + y) = \lambda T(x) + T(y)\,,$$

holding for all $\lambda \in \mathbb{R}$, $x, y \in \mathbb{R}^d$. This says that T is a *linear* mapping. It is immediate from (8.9) that T is then injective. The dimension of $T(\mathbb{R}^d) \subset \mathbb{R}^d$ is therefore d, so $T(\mathbb{R}^d) = \mathbb{R}^d$, and T is surjective. A motion T that is also a linear mapping, and the preceding deliberations show that this is equivalent to $T(\mathbf{0}) = \mathbf{0}$, is called an *orthogonal transformation*.

If T is any motion and we set $a := T(\mathbf{0})$, then the mapping $U := T - a = T_{-a} \circ T$ is a motion that fixes $\mathbf{0}$. Therefore by the above, every motion T is a composite $T_a \circ U$ of a translation and an orthogonal transformation, and is consequently a bijection of $\mathbb{R}^d$. From this and (8.9) it is clear that the mapping inverse to a motion is itself a motion, and that the set of motions is a group under composition, the motion group $\mathrm{Mot}(\mathbb{R}^d)$ of $\mathbb{R}^d$.

The translation-invariance of λ^d derived in 8.1 not only characterizes L-B measure but renders excellent service in the derivation of further invariance properties. We begin with the *motion-invariance* of λ^d, that is, with the proof that

$$(8.10) \qquad\qquad T(\lambda^d) = \lambda^d \qquad\qquad \text{for all } T \in \mathrm{Mot}(\mathbb{R}^d).$$

The reflection-invariance treated in Example 5 of §7 is contained in this as a special case.

8.3 Theorem. *Lebesgue–Borel measure λ^d is motion-invariant.*

Proof. Let a motion T of $\mathbb{R}^d$, about which we initially assume that $T(\mathbf{0}) = \mathbf{0}$, be given. Thus T is an orthogonal, linear transformation. Via the following considerations, we will quickly convince ourselves that $T(\lambda^d)$ is a translation-invariant measure on $\mathcal{B}^d$: Denoting as before by T_c the translation $x \mapsto x + c$, for each $c \in \mathbb{R}^d$, we consider any $a \in \mathbb{R}^d$, set $b := T^{-1}(a)$, and observe that

$$(8.11) \qquad\qquad T_a \circ T = T \circ T_b\,.$$

For every $x \in \mathbb{R}^d$, $T_a \circ T(x) = T(x) + a = T(x) + T(b) = T(x+b) = T \circ T_b(x)$, confirming (8.11). From this and the translation-invariance of λ^d we get $T_a(T(\lambda^d)) = T(T_b(\lambda^d)) = T(\lambda^d)$. As $a \in \mathbb{R}^d$ is arbitrary, this says that $\mu := T(\lambda^d)$ is a translation-invariant measure on $\mathcal{B}^d$. For the unit cube $W = [\mathbf{0}, \mathbf{1}[$ we have $\alpha := \mu(W) = \lambda^d(T^{-1}(W)) < +\infty$ by (6.2), since T is an isometry and therefore along with W the set $T^{-1}(W)$ is also bounded. Now Theorem 8.1 comes into action and guarantees that $T(\lambda^d) = \mu = \alpha\lambda^d$ holds. So what remains is to see that $\alpha = 1$. To this end we look at the compact ball $K := \{x \in \mathbb{R}^d : \varrho(\mathbf{0}, x) \le 1\}$ of radius 1 and center $\mathbf{0}$. Since T and T^{-1} are orthogonal transformations, they fix $\mathbf{0}$ and leave distances invariant (8.9). Hence $T^{-1}(K) = K$, and from $T(\lambda^d) = \alpha\lambda^d$ follows

$$\lambda^d(K) = \lambda^d(T^{-1}(K)) = T(\lambda^d)(K) = \alpha\lambda^d(K).$$

From this follows the desired $\alpha = 1$, because on the one hand $\lambda^d(K) < +\infty$ by (6.2) and on the other hand $\lambda^d(K) > 0$ because K contains a non-empty interval $I \in \mathscr{I}^d$, namely $I := [-t, t[$ with $t := (d^{-1/2}, \ldots, d^{-1/2})$. [In Exercise 6 of §23 we will compute $\lambda^d(K)$ explicitly.]

To handle the case of an arbitrary motion T, set $c := T(\mathbf{0})$ and $S := T_{-c} \circ T$, getting a motion that fixes $\mathbf{0}$, for which $\lambda^d = S(\lambda^d)$ by what was first proved. It follows finally from transitivity and $T = T_c \circ S$ that $T(\lambda^d) = T_c(S(\lambda^d)) = T_c(\lambda^d) = \lambda^d$. Thus the theorem is proved. □

Since with every motion T of $\mathbb{R}^d$ its inverse T^{-1} is also one, the motion-invariance can also be recorded in the following form: For every motion T of $\mathbb{R}^d$ and every Borel set $A \in \mathscr{B}^d$

$$(8.12) \qquad \lambda^d(T(A)) = \lambda^d(A).$$

In this form Theorem 8.3 just says that any two congruent Borel sets in $\mathbb{R}^d$ have the same d-dimensional Lebesgue measure. This however is the measure-theoretic formulation and refinement of the elementary geometric principle (A) enunciated in the introduction to the chapter. Via it L-B measure is seen in the final analysis to be a concept from euclidean geometry.

Examples. 2. *Every hyperplane $H \subset \mathbb{R}^d$ is an* L-B-*nullset*. This follows from Example 1 of §6 and the fact that there is a motion T which transforms a hyperplane of the kind considered in that example, say the hyperplane with equation $\xi_d = 0$, into H.

3.   Every closed or open box (meaning a parallelepiped with pairwise orthogonal edges) $Q \subset \mathbb{R}^d$ whose edge-lengths are $l_1, \ldots, l_d$ has Lebesgue measure $\lambda^d(Q) = l_1 \cdot \ldots \cdot l_d$. This follows analogously from Example 3 of §6.

The behavior of λ^d with respect to linear transformations of the vector space $\mathbb{R}^d$ into itself – and then too with respect to arbitrary affine mappings – can also be clarified using a slight modification of the preceding method of proof.

Linear mappings $T : \mathbb{R}^d \to \mathbb{R}^d$ are just those that with respect to the canonical basis in $\mathbb{R}^d$ (or indeed any basis) can be represented in the form $T(x) = Cx$, with C a $d \times d$ matrix and $x \in \mathbb{R}^d$ interpreted as a column vector. The determinant of T, in symbols $\det T$, is by definition that of C (and is independent of the choice of basis).

We will restrict ourselves to the case where T is non-singular, that is, $\det T \neq 0$, and consequently bijective. These are elements of the group $\mathrm{GL}(d, \mathbb{R})$ known as the *general linear group*. The mappings $T \in \mathrm{GL}(d, \mathbb{R})$ with $\det T = 1$ form a subgroup of $\mathrm{GL}(d, \mathbb{R})$, the *special linear group* $\mathrm{SL}(d, \mathbb{R})$. It is in fact the commutator subgroup of $\mathrm{GL}(d, \mathbb{R})$ and this fact is used by DIEROLF and SCHMIDT [1998] to give an alternative proof of our next theorem. (The behavior of λ^d with respect to linear mappings T with $\det T = 0$ is elucidated in Exercise 2 below.)

8.4 Theorem. *Every $T \in \mathrm{GL}(d, \mathbb{R})$ satisfies*

$$(8.13) \qquad\qquad T(\lambda^d) = \frac{1}{|\det T|} \lambda^d$$

or, equivalently

$$(8.14) \qquad\qquad \lambda^d(T(A)) = |\det T| \, \lambda^d(A)$$

for all $A \in \mathscr{B}^d$.

Proof. Consider first the elementary mappings $D_\alpha^{(i)}$ defined in Example 4 of §7 and $S^{(i,k)}$ defined in Example 1 of this section. It is obvious that $\det D_\alpha^{(i)} = \alpha$ and $\det S^{(i,k)} = 1$. Therefore (7.9) and (8.5) confirm (8.13) in the special case that T lies in the set

$$(8.15) \qquad\qquad \{D_\alpha^{(i)}, S^{(i,k)} : \alpha \in \mathbb{R} \setminus \{0\}, i, k \in \{1, \ldots, d\}, i \neq k\}.$$

Since $\det$ is a homomorphism of $\mathrm{GL}(d, \mathbb{R})$ into the multiplicative group $\mathbb{R}^\times := \mathbb{R} \setminus \{0\}$, it follows from this and (7.6) that in fact (8.13) holds for all T in the *subgroup* of $\mathrm{GL}(d, \mathbb{R})$ generated by the set (8.15). The proof of Theorem 8.4 is therefore completed by recalling the key fact, proved in every linear algebra text, that this subgroup is the whole group $\mathrm{GL}(d, \mathbb{R})$. Actually what is usually proved (see, e.g., Birkhoff and MacLane [1965], p. 217) is that $\mathrm{GL}(d, \mathbb{R})$ is generated by (8.15) together with the transformations $I^{(i,k)}$ which send every vector $x = (x_1, \ldots, x_d)$ to the vector whose coordinates are those of x but with x_i and x_k interchanged. However, every such transformation is already in the group generated by (8.15), for it is routine to confirm, by watching the behavior of the ith and the kth coordinates at each step, that for $i \neq k$

$$I^{(i,k)} = D_{-1}^{(i)} \circ S^{(k,i)} \circ D_{-1}^{(k)} \circ S^{(i,k)} \circ D_{-1}^{(k)} \circ S^{(k,i)}. \qquad \square$$

Theorems 8.3 and 8.4 taken together confirm an elementary fact from linear algebra, namely that $\det T = \pm 1$ for every orthogonal transformation T. And this means that 8.3 is contained in the following immediate consequence of 8.4:

8.5 Corollary. *The L-B measure λ^d is invariant under all transformations $T \in \mathrm{GL}(d, \mathbb{R})$ with $|\det T| = 1$, in particular, under mappings $T \in \mathrm{SL}(d, \mathbb{R})$.*

Remark. 1. As is known from the differential calculus in $\mathbb{R}^d$, a C_1-*diffeomorphism* $\varphi : G \to G'$ between two open subsets G and G' of $\mathbb{R}^d$ is approximable near each point $x \in G$ by a mapping $T_x \in \mathrm{GL}(d, \mathbb{R})$, namely by the derivative $T_x := \mathrm{D}\varphi(x)$. It should now not come as a surprise that the following transformation, involving the density concept from 17.2, relates the image of L-B measure on G to L-B measure on G':

$$(8.16) \qquad\qquad \varphi(\lambda_G^d) = \frac{1}{|\det \mathrm{D}\varphi| \circ \varphi^{-1}} \lambda_{G'}^d$$

or equivalently

$$(8.16') \qquad \varphi^{-1}(\lambda^d_{G'}) = |\det D\varphi|\, \lambda^d_G \,.$$

We will not go into this any further, but refer the reader to the textbook literature, e.g., STROMBERG [1981], or to VARBERG [1971].

We will conclude the chapter by proving the existence of non-Borel subsets of $\mathbb{R}^d$. A different approach is indicated in the prologue to Theorem 26.6.

8.6 Theorem. *For every dimension $d \in \mathbb{N}$, $\mathscr{B}^d \neq \mathscr{P}(\mathbb{R}^d)$.*

Proof. Let $\mathbb{Q}^d$ denote the set of points in $\mathbb{R}^d$ each of whose d coordinates is rational. This is a subgroup of the additive group $\mathbb{R}^d$, so congruence $x \sim y$ of points $x, y \in \mathbb{R}^d$ modulo $\mathbb{Q}^d$ is an equivalence relation; it is defined by $x \sim y$ if and only if $x - y \in \mathbb{Q}^d$. The space $\mathbb{R}^d$ decomposes into disjoint equivalence classes, each a set $x + \mathbb{Q}^d$ with $x \in \mathbb{R}^d$, the statement $x \sim y$ being equivalent to the equality $x + \mathbb{Q}^d = y + \mathbb{Q}^d$. Since to every real number η corresponds an integer n such that $n \leq \eta < n+1$, that is, such that $\eta - n \in [0, 1[$, every equivalence class contains a point $x \in [0, 1[$. Consequently, there is a set $K \subset [0, 1[$ which contains exactly one element from each equivalence class. (On the role of the Axiom of Choice from set theory in this existence claim see SOLOVAY [1970] and HALMOS [1974].) We have then

$$(8.17) \qquad \mathbb{R}^d = \bigcup_{k \in K} (k + \mathbb{Q}^d) = \bigcup_{y \in \mathbb{Q}^d} (y + K)$$

and

$$(8.18) \qquad y_1, y_2 \in \mathbb{Q}^d, \quad y_1 \neq y_2 \quad \Rightarrow \quad (y_1 + K) \cap (y_2 + K) = \emptyset \,.$$

(Otherwise there are $k, k' \in K$ with $y_1 + k = y_2 + k'$, that is, with $k \sim k'$, which by definition of K means that $k = k'$ and consequently also $y_1 = y_2$.) Let us now suppose that $K \in \mathscr{B}^d$. Since $\mathbb{Q}$ and therewith $\mathbb{Q}^d$ is countable, it follows from (8.17), (8.18) and the σ-additivity of λ^d that

$$(8.19) \qquad \sum_{y \in \mathbb{Q}^d} \lambda^d(y + K) = \lambda^d(\mathbb{R}^d) = +\infty \,.$$

Translation-invariance of λ^d says that

$$(8.20) \qquad \lambda^d(y + K) = \lambda^d(K) \qquad\qquad \text{for all } y \in \mathbb{Q}^d$$

and so in view of (8.19), $\lambda^d(K) > 0$. Now $K \subset [0, 1[$, and so

$$\bigcup_{y \in [0,1[\cap\mathbb{Q}^d} (y + K) \subset [0, 2[\,,$$

2 being the point in $\mathbb{R}^d$ each of whose coordinates equals 2. From this fact and (8.18) follows, again via σ-additivity of λ^d, that

$$\sum_{y \in [0,1[\cap\mathbb{Q}^d} \lambda^d(y + K) \leq \lambda^d([0, 2[) = 2^d < +\infty \,.$$

But then (8.20) means that we must have $\lambda^d(K) = 0$, contradicting (8.19). The assumption $K \in \mathscr{B}^d$ is what led to this contradiction, so we conclude that K is, after all, not a Borel set.   □

The following remarks serve to round out the foregoing and to provide a glimpse of some closely related issues.

Remarks. 2. The *"content-problem"* in $\mathbb{R}^d$ described in the introduction to this chapter, namely the problem of determining a d-dimensional volume for as large a class of subsets of $\mathbb{R}^d$ as possible, is quite satisfactorily solved by the Lebesgue–Borel measure λ^d, especially in view of its motion-invariance. More satisfactory still for a reader without preconceived notions would be a proof of the existence of a measure μ on the whole power set $\mathscr{P}(\mathbb{R}^d)$ which assigns mass 1 to the unit cube $[\mathbf{0}, \mathbf{1}[$ and is invariant under all motions of $\mathbb{R}^d$. According to Corollary 8.2 such a μ would have to be an extension of the Lebesgue–Borel measure λ^d. But it was shown by F. Hausdorff (1868–1942), cf. HAUSDORFF [1914], pp. 401–402, that no such measure μ on $\mathscr{P}(\mathbb{R}^d)$ exists, for any dimension $d \geq 1$; in fact, there is not even a σ-additive, motion-invariant *content* $\eta \neq 0$ on the ring of all bounded subsets of $\mathbb{R}^d$. For this result the reader is referred to the exposition in AUMANN [1969], pp. 275–276, which further exploits the ideas in the preceding proof of Theorem 8.6 and will consequently be mathematically accessible to him.

HAUSDORFF [1914], p. 469 further showed that the content-problem for bounded subsets of $\mathbb{R}^d$ does not even have a solution if the motion-invariant content $\eta \neq 0$ is only required to be finitely additive, and $d \geq 3$. The reader is referred for this to the presentation by STROMBERG [1979] or WAGON [1985] (in each of which a central role is played by the so-called Banach–Tarski paradox, discovered in 1924) and to the subsequent investigations of VON NEUMANN [1929] that introduced the idea of amenable groups.

S. Banach (1892–1945), cf. BANACH [1923], discovered that the *finitely-additive* content-problem mentioned above has a solution in dimensions $d = 1$ and $d = 2$. But such an η is not uniquely determined by the normalization $\eta([\mathbf{0}, \mathbf{1}[) = 1$ and for this very reason its further study has not seemed worthwhile. A remarkable generalization of Banach's result will be found on pp. 242–245 of HEWITT and ROSS [1979].

3. If $(\Omega, \mathscr{A}, \mu)$ is a measure space and A a μ-null set, then indeed because of isotoneity every subset of A that belongs to $\mathscr{A}$ is itself μ-null; nevertheless, not every subset of A need belong to $\mathscr{A}$. This phenomenon even occurs with the L-B measure, as the second part of Remark 4 will show. If $\mathscr{A}$ contains all subsets of each μ-null set, then μ is called a *complete measure*.

Exercise 7 of §5 describes how an arbitrary measure μ can be extended in a natural way to a complete measure by passage to its so-called *completion* μ_0. The completion of the Lebesgue–Borel measure in $\mathbb{R}^d$ is called *Lebesgue measure* in $\mathbb{R}^d$; the sets in the σ-algebra on which it is defined are called *Lebesgue measurable* and those of them having measure zero are called *Lebesgue-null sets* in $\mathbb{R}^d$.

In passing from Borel sets to Lebesgue measurable sets the important property of the former that they are determined only by the topology of $\mathbb{R}^d$ is lost. Because $\mathscr{B}^d$ is the defining σ-algebra for so many other important measures (for $d = 1$ Theorem 6.5 already attests to this), we will not dwell in detail on the transition from Lebesgue–Borel to Lebesgue measure; only the former will be employed in the sequel.

4. There exists a Borel set $B \in \mathscr{B}^2$ whose image $\pi_1(B)$ under the first projection map $\pi_1 : \mathbb{R}^2 \to \mathbb{R}$ (which sends every point $(x_1, x_2) \in \mathbb{R}^2$ to its first coordinate x_1) is not a Borel subset of $\mathbb{R}$. A proof of this will be found in SRIVASTAVA [1998], p. 130. Such a B can even be found which is G_δ-set, that is, the intersection of countably many open subsets of $\mathbb{R}^2$; see p. 36 of CHRISTENSEN [1974]. In particular, the continuous image of a Borel set need not be a Borel set. The system of all sets $\pi_1(B)$ with $B \in \mathscr{B}^2$ comprises rather the so-called *Souslin* or *analytic subsets* of $\mathbb{R}$. See SRIVASTAVA [1998] and CHOQUET [1969].

For any non-Borel set $A \subset \mathbb{R}$, Exercise 2 of §6 shows that $A \times \{0\}$ is a non-Borel subset of the λ^2-null set $\mathbb{R} \times \{0\}$.

5. Examples 4 and 5 of §7 as well as Theorem 8.4 illustrate that the L-B measure λ^d is not invariant with respect to all homeomorphisms $T : \mathbb{R}^d \to \mathbb{R}^d$ of $\mathbb{R}^d$ with itself. For such a homeomorphism T however, $\mu := T(\lambda^d)$ is always a measure on $\mathscr{B}^d$ with the following properties: (i) $\mu(K) < +\infty$ for every compact $K \subset \mathbb{R}^d$; (ii) $\mu(\{x\}) = 0$ for every $x \in \mathbb{R}^d$; (iii) $\mu(U) > 0$ for every non-empty open $U \subset \mathbb{R}^d$; (iv) $\mu(\mathbb{R}^d) = +\infty$. OXTOBY and ULAM [1941] showed that, conversely, every measure μ on $\mathscr{B}^d$ enjoying properties (i)–(iv) has the form $\mu = T(\lambda^d)$ for some homeomorphism $T : \mathbb{R}^d \to \mathbb{R}^d$. A simpler treatment of their result was later provided by GOFFMAN and PEDRICK [1975].

Exercises.

1. Let $T : (\Omega, \mathscr{A}) \to (\Omega', \mathscr{A}')$ be a measurable mapping, μ a measure on the σ-algebra $\mathscr{A}$, and $\mu' := T(\mu)$ its image under this mapping. $(\Omega, \mathscr{A}_0, \mu_0)$ and $(\Omega', \mathscr{A}_0', \mu_0')$ will denote the completions of these measure spaces (Exercise 7, §5). Show that the mapping T is also $\mathscr{A}_0$-$\mathscr{A}_0'$-measurable and that $T(\mu_0) = \mu_0'$. From this it follows that Lebesgue measure in $\mathbb{R}^d$ is also motion-invariant.

2. Let T be a linear mapping of $\mathbb{R}^d$ into itself with $\det T = 0$. Show that for every $A \in \mathscr{B}^d$, $T(A)$, although it may fail to be a Borel set (as noted in Remark 4) is at least a Lebesgue-null set, thus a subset of an L-B-null set, namely the linear subspace $T(\mathbb{R}^d)$ of $\mathbb{R}^d$. In this sense equality (8.14) retains its validity for linear transformation $T : \mathbb{R}^d \to \mathbb{R}^d$ with $\det T = 0$, i.e., (8.14) is valid for every linear transformation T of $\mathbb{R}^d$ into itself.

3. Show that the set K constructed in the proof of Theorem 8.6 is not even Lebesgue measurable.

4. In the section entitled "Fallacies, Flaws and Flimflam", p. 39, vol. 22, no. 1 (1991) of the *College Mathematics Journal* the following short "proof" of Theorem 8.6 is offered: Suppose that $\lambda^1(X)$ is defined for every subset X of $[0, 1]$. By isotoneity it is a number in $[0, 1]$. Consider the set B defined as $\{\lambda^1(X) : X \in$

$\mathscr{P}([0,1]), \lambda^1(X) \notin X\}$. It is a subset of $[0,1]$ and upon testing the number $\lambda^1(B)$ for membership in B we find that the statements $\lambda^1(B) \in B$ and $\lambda^1(B) \notin B$ are equivalent, a contradiction. What is the error in this reasoning, or is it perhaps a legitimate proof of Theorem 8.6?

Chapter II

Integration Theory

A measure space $(\Omega, \mathscr{A}, \mu)$ is given. We pose the problem of assigning to each function on Ω from as large a class as possible an integral, that is, a "mean value" constructed with respect to μ. After the introduction in §9 of the property of measurability, which is fundamental, this problem will be resolved step by step in §10–§12. The later sections of this chapter are devoted to erecting the theory and exploring the applications of the integration procedure thus defined.

§9. Measurable numerical functions

On the number line $\mathbb{R}$ we have defined the σ-algebra $\mathscr{B}^1$ of Borel sets. If we compactify $\mathbb{R}$ to $\overline{\mathbb{R}}$ in the customary way by adjoining the "ideal" points $-\infty$ and $+\infty$, the sets $A \subset \overline{\mathbb{R}}$ for which $A \cap \mathbb{R} \in \mathscr{B}^1$ are called *Borel* in $\overline{\mathbb{R}}$. Constructively, the Borel sets in $\overline{\mathbb{R}}$ are precisely all sets B, $B \cup \{-\infty\}$, $B \cup \{+\infty\}$, $B \cup \{-\infty, +\infty\}$ with $B \in \mathscr{B}^1$. The system $\overline{\mathscr{B}}^1$ of these sets is obviously a σ-algebra in $\overline{\mathbb{R}}$ whose trace in $\mathbb{R}$ is $\mathscr{B}^1$:

$$(9.1) \qquad \qquad \mathbb{R} \cap \overline{\mathscr{B}}^1 = \mathscr{B}^1 .$$

If now $(\Omega, \mathscr{A})$ is a measurable space, the $\mathscr{A}$-$\overline{\mathscr{B}}^1$-measurability of functions $f : \Omega \to \overline{\mathbb{R}}$ is defined. Such functions will henceforth be called $(\mathscr{A}\text{-})$*measurable numerical functions* on Ω. Real functions $f : \Omega \to \mathbb{R}$ are special numerical functions; in view of (9.1) the $\mathscr{A}$-$\overline{\mathscr{B}}^1$-measurability of such a function is just the same as its $\mathscr{A}$-$\mathscr{B}^1$-measurability.

Examples. 1. Let $(\Omega, \mathscr{A})$ be a measurable space, A a subset of Ω. The function

$$(9.2) \qquad \qquad 1_A(\omega) := \begin{cases} 1 & \text{if } \omega \in A \\ 0 & \text{if } \omega \in \Omega \setminus A \end{cases}$$

is called the *indicator function* (sometimes also the *characteristic function*) of A. This real function on Ω is $\mathscr{A}$-measurable just if $A \in \mathscr{A}$, because for every $B \subset \overline{\mathbb{R}}$ the set $(1_A)^{-1}(B)$ must be one of the four $\Omega, A, \Omega \setminus A, \emptyset$.

Thus sets and their indicator functions correspond biuniquely. The following calculation rules, in which $A, B \subset \Omega$ and $A_i \subset \Omega$ for $i \in I$, are often used, and their validity is immediately perceived:

$$A \subset B \quad \Leftrightarrow \quad 1_A \leq 1_B ;$$

$$1_{\complement A} = 1 - 1_A\,; \quad 1_{A \triangle B} = |1_A - 1_B|\,; \quad 1_{A \cap B} = 1_A \cdot 1_B\,;$$
$$1_{\bigcup_{i \in I} A_i} = \sup_{i \in I} 1_{A_i}\,; \qquad 1_{\bigcap_{i \in I} A_i} = \inf_{i \in I} 1_{A_i}\,.$$

2.   For an arbitrary subset Q of $\mathbb{R}^d$ consider the measurable space $(Q, Q \cap \mathscr{B}^d)$. The corresponding measurable numerical functions on Q will be called *Borel measurable functions* or *Borel functions* on Q. *Every continuous numerical function f on Q is such a Borel measurable function.* Indeed, for every $\alpha \in \mathbb{R}$ the set Q_α of all $x \in Q$ with $f(x) \geq \alpha$ is a relatively closed subset of Q, that is, of the form $Q \cap F$ for a set F which is closed in $\mathbb{R}^d$. (Such an F would be, e.g., the closure of Q_α in $\mathbb{R}^d$.) Since $F \in \mathscr{B}^d$, this intersection lies in the trace σ-algebra $Q \cap \mathscr{B}^d$. The claim therefore follows from the next theorem. (Cf. also Example 2 of §7.)

9.1 Theorem. *A numerical function f on Ω is $\mathscr{A}$-measurable if and only if*

(9.3) $\{\omega \in \Omega : f(\omega) \geq \alpha\} \in \mathscr{A}$ *for all $\alpha \in \mathbb{R}$.*

Proof. According to 7.2 we have only to show that the system $\overline{\mathscr{E}}$ of all intervals $[\alpha, +\infty]$ with $\alpha \in \mathbb{R}$ generates the σ-algebra $\overline{\mathscr{B}}^1$ in $\overline{\mathbb{R}}$. Since $[\alpha, +\infty] \in \overline{\mathscr{B}}^1$ for every $\alpha \in \mathbb{R}$, we have at any rate that $\overline{\mathscr{Q}} \subset \overline{\mathscr{B}}^1$ for the σ-algebra $\overline{\mathscr{Q}}$ generated by $\overline{\mathscr{E}}$. Because $[\alpha, \beta[= [\alpha, +\infty[\setminus [\beta, +\infty[$, the intervals $[\alpha, \beta[$ with $\alpha, \beta \in \mathbb{R}$ and $\alpha \leq \beta$ all lie in $\mathbb{R} \cap \overline{\mathscr{Q}}$. From 6.1 therefore follows that $\mathscr{B}^1 \subset \mathbb{R} \cap \overline{\mathscr{Q}}$. Now the single-element sets

$$\{-\infty\} = \bigcap_{n \in \mathbb{N}} \complement[-n, +\infty] \quad \text{and} \quad \{+\infty\} = \bigcap_{n \in \mathbb{N}} [n, +\infty]$$

both lie in $\overline{\mathscr{Q}}$. Consequently, along with each $Q \in \overline{\mathscr{Q}}$, the set $\mathbb{R} \cap Q$ is also in $\overline{\mathscr{Q}}$. In other words, $\mathbb{R} \cap \overline{\mathscr{Q}} \subset \overline{\mathscr{Q}}$ and therewith $\mathscr{B}^1 \subset \overline{\mathscr{Q}}$. This fact together with $\{-\infty\}, \{+\infty\} \in \overline{\mathscr{Q}}$ and the remarks preceding (9.1) make it clear that $\overline{\mathscr{B}}^1 \subset \overline{\mathscr{Q}}$, so that finally we have $\overline{\mathscr{Q}} = \overline{\mathscr{B}}^1$. $\square$

We now introduce some popular short-hand notation: For numerical functions f and g on Ω

(9.4) $\{f \leq g\} := \{\omega \in \Omega : f(\omega) \leq g(\omega)\}$

and the sets $\{f < g\}$, $\{f = g\}$, $\{f \neq g\}$, etc., are defined analogously. Condition (9.3) in this language reads: $\{f \geq \alpha\} \in \mathscr{A}$ for all $\alpha \in \mathbb{R}$.

That we can just as well employ the sets $\{f > \alpha\}$, $\{f \leq \alpha\}$, etc., in the preceding characterization is the content of

9.2 Theorem. *Each of the following conditions is equivalent to the $\mathscr{A}$-measurability of the numerical function f on Ω:*

(a) $\{f \geq \alpha\} \in \mathscr{A}$ *for all $\alpha \in \mathbb{R}$;*
(b) $\{f > \alpha\} \in \mathscr{A}$ *for all $\alpha \in \mathbb{R}$;*

(c) $\qquad\qquad\qquad\qquad \{f \leq \alpha\} \in \mathscr{A} \qquad\qquad\qquad$ *for all $\alpha \in \mathbb{R}$;*

(d) $\qquad\qquad\qquad\qquad \{f < \alpha\} \in \mathscr{A} \qquad\qquad\qquad$ *for all $\alpha \in \mathbb{R}$.*

Proof. All that has to be shown is the equivalence of these four assertions, and that results from the validity, for all $\alpha \in \mathbb{R}$, of the equations

$$\{f > \alpha\} = \bigcup_{n \in \mathbb{N}} \{f \geq \alpha + n^{-1}\}; \qquad \{f \leq \alpha\} = \complement\{f > \alpha\};$$

$$\{f < \alpha\} = \bigcup_{n \in \mathbb{N}} \{f \leq \alpha - n^{-1}\}; \qquad \{f \geq \alpha\} = \complement\{f < \alpha\}. \quad \square$$

It may be noted that the four related assertions in which quantification is over all $\alpha \in \overline{\mathbb{R}}$ are also equivalent.

A plethora of assertions about calculating with measurable numerical functions now presents itself.

9.3 Theorem. *For any $\mathscr{A}$-measurable functions $f, g : \Omega \to \overline{\mathbb{R}}$ the sets $\{f < g\}$, $\{f \leq g\}$, $\{f = g\}$ and $\{f \neq g\}$ lie in $\mathscr{A}$.*

Proof. Because the set $\mathbb{Q}$ of rational numbers is countable, the claims follow (with the help of 9.2) from the equalities

$$\{f < g\} = \bigcup_{\varrho \in \mathbb{Q}} \{f < \varrho\} \cap \{\varrho < g\};$$

$$\{f \leq g\} = \complement\{f > g\}; \quad \{f = g\} = \{f \leq g\} \cap \{g \leq f\};$$

$$\{f \neq g\} = \complement\{f = g\}. \quad \square$$

9.4 Theorem. *Along with $f, g : \Omega \to \overline{\mathbb{R}}$, the function $f \cdot g$ and, if everywhere defined, the functions $f + g$ and $f - g$ are also $\mathscr{A}$-measurable.*

Proof. First of all, along with g, $\sigma + \tau g$ is measurable for all $\sigma, \tau \in \mathbb{R}$. This follows from 9.2 because $\{\sigma + \tau g \geq \alpha\}$ is $\{g \geq (\alpha - \sigma)/\tau\}$ if $\tau > 0$ and is $\{g \leq (\alpha - \sigma)/\tau\}$ if $\tau < 0$, the case $\tau = 0$ being trivial. This preliminary remark takes care of the passage from g to $-g$ and reduces the case $f - g$ to the case $f + g$. Furthermore, together with the remark following 9.2 and the equalities

$$\{f + g \geq \alpha\} = \{f \geq \alpha - g\} \qquad\qquad\qquad (\alpha \in \mathbb{R})$$

it yields the measurability of $f + g$.

In investigating fg we will first suppose both functions are real-valued. Then the identity

$$fg = \frac{1}{4}(f + g)^2 - \frac{1}{4}(f - g)^2$$

reduces the product question to the case $g = f$. But $\{f^2 \geq \alpha\}$ is Ω if $\alpha \leq 0$ and is $\{f \geq \sqrt{\alpha}\} \cup \{f \leq -\sqrt{\alpha}\}$ if $\alpha > 0$, which shows that the measurability of f^2 follows from that of f.

If finally f and g are numerical functions, introduce $\Omega_1 := \{fg = +\infty\}$, $\Omega_2 := \{fg = -\infty\}$, $\Omega_3 := \{fg = 0\}$ and $\Omega_4 := \complement(\Omega_1 \cup \Omega_2 \cup \Omega_3)$. Using 9.3 and the measurability of constant functions we check that these four pairwise disjoint sets lie in $\mathscr{A}$. The restrictions f', g' of f, g to Ω_4 are $\Omega_4 \cap \mathscr{A}$-measurable and real-valued. The product $f'g'$ is therefore $\Omega_4 \cap \mathscr{A}$-measurable. From this follows immediately the $\mathscr{A}$-measurability of fg. $\quad\square$

A useful special case of 9.4, isolated already in the course of its proof, is that αf is $\mathscr{A}$-measurable whenever f is and $\alpha \in \mathbb{R}$.

9.5 Theorem. *Let $(f_n)_{n \in \mathbb{N}}$ be a sequence of $\mathscr{A}$-measurable numerical functions on Ω. Then each of the following numerical functions is $\mathscr{A}$-measurable:*

$$\inf_{n \in \mathbb{N}} f_n, \quad \sup_{n \in \mathbb{N}} f_n, \quad \liminf_{n \in \mathbb{N}} f_n, \quad \limsup_{n \in \mathbb{N}} f_n.$$

Proof. The function $s := \sup f_n$ is measurable because

$$\{s \leq \alpha\} = \bigcap_{n \in \mathbb{N}} \{f_n \leq \alpha\} \qquad\qquad \text{for all } \alpha \in \mathbb{R}.$$

Due to 9.4, $\inf f_n = -\sup(-f_n)$ is then also measurable. By definition we have

$$\liminf_{n \to \infty} f_n = \sup_{n \in \mathbb{N}} \inf_{m \geq n} f_m,$$

$$\limsup_{n \to \infty} f_n = \inf_{n \in \mathbb{N}} \sup_{m \geq n} f_m.$$

By what has already been proved, each of these functions is measurable. $\quad\square$

9.6 Corollary 1. *For every finitely many $\mathscr{A}$-measurable numerical functions $f_1, \ldots, f_n$ on Ω, their lower and upper envelopes*

$$f_1 \wedge \ldots \wedge f_n \quad \text{and} \quad f_1 \vee \ldots \vee f_n$$

are $\mathscr{A}$-measurable.

Proof. Apply 9.5 to the ultimately constant sequence $f_1, \ldots, f_{n-1}, f_n, f_n, f_n \ldots$. $\quad\square$

9.7 Corollary 2. *If a sequence $(f_n)_{n \in \mathbb{N}}$ of $\mathscr{A}$-measurable functions converges pointwise throughout Ω, that is, if $\lim_{n \to \infty} f_n(\omega)$ exists on $\overline{\mathbb{R}}$ for every $\omega \in \Omega$, then the limit function $\lim_{n \to \infty} f_n$ is $\mathscr{A}$-measurable.*

This is immediate from

$$\lim_{n \to \infty} f_n = \liminf_{n \to \infty} f_n = \limsup_{n \to \infty} f_n.$$

To every numerical function $f : \Omega \to \overline{\mathbb{R}}$ three other functions on Ω are associated (cf. the section "Notations"): the *absolute value*

(9.5) $$|f| := f \vee (-f),$$

the *positive part*

(9.6)
$$f^+ := f \vee 0,$$

and the *negative part* of f

(9.7)
$$f^- := (-f)^+ = -(f \wedge 0).$$

Thus $f^+(\omega) = f(\omega)$ in case $f(\omega) \geq 0$ and $f^+(\omega) = 0$ in case $f(\omega) \leq 0$. Observe that not only $f^+ \geq 0$, but also $f^- \geq 0$. The important equalities

(9.8)
$$f = f^+ - f^- \quad \text{and} \quad |f| = f^+ + f^-$$

are immediate.

From 9.4 and 9.6 we effortlessly infer our concluding result:

9.8 Theorem. *A numerical function f on Ω is $\mathscr{A}$-measurable if and only if both its positive part f^+ and its negative part f^- are each $\mathscr{A}$-measurable. Furthermore, along with f, its absolute value $|f|$ is always $\mathscr{A}$-measurable.*

Exercises.

1. Let $(\Omega, \mathscr{A})$ be a measurable space, D a dense subset of $\mathbb{R}$ (e.g., $\mathbb{Q}$). Show that a numerical function f on Ω is $\mathscr{A}$-measurable if the analog for all $\alpha \in D$ of one of (a)–(d) in Theorem 9.2 holds.

2. Let $(f_n)_{n \in \mathbb{N}}$ be a sequence of $\mathscr{A}$-measurable numerical functions on a measurable space $(\Omega, \mathscr{A})$. Why is the set of all $\omega \in \Omega$ for which the sequence $(f_n(\omega))_{n \in \mathbb{N}}$ converges in $\overline{\mathbb{R}}$, and that for which it converges in $\mathbb{R}$, $\mathscr{A}$-measurable?

3. The real function $f : \Omega \to \mathbb{R}$ is measurable on the measurable space $(\Omega, \mathscr{A})$. Are $\exp f$ and $\sin f$, that is, the function $\omega \mapsto e^{f(\omega)}$ and $\omega \mapsto \sin f(\omega)$, $\mathscr{A}$-measurable?

4. With the aid of Theorem 9.1 show that the real function defined on $\mathbb{R}^2$ by $(x, y) \mapsto \max\{x, y\}$ is $\mathscr{B}^2$-measurable. Deduce from this another proof of Corollary 9.6.

5. Show via an example that the measurability of a numerical function f is not always a consequence of the measurability of $|f|$.

§10. Elementary functions and their integral

Our path to the integral proceeds via the set

$$E = E(\Omega, \mathscr{A})$$

of $\mathscr{A}$-*elementary functions on* Ω, which we define as follows:

10.1 Definition. A real function on Ω is called an $(\mathscr{A}$-$)$*elementary function* (or a *non-negative step function*) if it is non-negative, $\mathscr{A}$-measurable, and assumes only finitely many different values.

If $\{\alpha_1, \ldots, \alpha_n\}$ is the set of distinct values of a function $u \in E$, then the sets $A_i := u^{-1}(\alpha_i)$, $i = 1, \ldots, n$, are pairwise disjoint, and as pre-images of the Borel sets $\{\alpha_i\}$ they each lie in $\mathscr{A}$. Using the notation for indicator functions introduced in (9.2), we have then

$$(10.1) \qquad\qquad u = \sum_{i=1}^{n} \alpha_i 1_{A_i} \,.$$

If conversely, numbers $\alpha_1, \ldots, \alpha_n \in \mathbb{R}_+$ and sets $A_1, \ldots, A_n \in \mathscr{A}$ are given ($n \in \mathbb{N}$) and we define u via (10.1), then u is an elementary function, because by 9.4 it is measurable. Thus E is the set of all functions having a representation of the form (10.1), with $n \in \mathbb{N}$, coefficients α_i in $\mathbb{R}_+$ and sets A_i from $\mathscr{A}$.

From Definition 10.1 and the results of §9 the following further properties of E are immediate:

$$(10.2) \qquad u, v \in E, \alpha \in \mathbb{R}_+ \quad \Rightarrow \quad \alpha u, \ u + v, \ u \cdot v, \ u \vee v, \ u \wedge v \in E \,.$$

The derivation of (10.1) shows moreover that every function $u \in E$ has a representation of the form (10.1) in which the sets $A_i \in \mathscr{A}$ are pairwise disjoint and cover Ω, that is, constitute a *decomposition* of Ω. Such representations will henceforth be called *normal representations* of u.

It is easy to see that generally functions $u \in E$ can have several different normal representations. However, for $u \neq 0$ there is only one representation in which the coefficients are the distinct non-zero values taken by u. Anyway, for purposes of integration non-uniqueness of normal representations is not an issue, as the next lemma shows.

10.2 Lemma. *Let $(\Omega, \mathscr{A}, \mu)$ be a measure space. For any normal representations*

$$u = \sum_{i=1}^{m} \alpha_i 1_{A_i} = \sum_{j=1}^{n} \beta_j 1_{B_j}$$

of an elementary function $u \in E$ we have

$$\sum_{i=1}^{m} \alpha_i \mu(A_i) = \sum_{j=1}^{n} \beta_j \mu(B_j)$$

(bearing in mind the conventions for calculating with $+\infty$).

Proof. From

$$\Omega = A_1 \cup \ldots \cup A_m = B_1 \cup \ldots \cup B_n$$

follows

$$A_i = \bigcup_{j=1}^{n} (A_i \cap B_j) \quad \text{and} \quad B_j = \bigcup_{i=1}^{m} (A_i \cap B_j)$$

in which the sets $A_i \cap B_j$ are pairwise disjoint. The finite additivity of μ therefore supplies the equalities

$$\mu(A_i) = \sum_{j=1}^{n} \mu(A_i \cap B_j) \quad \text{and} \quad \mu(B_j) = \sum_{i=1}^{m} \mu(A_i \cap B_j),$$

the first for all $i \in \{1, \ldots, m\}$, the second for all $j \in \{1, \ldots, n\}$. After further summation

$$\sum_{i=1}^{m} \alpha_i \mu(A_i) = \sum_{i,j} \alpha_i \mu(A_i \cap B_j) \quad \text{and} \quad \sum_{j=1}^{n} \beta_j \mu(B_j) = \sum_{i,j} \beta_j \mu(A_i \cap B_j).$$

From these two equalities the claim follows when we observe the following fact: Because we started with normal representations of u, $\alpha_i = \beta_j$ for every index pair (i, j) such that $A_i \cap A_j \neq \emptyset$, in particular, for every pair (i, j) such that $\mu(A_i \cap A_j) \neq 0$. $\square$

Thanks to the preceding our next definition is sound:

10.3 Definition. Let u be an elementary function. The number

$$(10.3) \qquad \int u \, d\mu := \sum_{i=1}^{n} \alpha_i \mu(A_i),$$

which is independent of the special choice of normal representation

$$u = \sum_{i=1}^{n} \alpha_i 1_{A_i}$$

of u, is called the $(\mu\text{-})integral$ of u (over Ω).

Thus $u \mapsto \int u \, d\mu$ defines a mapping from E into $\overline{\mathbb{R}}_+$. Clearly it is a mapping in $\mathbb{R}_+$ just if μ is finite. The most important properties of this mapping are summarized in:

$$(10.4) \qquad \int 1_A \, d\mu = \mu(A) \qquad\qquad \text{for all } A \in \mathscr{A};$$

$$(10.5) \qquad \int (\alpha u) \, d\mu = \alpha \int u \, d\mu \qquad \text{for all } u \in E, \, \alpha \in \mathbb{R}_+;$$

$$(10.6) \qquad \int (u + v) \, d\mu = \int u \, d\mu + \int v \, d\mu \qquad \text{for all } u, v \in E;$$

$$(10.7) \qquad u \leq v \quad \Rightarrow \quad \int u \, d\mu \leq \int v \, d\mu \qquad \text{for all } u, v \in E.$$

Properties (10.4) and (10.5) are immediate from 10.3. The next property in the list is confirmed thus: Start with normal representations

$$u = \sum_{i=1}^{m} \alpha_i 1_{A_i} \quad \text{and} \quad v = \sum_{j=1}^{n} \beta_j 1_{B_j}$$

of the functions u and v in E. As before

$$A_i = \bigcup_{j=1}^{n}(A_i \cap B_j) \quad \text{and} \quad B_j = \bigcup_{i=1}^{m}(A_i \cap B_j);$$

and because the sets $A_i \cap B_j$ are pairwise disjoint, these equations entail

$$1_{A_i} = \sum_{j=1}^{n} 1_{A_i \cap B_j} \quad \text{and} \quad 1_{B_j} = \sum_{i=1}^{m} 1_{A_i \cap B_j},$$

the first for all $i \in \{1, \dots, m\}$, the second for all $j \in \{1, \dots, n\}$, from which in turn new normal representations

$$u = \sum_{i,j} \alpha_i 1_{A_i \cap B_j}, \quad v = \sum_{i,j} \beta_i 1_{A_i \cap B_j} \quad \text{and} \quad u + v = \sum_{i,j} (\alpha_i + \beta_j) 1_{A_i \cap B_j}$$

emerge. Using them to compute all the integrals,

$$\int u \, d\mu = \sum_{i,j} \alpha_i \mu(A_i \cap B_j), \quad \int v \, d\mu = \sum_{i,j} \beta_j \mu(A_i \cap B_j),$$

$$\int (u + v) \, d\mu = \sum_{i,j} (\alpha_i + \beta_j) \mu(A_i \cap B_j)$$

makes clear the validity of (10.6).

These deliberations have shown that every $u, v \in E$ admit normal representations

$$u = \sum_{i=1}^{k} \gamma_i 1_{C_i} \quad \text{and} \quad v = \sum_{i=1}^{k} \delta_i 1_{C_i}$$

involving the same sets $C_1, \dots, C_k \in \mathscr{A}$. In case $u \le v$, it then follows that $\gamma_i \le \delta_i$ for each $i \in \{1, \dots, k\}$ such that $C_i \ne \emptyset$, and from this we have (10.7).

Now let $u = \sum_{i=1}^{n} \alpha_i 1_{A_i}$ be an arbitrary representation of an elementary function $u \in E$ with coefficients $\alpha_i \in \mathbb{R}_+$ and sets $A_i \in \mathscr{A}$, but not necessarily a normal representation. From (10.4)–(10.6) it follows that

$$\int u \, d\mu = \sum_{i=1}^{n} \alpha_i \mu(A_i).$$

For normal representations this equation served as the definition of $\int u \, d\mu$. Its validity without this restriction, which we now perceive, indicates that the introduction of normal representations was simply a technique of proof.

Exercises.

1. Let $(\Omega, \mathscr{A}, \mu)$ be a measure space and $(\Omega, \mathscr{A}_0, \mu_0)$ its completion. Prove that for every $\mathscr{A}_0$-elementary function u there are $\mathscr{A}$-elementary functions u_1, u_2 such that $u_1 \le u \le u_2$ and $\mu(\{u_1 \ne u_2\}) = 0$. For every such pair, $\int u_1 \, d\mu = \int u_2 \, d\mu = \int u \, d\mu_0$. (Cf. Exercise 7(d) in §5.)

2. The function $1_\mathbb{Q}$ on $\mathbb{R}$ has long been known as *Dirichlet's jump function*. Is it a $\mathscr{B}^1$-elementary function?

§11. The integral of non-negative measurable functions

Further progress hinges on the following result:

11.1 Theorem. *For every isotone sequence $(u_n)_{n\in\mathbb{N}}$ of functions from E and every $u \in E$*

$$(11.1) \qquad u \le \sup_{n\in\mathbb{N}} u_n \quad \Rightarrow \quad \int u\,d\mu \le \sup_{n\in\mathbb{N}} \int u_n\,d\mu .$$

Proof. Choose a representation

$$u = \sum_{j=1}^{m} \alpha_j 1_{A_j}$$

of u with sets $A_j \in \mathscr{A}$ and coefficients $\alpha_j \in \mathbb{R}_+$, and let α be any number in $]0, 1[$. Then because of measurability the set

$$B_n := \{u_n \ge \alpha u\}$$

lies in $\mathscr{A}$ for each $n \in \mathbb{N}$. From this definition follows on the one hand that $u_n \ge \alpha u 1_{B_n}$ and consequently by (10.5) and (10.7)

$$\int u_n\,d\mu \ge \alpha \int u 1_{B_n}\,d\mu$$

for every $n \in \mathbb{N}$. Since the sequence (u_n) is isotone and $u \le \sup u_n$, it follows on the other hand that $B_n \uparrow \Omega$, and so $A_j \cap B_n \uparrow A_j$ for each $j \in \{1, \ldots, m\}$ and consequently, because μ is continuous from below

$$\int u\,d\mu = \sum_{j=1}^{m} \alpha_j \mu(A_j) = \lim_{n\to\infty} \sum_{j=1}^{m} \alpha_j \mu(A_j \cap B_n) = \lim_{n\to\infty} \int u 1_{B_n}\,d\mu .$$

Then

$$\sup_{n\in\mathbb{N}} \int u_n\,d\mu \ge \sup_{n\in\mathbb{N}} \alpha \int u 1_{B_n}\,d\mu$$

$$= \alpha \lim_{n\to\infty} \int u 1_{B_n}\,d\mu = \alpha \int u\,d\mu .$$

where the first step follows from $\int u_n\,d\mu \ge \alpha \int u 1_{B_n}\,d\mu$. Since $\alpha \in\ ]0, 1[$ is arbitrary here, the claim follows. $\square$

11.2 Corollary. *For any sequences* $(u_n)_{n\in\mathbb{N}}$, $(v_n)_{n\in\mathbb{N}}$ *of functions from E*

$$(11.2) \qquad \sup_{n\in\mathbb{N}} u_n = \sup_{n\in\mathbb{N}} v_n \quad \Rightarrow \quad \sup_{n\in\mathbb{N}} \int u_n \, d\mu = \sup_{n\in\mathbb{N}} \int v_n \, d\mu \,.$$

Proof. For every $m \in \mathbb{N}$, $v_m \leq \sup_n u_n$ and $u_m \leq \sup_n v_n$, from which inequalities and 11.1 follow

$$\int v_m \, d\mu \leq \sup_{n\in\mathbb{N}} \int u_n \, d\mu \quad \text{and} \quad \int u_m \, d\mu \leq \sup_{n\in\mathbb{N}} \int v_n \, d\mu \,.$$

Claim (11.2) is immediate from the validity of these inequalities for all $m \in \mathbb{N}$. □

Now let

$$(11.3) \qquad\qquad E^* = E^*(\Omega, \mathscr{A})$$

designate the set of all non-negative numerical functions f on Ω for which an isotone sequence (u_n) of functions from E can be found satisfying

$$\sup_{n\in\mathbb{N}} u_n = f \,.$$

Then according to (11.2) the number

$$\sup_{n\in\mathbb{N}} \int u_n \, d\mu \in \overline{\mathbb{R}}_+$$

depends only on f and not on the special representating sequence (u_n) of f used to compute it. We're in a position similar to that of 10.3. Therefore we make the

11.3 Definition. Let f be a function in E^*, represented as the upper envelope $f = \sup_n u_n$ of an isotone sequence $(u_n)_{n\in\mathbb{N}}$ for elementary functions. Then the number

$$(11.4) \qquad\qquad \int f \, d\mu := \sup_{n\in\mathbb{N}} \int u_n \, d\mu \in \overline{\mathbb{R}}_+ \,,$$

shown above to be independent of the special representing (u_n), is called the $(\mu\text{-})$*integral of* f (over Ω).

Evidently $E \subset E^*$, because every $u \in E$ satisfies $u = \sup_n u_n$ for the constant sequence $u_n := u$. Moreover, using this sequence (as we may) in (11.4), we see that in case $f = u \in E$, that definition of the integral coincides with the earlier one. The mapping $f \mapsto \int f \, d\mu$ initially defined only on E is thereby extended to a mapping of E^* into $\overline{\mathbb{R}}_+$. That in this extension process the known properties of the integral persist, will now be confirmed.

The analogs of (10.2) and of (10.5)–(10.7) are

$$(11.5) \quad f, g \in E^*, \alpha \in \mathbb{R}_+ \quad \Rightarrow \quad \alpha f, \ f + g, \ f \cdot g, \ f \vee g, \ f \wedge g \in E^* \,;$$

$$(11.6) \qquad\qquad \int (\alpha f) \, d\mu = \alpha \int f \, d\mu \qquad \text{for all } f \in E^*, \alpha \in \mathbb{R}_+ \,;$$

(11.7)
$$\int (f + g)\, d\mu = \int f\, d\mu + \int g\, d\mu \qquad \text{for all } f, g \in E^*;$$

(11.8)
$$f \leq g \quad \Rightarrow \quad \int f\, d\mu \leq \int g\, d\mu \qquad \text{for all } f, g \in E^*.$$

Proof. From the definition of E^* and from (10.2) follows (11.5). One only has to note that $\sup_n u_n = \lim_n u_n$ for isotone sequences (u_n). The earlier proofs carry over almost verbatim to (11.6) and (11.7). We'll do (11.7) and leave (11.6) for the reader: Let $f = \sup_n u_n$, $g = \sup_n v_n$ be representations of $f, g \in E^*$ by means of isotone sequences of elementary functions. Then by definition

$$\int f\, d\mu = \sup_n \int u_n\, d\mu \quad \text{and} \quad \int g\, d\mu = \sup_n \int v_n\, d\mu,$$
$$\int (f + g)\, d\mu = \sup_n \int (u_n + v_n)\, d\mu.$$

From this and (10.6) we get (11.7), since due to isotoneity

$$\sup_n \int (u_n + v_n)\, d\mu = \lim_n \left(\int u_n\, d\mu + \int v_n\, d\mu \right) = \int f\, d\mu + \int g\, d\mu.$$

If in addition we assume that $f \leq g$, then $u_m \leq \sup_n v_n$ for every $m \in \mathbb{N}$. (11.8) therefore follows from 11.1 □

Properties (11.6)–(11.8) say that the integral is a *positively-homogeneous, additive and isotone* function on E^*.

Finally, it turns out that Theorem 11.1, which is so critical for our program, is valid also in E^*. This is the content of a theorem which goes back to B. LEVI (1875–1961):

11.4 Theorem (on monotone convergence). *For every isotone sequence $(f_n)_{n \in \mathbb{N}}$ of functions from E^**

$$\sup_{n \in \mathbb{N}} f_n \in E^* \quad \text{and} \quad \int \sup_{n \in \mathbb{N}} f_n\, d\mu = \sup_{n \in \mathbb{N}} \int f_n\, d\mu.$$

Proof. Set $f := \sup_n f_n$. It suffices to find an isotone sequence (v_n) of functions from E which satisfy

$$\sup_{n \in \mathbb{N}} v_n = f \quad \text{and} \quad v_n \leq f_n \qquad \text{for every } n \in \mathbb{N}.$$

For then $f \in E^*$ and $\int f\, d\mu = \sup_n \int v_n\, d\mu$ by definition of the integral in E^*, while $\int v_n\, d\mu \leq \int f_n\, d\mu$ by (11.8). Consequently, $\int f\, d\mu \leq \sup_n \int f_n\, d\mu$ and therewith the equality claimed by the theorem follows, since the other inequality $\sup_n \int f_n\, d\mu \leq \int f\, d\mu$ is immediate from (11.8) and the fact that $f_n \leq f$ for all n.

The sequence (v_n) is gotten thus: For each f_n there is by definition an isotone sequence $(u_{mn})_{m\in\mathbb{N}}$ of functions from E with $\sup\limits_{m\in\mathbb{N}} u_{mn} = f_n$. According to (10.2) the functions

$$v_m := u_{m1} \vee \ldots \vee u_{mm}$$

lie in E (for each $m \in \mathbb{N}$). The isotoneity of each sequence $(u_{mn})_{m\in\mathbb{N}}$ clearly entails that of the sequence $(v_m)_{m\in\mathbb{N}}$. From the isotoneity of $(f_m)_{m\in\mathbb{N}}$ follows $v_m \leq f_m$ for all m, and thus $\sup\limits_{m} v_m \leq f$. For all $m \geq n$ we have $u_{mn} \leq v_m$ and so

$$\sup_{m\in\mathbb{N}} u_{mn} = f_n \leq \sup_{m\in\mathbb{N}} v_m \qquad\qquad \text{for every } n \in \mathbb{N}.$$

Together with the preceding this gives finally $\sup\limits_{m} v_m = f$. Therefore (v_n) is a sequence with the needed properties $\square$

11.5 Corollary. *For every sequence* $(f_n)_{n\in\mathbb{N}}$ *of functions from* E^*

$$\sum_{n=1}^{\infty} f_n \in E^* \quad\text{and}\quad \int\left(\sum_{n=1}^{\infty} f_n\right) d\mu = \sum_{n=1}^{\infty} \int f_n \, d\mu\,.$$

Proof. Apply 11.4 to the sequence $(f_1 + \ldots + f_n)_{n\in\mathbb{N}}$ and recall (11.7). $\square$

In analogy with the device of writing $A_n \uparrow A$, $A_n \downarrow A$ for sets, introduced in §3, we will from now on write

$$f_n \uparrow f, \quad f_n \downarrow f$$

for numerical function $f, f_1, f_2, \ldots$ on the set Ω to signal that $f_n(\omega) \uparrow f(\omega)$ for every $\omega \in \Omega$, or $f_n(\omega) \downarrow f(\omega)$ for every $\omega \in \Omega$; that is, the notations mean (f_n) is an isotone sequence and f is its upper envelope, or (f_n) is an antitone sequence and f is its lower envelope. Obviously for a sequence (A_n) of subsets of Ω

$$A_n \uparrow A \quad \Leftrightarrow \quad 1_{A_n} \uparrow 1_A \quad\text{and}\quad A_n \downarrow A \quad \Leftrightarrow \quad 1_{A_n} \downarrow 1_A\,.$$

Examples. 1. Let $(\Omega, \mathscr{A})$ be an arbitrary measurable space and ε_ω the measure defined on $\mathscr{A}$ by unit mass at the point $\omega \in \Omega$ (cf. Example 5 in §3). Then

$$\int f \, d\varepsilon_\omega = f(\omega)$$

for every $f \in E^*$. Due to 11.3 we can at once assume that $f \in E$.

If, however, $f = \sum \alpha_i 1_{A_i}$ is a normal representation of f, then ω lies in exactly one of the sets A_i, say in A_{i_0}. Then $\int f \, d\varepsilon_\omega = \sum \alpha_i \varepsilon_\omega(A_i) = \alpha_{i_0} = f(\omega)$.

2.   Consider $\Omega := \mathbb{N}$ and $\mathscr{A} := \mathscr{P}(\mathbb{N})$. The σ-additivity requirement means that a measure μ on $\mathscr{A}$ is uniquely defined whenever numbers $d_n = \mu(\{n\}) \in \overline{\mathbb{R}}_+$ are specified for each $n \in \mathbb{N}$. E^* consists of all numerical function $f \geq 0$ on Ω. Indeed, one sets $f_n := f(n)1_{\{n\}}$ for each $n \in \mathbb{N}$ and then $f_n \in E^*$, and in case $f(n) < +\infty$,

$f_n \in E$. Since

$$f = \sum_{n=1}^{\infty} f_n,$$

it follows from 11.5 that $f \in E^*$ and

$$\int f \, d\mu = \sum_{n=1}^{\infty} f(n)\alpha_n.$$

3.   Let $(\Omega, \mathscr{A})$ be a measurable space, $(\mu_n)_{n \in \mathbb{N}}$ a sequence of measures on $\mathscr{A}$ and $\mu := \sum\limits_{n=1}^{\infty} \mu_n$ (cf. Example 4 in §3). Then for every $f \in E^*$

$$\int f \, d\mu = \sum_{n=1}^{\infty} \int f \, d\mu_n.$$

This is evidently true of indicator functions f, so the claimed equality holds for all elementary functions. Transition to an arbitrary $f \in E^*$ is accomplished thus: Let (u_n) be a sequence in E with $u_n \uparrow f$. Then the double sequence

$$\alpha_{mn} := \sum_{i=1}^{n} \int u_m \, d\mu_i \qquad\qquad (m, n \in \mathbb{N})$$

satisfies

$$\sup_{m \in \mathbb{N}} \left(\sup_{n \in \mathbb{N}} \alpha_{mn} \right) = \sup_{n \in \mathbb{N}} \left(\sup_{m \in \mathbb{N}} \alpha_{mn} \right) \quad \left(= \sup_{m,n \in \mathbb{N}} \alpha_{mn} \right),$$

which confirms the assertion.

Now that E^* is seen as a natural generalization of E, we might ask for a more workable characterization of it. A surprisingly simple one exists which brings us back to the measurability concept in §9.

11.6 Theorem. E^* *is the set of non-negative, $\mathscr{A}$-measurable, numerical functions on Ω.*

Proof. Every elementary function is measurable and so therefore is every function in E^*, by 9.5. Suppose conversely that f is a non-negative, measurable, numerical function on Ω. The sets

$$A_{in} := \begin{cases} \{f \geq i2^{-n}\} \cap \{f < (i+1)2^{-n}\}, & i = 0, 1, \ldots, n2^n - 1 \\ \{f \geq n\}, & i = n2^n \end{cases}$$

all lie in $\mathscr{A}$, and for each fixed $n \in \mathbb{N}$ the $n2^n$ sets are a decomposition of Ω. Consequently, for each n

$$u_n := \sum_{i=1}^{n2^n} i2^{-n} 1_{A_{in}}$$

is a normal representation of a function in E. On the set A_{in} the function u_{n+1} can take only the values $(2i)2^{-n-1}$ and $(2i+1)2^{-n-1}$ if $i \in \{0, \ldots, n2^n - 1\}$, and only values $\geq n$ when $i = n2^n$. Therefore the sequence (u_n) is isotone. It satisfies $\sup_n u_n = f$, because for any $\omega \in \Omega$ either $f(\omega) = +\infty$, in which case $u_n(\omega) = n$ for every n, or $f(\omega) < +\infty$, in which case $u_n(\omega) \leq f(\omega) < u_n(\omega) + 2^{-n}$ for all $n > f(\omega)$. Thus f lies in E^*. $\square$

Example. 4. Let Ω be an uncountable set, $\mathscr{A}$ the σ-algebra in Ω comprised of all sets which are either countable or have countable complement (introduced in Example 2 of §1). We claim that a numerical function f on Ω is $\mathscr{A}$-measurable just if there is a countable set A in the complement of which f is constant. This constant $\alpha(f)$ does not depend on the particular set A, because if B is another such, $\complement A \cap \complement B$, being the complement in uncountable Ω of the countable set $A \cup B$, is not empty. That this condition really implies the $\mathscr{A}$-measurability of f follows from Theorem 9.1, because for every $a \in \mathbb{R}$ either $\{f \geq a\} \subset A$ or $\complement A \subset \{f \geq a\}$. In proving the converse we can, thanks to (9.8), assume that $f \geq 0$. The claim is then true for elementary functions $f \in E(\Omega, \mathscr{A})$, because among finitely many pairwise disjoint sets whose union is Ω, exactly one has a countable complement. For arbitrary $f \in E^*(\Omega, \mathscr{A})$ let (u_n) be a sequence of elementary functions with $u_n \uparrow f$. Each function u_n is constantly $\alpha(u_n)$ in the complement of some countable set A_n. But then $f(\omega)$ has the constant value $\sup_n \alpha(u_n)$ for all $\omega \in \bigcap_{n \in \mathbb{N}} \complement A_n = \complement\left(\bigcup_{n \in \mathbb{N}} A_n\right)$. As the set $\bigcup_{n \in \mathbb{N}} A_n$ is countable, this proves that f has the asserted property and that moreover $\alpha(f) = \sup_n \alpha(u_n)$.

If now μ is the measure defined in Examples 2 and 7 of §3 which takes only the values 0 and 1, then it follows from the preceding deliberations that

$$\int f \, d\mu = \alpha(f) \qquad \qquad \text{for all } f \in E^*(\Omega, \mathscr{A}).$$

In closing we will use Theorem 11.6 to derive a factorization lemma, due to J.L. Doob, which is interesting in its own right and quite important for its applications in probability theory.

11.7 Factorization lemma. *Let $T : \Omega \to \Omega'$ be a mapping of a set Ω into a measurable space $(\Omega', \mathscr{A}')$ and $f : \Omega \to \overline{\mathbb{R}}$ a numerical function on Ω. The function f is measurable with respect to the σ-algebra $\sigma(T) = T^{-1}(\mathscr{A}')$ in Ω generated by T if and only if there exists a measurable numerical function g on $(\Omega', \mathscr{A}')$ such that*

(11.9) $$f = g \circ T.$$

In case f is $\sigma(T)$-measurable and real (resp., non-negative)-valued, then there is such a g which is real (resp., non-negative)-valued.

Proof. If f has the form $f = g \circ T$ as specified, then it is the composite of a $\boldsymbol{\sigma}(T)$-$\mathscr{A}'$-measurable with an $\mathscr{A}'$-$\overline{\mathscr{B}}^1$-measurable mapping, making it $\boldsymbol{\sigma}(T)$-$\overline{\mathscr{B}}^1$-measurable. For the proof of the converse we distinguish three cases:

1. Let $f = \sum\limits_{i=1}^{n} \alpha_i 1_{A_i}$ be a $\boldsymbol{\sigma}(T)$-elementary function; so $A_i \in \boldsymbol{\sigma}(T)$ and $\alpha_i \in \mathbb{R}_+$ for $i = 1, \ldots, n$. For each A_i there is a set $A_i' \in \mathscr{A}'$ with $A_i = T^{-1}(A_i')$, by definition of $\boldsymbol{\sigma}(T)$. Therefore the function $g := \sum\limits_{i=1}^{n} \alpha_i 1_{A_i'}$ does what is wanted.

2. Let $f \geq 0$. According to Theorem 11.6 there is an isotone sequence $(u_n)_{n \in \mathbb{N}}$ of $\boldsymbol{\sigma}(T)$-elementary functions with $f = \sup\limits_{n} u_n$, and by the proof just given, there are $\mathscr{A}'$-elementary functions g_n such that $u_n = g_n \circ T$. The function $g := \sup\limits_{n} g_n$ then does what is wanted in this case.

3. An arbitrary $\boldsymbol{\sigma}(T)$-measurable $f : \Omega \to \overline{\mathbb{R}}$ decomposes into its positive part f^+ and its negative part f^-. From 2. we get $\mathscr{A}'$-measurable $g_0' \geq 0$ and $g_0'' \geq 0$ on Ω' for which $f^+ = g_0' \circ T$ and $f^- = g_0'' \circ T$. For ω' in the set $U' := \{g_0' = +\infty\} \cap \{g_0'' = +\infty\}$ the difference $g_0'(\omega') - g_0''(\omega')$ is not defined. But the set $T(\Omega)$ is disjoint from U', because $g_0'(T(\omega)) = +\infty$ always entails that $g_0''(T(\omega)) = f^-(\omega) = 0$. Therefore if we set

$$g' := 1_{\complement U'} g_0' \quad \text{and} \quad g'' := 1_{\complement U'} g_0'',$$

then $g := g' - g''$ will do the desired job.

4. If f is real, 3. supplies a numerical $\mathscr{A}'$-measurable function g_0 on Ω' such that $f = g_0 \circ T$. If we set $U := \{|g_0| = +\infty\}$, then $U \cap T(\Omega) = \emptyset$ since f takes only real values, and so the real function $g := 1_{\complement U} g_0$ does what is wanted. □

Remark. The restriction of g to $T(\Omega)$ is uniquely determined by f and (11.9). Specifically, for each $\omega' \in T(\Omega)$, $g(\omega') = f(\omega)$ for every $\omega \in T^{-1}(\omega')$. On $T(\Omega)$ one therefore has no other choice than to set $g(T(\omega)) := f(\omega)$. In case $T(\Omega) \in \mathscr{A}'$, in particular when $T(\Omega) = \Omega'$, the existence of g can thus be secured without recourse to 11.7 – cf. Exercise 3 below.

The factorization lemma is therefore noteworthy only in so far as it allows the measurability of $T(\Omega)$ to be dispensed with. And in doing that the special structure of $(\overline{\mathbb{R}}, \overline{\mathscr{B}}^1)$ is critical. Remark 4 in §8 shows how we are sometimes forced to do without the measurability of $T(\Omega)$.

Exercises.

1. Show that every bounded, $\mathscr{A}$-measurable, non-negative real-valued function on a measurable space $(\Omega, \mathscr{A})$ is the uniform limit of an isotone sequence of $\mathscr{A}$-measurable elementary functions.

2. Let $(\Omega, \mathscr{A}, \mu)$ be a measurable space with a *finite* measure μ. Further, let $f, f_1, f_2, \ldots$ be measurable numerical functions on Ω. Prove the equivalence of

the two assertions:

(i)
$$\lim_{n\to\infty} \mu\left(\bigcup_{m\geq n} \{f_m > f + \varepsilon\}\right) = 0 \qquad \text{for every } \varepsilon > 0;$$

(ii) for every $\delta > 0$ there exists an $A_\delta \in \mathscr{A}$ with $\mu(A_\delta) < \delta$ such that for every $\varepsilon > 0$, $f_n(\omega) \leq f(\omega) + \varepsilon$ holds for all $\omega \in \complement A_\delta$ and all sufficiently large $n \in \mathbb{N}$. [*Hints*: Note that (i) is also equivalent to the statement that for every $\varepsilon > 0$ and $\delta > 0$ there exists an $A_{\delta,\varepsilon} \in \mathscr{A}$ with $\mu(A_{\delta,\varepsilon}) < \delta$ and an $N_{\delta,\varepsilon} \in \mathbb{N}$ such that $f_n(\omega) \leq f(\omega) + \varepsilon$ for all $\omega \in \complement A_{\delta,\varepsilon}$ and $n \geq N_{\delta,\varepsilon}$.] Why does (i) hold, given the sequence $(f_n)_{n\in\mathbb{N}}$, for every measurable function f which satisfies $f \geq \limsup_{n\to\infty} f_n$?

3. With the hypotheses and notation of the factorization lemma, show that for any $\omega_1, \omega_2 \in \Omega$ with $T(\omega_1) = T(\omega_2)$, and every $C \in \sigma(T)$, either $\omega_1, \omega_2 \in C$ or $\omega_1, \omega_2 \in \complement C$. (That is, ω_1 and ω_2 cannot be "separated" by any set in $\sigma(T)$.) From this fact infer that a $\sigma(T)$-measurable f satisfies $f(\omega_1) = f(\omega_2)$ whenever $T(\omega_1) = T(\omega_2)$. In case $T(\Omega) \in \mathscr{A}'$, deduce the existence of a $\sigma(T)$-measurable mapping $g : \Omega' \to \overline{\mathbb{R}}$ with $f = g \circ T$. [*Hint*: Consider the system $\mathscr{C}$ of all $C \subset \Omega$ which have this two-point property and conclude that $\sigma(T) \subset \mathscr{C}$. Further, take note of the equality $T(T^{-1}(A')) = A' \cap T(\Omega)$ for $A' \subset \Omega'$.]

§12. Integrability

By now the integral $\int f \, d\mu$ is defined for all non-negative $\mathscr{A}$-measurable numerical functions on Ω, as a result of 11.4 and 11.6 together. In a third and final step $\int f \, d\mu$ will now be defined for certain numerical functions f which are not of constant sign.

According to Theorem 9.8, f is measurable just if both its positive part f^+ and its negative part f^- are measurable. This remark prompts the following definition:

12.1 Definition. A numerical function f on the measure space $(\Omega, \mathscr{A}, \mu)$ is called (μ-)*integrable* if it is $\mathscr{A}$-measurable and the integrals $\int f^+ \, d\mu$, $\int f^- \, d\mu$ are real numbers. Then

$$\int f \, d\mu := \int f^+ \, d\mu - \int f^- \, d\mu.$$

is called the (μ-)*integral* of f (over Ω).

If for some reason one wants to put the variable $\omega \in \Omega$ into evidence, he also writes

$$\int f(\omega)\mu(d\omega) \quad \text{or} \quad \int f(\omega) \, d\mu(\omega).$$

Remarks. 1. The right side of (12.1) is meaningful for measurable f if at least one of f^+, f^- has a real integral. One says that then f is *quasi-integrable* or that the *integral of f exists* and one uses (12.1) to define $\int f \, d\mu \in \overline{\mathbb{R}}$. Only occasionally will we be concerned with this obvious generalization.

2. In the special case $\mu = \lambda^d$ we speak of *Lebesgue integrable functions* (on $\mathbb{R}^d$) and of their *Lebesgue integrals*. If a Borel measure μ_F on $\mathbb{R}^d$ is described with the help of a measure-generating function F on $\mathbb{R}^d$ (cf. §6), the μ_F-integrable functions f on $\mathbb{R}^d$ are called *Lebesgue–Stieltjes integrable* (or *Stieltjes integrable*) with respect to F. One speaks of its *(Lebesgue-)Stieltjes integral* and writes $\int f \, dF$ instead of $\int f \, d\mu_F$. The general theory of measure and integration has however displaced this terminology and the notation $\int f \, dF$, despite their historical significance.

Let us now summarize the most important properties of the conceptual edifice just built:

12.2 Theorem. *Each of the following four statements is equivalent to the integrability of the measurable numerical function f on Ω:*
(a) *f^+ and f^- are integrable.*
(b) *There are integrable functions $u \geq 0$, $v \geq 0$ such that $f = u - v$. (Note that the last equality entails that $u(\omega) - v(\omega)$ is defined (in $\overline{\mathbb{R}}$) for every $\omega \in \Omega$.)*
(c) *There is an integrable function g with $|f| \leq g$.*
(d) *$|f|$ is integrable.*

From (b) follows: $\int f \, d\mu = \int u \, d\mu - \int v \, d\mu$.

Proof. What has to be shown is the equivalence of (a) through (d), since (a) constitutes the definition of f being integrable.

(a)$\Rightarrow$(b): According to (9.8), $u := f^+$ and $v := f^-$ do the job required in (b).

(b)$\Rightarrow$(c): Because the integral is additive on E^*, along with u and v, $u + v$ is also integrable. Since $f = u - v \leq u \leq u + v$ and $-f = v - u \leq v \leq u + v$, the function $g := u + v$ is as required.

(c)$\Rightarrow$(d): This follows from the isotoneity of the integral on E^* and the fact that $|f| \in E^*$ (Theorems 11.6 and 9.8): $\int |f| \, d\mu \leq \int g \, d\mu < +\infty$.

(d)$\Rightarrow$(a): Upon recalling that $f^+ \leq |f|$ and $f^- \leq |f|$, this too follows from the isotoneity of the integral on E^*.

In (b), $f = u - v = f^+ - f^-$ and so $u + f^- = v + f^+$, which via (11.7) yields $\int u \, d\mu + \int f^- \, d\mu = \int v \, d\mu + \int f^+ \, d\mu$ and therewith the last assertion of the theorem, since all the integrals here are finite. □

12.3 Theorem. *Let f and g be integrable numerical functions on Ω, $\alpha \in \mathbb{R}$. Then the functions αf and, if it is everywhere defined on Ω, $f + g$ are integrable, and satisfy*

$$(12.2) \qquad \int (\alpha f) \, d\mu = \alpha \int f \, d\mu \quad and \quad \int (f + g) \, d\mu = \int f \, d\mu + \int g \, d\mu.$$

Furthermore, the functions

$$f \vee g \quad and \quad f \wedge g$$

are integrable.

Proof. The claims regarding αf follow from (11.6), since

$$(\alpha f)^+ = \alpha f^+, \qquad (\alpha f)^- = \alpha f^- \qquad \text{if } \alpha \geq 0, \text{ and}$$
$$(\alpha f)^+ = |\alpha| f^-, \qquad (\alpha f)^- = |\alpha| f^+ \qquad \text{if } \alpha < 0.$$

Regarding $f + g$, we argue as follows: from $f = f^+ - f^-$ and $g = g^+ - g^-$ follow $f + g = f^+ + g^+ - (f^- + g^-)$. (11.7) insures that $u := f^+ + g^+$ and $v := f^- + g^-$ are integrable. Then the claims about $f + g$ follow from the equality $f + g = u - v$ via 12.2. Finally, $|f \vee g| \leq |f| + |g|$ and $|f \wedge g| \leq |f| + |g|$, and we know that $|f| + |g|$ is integrable. The integrability of the measurable functions $f \vee g$ and $f \wedge g$ follows then from these inequalities and part (c) of 12.2.   □

12.4 Theorem. *For any integrable numerical functions f and g on Ω*

$$(12.3) \qquad\qquad f \leq g \quad \Rightarrow \quad \int f \, d\mu \leq \int g \, d\mu;$$

$$(12.4) \qquad\qquad \left| \int f \, d\mu \right| \leq \int |f| \, d\mu.$$

Proof. From $f \leq g$ follows $f^+ \leq g^+$ and $f^- \geq g^-$, and from these inequalities and the isotoneity of the integral on E^* follows (12.3). Because $f \leq |f|$ and $-f \leq |f|$, (12.4) follows from the first equality in (12.2) and from (12.3), with $|f|$ in the role of g there.   □

The relevant properties are particularly clearly perceptible when we consider only real-valued integrable functions. To aid in that we define

$$(12.5) \qquad \mathscr{L}^1(\mu) := \text{ the set of all } \mu\text{-integrable } \textit{real} \text{ functions on } \Omega.$$

Using this widespread notation it follows immediately from Theorem 12.3 and from (12.3) that: With respect to the operations

$$(f + g)(\omega) := f(\omega) + g(\omega) \quad \text{and} \quad (\alpha f)(\omega) := \alpha f(\omega) \qquad\qquad \omega \in \Omega$$

of pointwise addition and multiplication by scalars $\alpha \in \mathbb{R}$, $\mathscr{L}^1(\mu)$ is a *vector space* over $\mathbb{R}$, and on it $f \mapsto \int f \, d\mu$ is an (isotone) *linear form.*

Examples. 1. Let $(\Omega, \mathscr{A})$ be any measurable space, ε_ω the measure on $\mathscr{A}$ defined by unit mass at $\omega \in \Omega$. According to Example 1 of §11, the ε_ω-integrable functions are just the $\mathscr{A}$-measurable numerical function f on Ω with $|f(\omega)| < +\infty$. For them

$$\int f \, d\varepsilon_\omega = f(\omega).$$

2.   Let $(\Omega, \mathscr{A}, \mu)$ be the measure space defined in Example 2 of §11, $\mu(\{n\}) = \alpha_n$ for $n \in \mathbb{N}$. From what was shown there it follows that the μ-integrable functions

$f : \Omega \to \overline{\mathbb{R}}$ are precisely those for which

$$\sum_{n=1}^{\infty} |f(n)| \, \alpha_n < +\infty$$

and for such an f

$$\int f \, d\mu = \sum_{n=1}^{\infty} f(n) \alpha_n \,.$$

3. Let $(\Omega, \mathscr{A}, \mu)$ be the measure space defined in Examples 2 and 1 of §3. A function $f : \Omega \to \overline{\mathbb{R}}$ is then μ-integrable if and only if it is equal to a *real* constant α throughout the complement of some countable subset of Ω. From Example 4 of §11 we have $\int f \, d\mu = \alpha$ for such an f.

4. Let $(\Omega, \mathscr{A}, \mu)$ be a measure space with $\mu(\Omega) < +\infty$. Then every constant real function, and consequently after 12.2, every bounded, measurable real function on Ω is μ-integrable.

5. Let μ and ν be measures on a σ-algebra $\mathscr{A}$ in Ω. A numerical function f on Ω is $(\mu + \nu)$-integrable if and only if it is both μ- and ν-integrable, and in this case

$$\int f \, d(\mu + \nu) = \int f \, d\mu + \int f \, d\nu \,.$$

In fact: For every non-negative $\mathscr{A}$-measurable function g on Ω, $\int g \, d(\mu + \nu) = \int g \, d\mu + \int g \, d\nu$ holds by Example 3 in §11. Applied to $g := |f|$ this and 12.2 prove the integrability claim, and applied to $g := f^+$ and $g := f^-$ it implies the claimed equality. In particular

$$\mathscr{L}^1(\mu + \nu) = \mathscr{L}^1(\mu) \cap \mathscr{L}^1(\nu)$$

is valid.

We can now free ourselves of the restriction that functions always be integrated over the whole Ω. (11.5) insures that along with any pair of functions from $E^* = E^*(\Omega, \mathscr{A})$ their product is also in E^*. So from $f \in E^*$ and $A \in \mathscr{A}$ follows $1_A f \in E^*$. If f is an integrable numerical function on Ω, then so is $1_A f$, for every $A \in \mathscr{A}$: Because of the trivial inequality $|1_A f| \le |f|$, this is immediate from 12.2 (and 9.4). In the light of this the following seems natural:

12.4 Definition. If f is a numerical function which is defined on Ω and either belongs to E^* or is μ-integrable, we set

(12.6)
$$\int_A f \, d\mu := \int 1_A f \, d\mu$$

for every $A \in \mathscr{A}$ and call it the μ-integral of f over A.

As a special case of this notation

(12.7)
$$\int_{\Omega} f \, d\mu = \int f \, d\mu \,.$$

The following rules of calculation are evident, for all f, g which either lie in E^* or are integrable:

$$(12.8) \qquad \int_{A \cup B} f \, d\mu + \int_{A \cap B} f \, d\mu = \int_A f \, d\mu + \int_B f \, d\mu \quad \text{for all } A, B \in \mathscr{A}$$

and, as a special case,

$$(12.8') \qquad \int_{A \cup B} f \, d\mu = \int_A f \, d\mu + \int_B f \, d\mu \qquad \text{for all disjoint } A, B \in \mathscr{A};$$

(12.9) $A \in \mathscr{A}$ and $f(\omega) \leq g(\omega)$ for all $\omega \in A$ $\Rightarrow$ $\displaystyle\int_A f \, d\mu \leq \int_A g \, d\mu$.

One merely has to reflect on the definitions involved. Moreover, pursuant to the discussion after (12.5),

$$(12.10) \qquad f \mapsto \int_A f \, d\mu \quad \text{is a linear form on } \mathscr{L}^1(\mu), \qquad \text{for each } A \in \mathscr{A}.$$

But we can get at integrals over sets in $\mathscr{A}$ in a different way, namely by considering the restriction μ_A of the given measure μ to the trace σ-algebra $A \cap \mathscr{A}$. That one is thereby led to the same result is the content of

12.5 Lemma. *Let $A \in \mathscr{A}$ and for every function f on Ω which either lies in E^* or is μ-integrable let f' denote the restriction of f to A, and μ_A the restriction of μ to $A \cap \mathscr{A}$. Then*

$$(12.11) \qquad \int f' \, d\mu_A = \int_A f \, d\mu.$$

Proof. First consider $f \in E^*(\Omega, \mathscr{A})$. Then $f' \in E^*(A, A \cap \mathscr{A})$ since

$$(f')^{-1}(B) = A \cap f^{-1}(B)$$

holds for all Borel sets B in $\overline{\mathbb{R}}$ (cf. 11.6). For the function $1_A f \in E^*$ there is a sequence (u_n) of $\mathscr{A}$-elementary functions satisfying $u_n \uparrow 1_A f$. The sequence (u'_n) of restrictions to A obviously consists of $A \cap \mathscr{A}$-elementary functions that satisfy $u'_n \uparrow f'$, from all of which follows that

$$(12.12) \qquad \int_A f \, d\mu = \sup_{n \in \mathbb{N}} \int u_n \, d\mu \quad \text{and} \quad \int f' \, d\mu_A = \sup_{n \in \mathbb{N}} \int u'_n \, d\mu_A.$$

Since $0 \leq u_n \leq 1_A f$, $u_n = 0$ in $\complement A$, so $u_n = 1_A u_n$ and consequently

$$u_n = \sum_{i=1}^{k_n} \alpha_i 1_{A_i}$$

for appropriate (depending on n) sets $A_i \in \mathscr{A}$ which are all subsets of A, and appropriate (also n-dependent) real coefficients $\alpha_i \geq 0$, and $k_n \in \mathbb{N}$. It follows

that

$$u'_n = \sum_{i=1}^{k_n} \alpha_i 1'_{A_i} .$$

(Notice that for $Q \subset A$, the restriction $1'_Q$ coincides with the indicator function with respect to A of Q.) From the last two equalities we see that

$$\int u_n \, d\mu = \int u'_n \, d\mu_A \qquad \text{for all } n \in \mathbb{N},$$

because each integral equals $\sum_{i=1}^{k_n} \alpha_i \mu(A_i)$, and from these equalities and (12.12) follows (12.11) for $f \in E^*(\Omega, \mathscr{A})$.

If f is μ-integrable the preceding can be applied to both f^+ and f^-. All integrals are finite and it is obvious that $(f')^+ = (f^+)'$, $(f')^- = (f^-)'$, so (12.11) follows from linearity of the integral. □

In a final step of generalization let us note that we can conversely proceed from an $A \in \mathscr{A}$ and a function f on A which either lies in $E^*(A, A \cap \mathscr{A})$ or is μ_A-integrable to define the μ-integral of f over A via

(12.13) $$\int_A f \, d\mu := \int f \, d\mu_A$$

and in the second case to say that f is also μ-integrable over A. With the aid of Lemma 12.5 we thereby get:

12.6 Corollary. *A numerical function f defined on a set $A \in \mathscr{A}$ is μ-integrable over A if the function defined on the whole of Ω by*

$$f_A(\omega) := \begin{cases} f(\omega) & \text{if } \omega \in A \\ 0 & \text{if } \omega \in \Omega \setminus A \end{cases}$$

is μ-integrable. In this case

$$\int_A f \, d\mu = \int f_A \, d\mu = \int f \, d\mu_A .$$

From this discussion we see that a μ-integral over a set $A \in \mathscr{A}$ is nothing other than a μ_A-integral over the new base space A. It can also be thought of as a μ-integral over Ω employing the integrands f_A.

Exercises.

1. Characterize the functions $u \in E(\Omega, \mathscr{A})$ which are μ-integrable.

2. Let $(\Omega, \mathscr{A}, \mu)$ be a measure space. The indicator function 1_A of a set $A \in \mathscr{A}$ is μ-integrable just when $\mu(A) < +\infty$. Such sets are called μ-integrable, and $\mathscr{R}$ will denote their totality. Show that $\mathscr{R}$ is an ideal in the ring $\mathscr{A}$ (cf. Exercise 4 in §1); in particular, $R \in \mathscr{R}$ and $A \in \mathscr{A} \Rightarrow A \cap R \in \mathscr{R}$. For a σ-finite measure μ a converse also holds: $A \subset \Omega$ and $A \cap R \in \mathscr{R}$ for all $R \in \mathscr{R}$ implies $A \in \mathscr{A}$.

3. Is the Dirichlet jump function from Exercise 2, §10 λ^1-integrable?

4. Consider the measurable space $(\Omega, \mathscr{A})$ from Example 4 of §11. On $\mathscr{A}$ is defined the measure μ which assigns countable sets the value 0, uncountable sets (from $\mathscr{A}$) the value $+\infty$. Determine all the μ-integrable functions and their integrals.

5. Let $(\Omega, \mathscr{A}, \mu)$ be any measure space, (A_n) a sequence of pairwise disjoint sets from $\mathscr{A}$, A their union, f a numerical function on A. Show that f is μ-integrable over A if and only if it is μ-integrable over each A_n and $\sum\limits_{n=1}^{\infty} \int_{A_n} |f| \, d\mu < +\infty$.

6. Let $(\Omega, \mathscr{A}, \mu)$ be a measure space with μ finite. Show that every real function f on Ω which is the uniform limit of a sequence (f_n) in $\mathscr{L}^1(\mu)$ itself belongs to $\mathscr{L}^1(\mu)$. Why does this conclusion fail for every non-finite μ which is σ-finite? [*Hint*: Construct a sequence (g_n) in $\mathscr{L}^1(\mu)$ with $0 \le g_n \le 1$ and $\int g_n \, d\mu \ge n^2$ for each $n \in \mathbb{N}$ and then consider $f_n := \sum\limits_{j=1}^{n} j^{-2} g_j$.]

§13. Almost everywhere prevailing properties

For the further construction of the theory the concept of a negligible set, already frequently mentioned in Chapter I, will now play an important role. We recall: $N \subset \Omega$ is called a $(\mu\text{-})nullset$ if $N \in \mathscr{A}$ and $\mu(N) = 0$. The union of every sequence of μ-nullsets is again one (3.10), as is every set in $\mathscr{A}$ which is contained in a μ-nullset, thanks to isotoneity (cf. Exercise 5 in §3).

13.1 Definition. Let η be a property of points in Ω: every $\omega \in \Omega$ either enjoys property η or does not. We say that "$(\mu\text{-})almost\ all\ points\ of\ \Omega\ have\ property\ \eta$" or "$\eta\ prevails\ (\mu\text{-})almost\ everywhere\ in\ \Omega$" if there is a μ-nullset N such that all points of $\complement N$ enjoy property η.

Be careful: It is not required that the set N_η of all $\omega \in \Omega$ which enjoy property η be a μ-nullset. Indeed, generally N_η may not belong to $\mathscr{A}$. For example, if A is a subset of Ω which does not belong to $\mathscr{A}$ and η is the property "ω is a point of A", then $N_\eta = \complement A$ is not in $\mathscr{A}$.

Examples of properties η which will come up in the sequel are: Equality of the values at a point $\omega \in \Omega$ of two functions f and g which are defined on Ω, finiteness of the value at $\omega \in \Omega$ of a function f, etc. Corresponding to these we have the following modes of speaking: f and g are $(\mu\text{-})$ *almost everywhere equal* on Ω, in symbols

$$f = g \qquad (\mu\text{-})\text{almost everywhere};$$

f is $(\mu\text{-})$ *almost everywhere finite*, in symbols

$$|f| < +\infty \qquad (\mu\text{-})\text{almost everywhere};$$

f is $(\mu\text{-})$*almost everywhere bounded*, meaning that for some $\alpha \in \mathbb{R}$

$$|f| \leq \alpha \qquad (\mu\text{-})\text{almost everywhere},$$

etc.

The theorems that follow explicate the significance that this new concept has for integration theory:

13.2 Theorem. *For every $f \in E^*(\Omega, \mathscr{A})$, that is, (cf. 11.6) for every $\mathscr{A}$-measurable, non-negative numerical function f*

$$\int f \, d\mu = 0 \quad \Leftrightarrow \quad f = 0 \quad \mu\text{-almost everywhere.}$$

Proof. Since f is measurable, the set

$$N := \{f \neq 0\} = \{f > 0\}$$

lies in $\mathscr{A}$. What has to be shown is that

$$\int f \, d\mu = 0 \quad \Leftrightarrow \quad \mu(N) = 0.$$

Suppose $\int f \, d\mu = 0$. For each $n \in \mathbb{N}$ the set $A_n := \{f \geq n^{-1}\}$ also lies in $\mathscr{A}$ and $A_n \uparrow N$, so that $\mu(N) = \lim_{n \to \infty} \mu(A_n)$ and it is enough to show that $\mu(A_n) = 0$ for every n. But obviously $f \geq n^{-1} 1_{A_n}$, entailing that $0 = \int f \, d\mu \geq n^{-1} \mu(A_n) \geq 0$, that is, $\mu(A_n) = 0$, as wanted.

Suppose conversely that $\mu(N) = 0$. Each of the functions $u_n := n 1_N$ $(n \in \mathbb{N})$ lies in $E(\Omega, \mathscr{A})$ and satisfies $\int u_n \, d\mu = 0$. Setting $g := \sup_n u_n$ gives a function $g \in E^*(\Omega, \mathscr{A})$ such that $u_n \uparrow g$, so $\int g \, d\mu = \sup_n \int u_n \, d\mu = 0$. Finally, since evidently $f \leq g$, $0 \leq \int f \, d\mu \leq \int g \, d\mu = 0$ gives the desired equality $\int f \, d\mu = 0$. $\square$

13.3 Corollary. *Every $\mathscr{A}$-measurable numerical function f on Ω is integrable over every μ-nullset N, and*

$$\int_N f \, d\mu = 0.$$

Proof. If $f \geq 0$, this claim follows from the theorem, because each function $1_N f$ lies in $E^*(\Omega, \mathscr{A})$ and is almost everywhere 0. In turn, application of this to f^+ and f^- delivers the full claim. $\square$

13.4 Theorem. *Let f, g be $\mathscr{A}$-measurable numerical functions on Ω which are μ-almost everywhere equal on Ω. Then*

(a) $$f \geq 0, g \geq 0 \quad \Rightarrow \quad \int f \, d\mu = \int g \, d\mu;$$

(b) $$f \text{ integrable} \quad \Rightarrow \quad g \text{ integrable and } \int f \, d\mu = \int g \, d\mu.$$

Proof. (a): By hypothesis (and 9.3) $N := \{f \neq g\}$ is a μ-nullset. From 13.3 then

$$\int_N f \, d\mu = \int_N g \, d\mu = 0 \, .$$

On the other hand, for $M = \complement N$ we have $1_M f = 1_M g$ due to the definition of N, and so by (12.6)

$$\int_M f \, d\mu = \int_M g \, d\mu \, .$$

Adding integrals and using (12.8′) leads to the conclusion in (a).

(b): The almost everywhere equality hypothesis entails that

$$f^+ = g^+ \text{ almost everywhere and } f^- = g^- \text{ almost everywhere.}$$

From (a) then

$$\int f^+ \, d\mu = \int g^+ \, d\mu \quad \text{and} \quad \int f^- \, d\mu = \int g^- \, d\mu \, .$$

Because f is integrable, what we have here are non-negative real numbers, showing that g is integrable (part (a) of 12.2) and, upon subtracting the second equality from the first, we get the equality claimed in (b). □

Since, roughly speaking, all this shows that integrability and the integral of a function are insensitive to (measurable) changes of the function on nullsets, results proved earlier can easily be reformulated somewhat more sharply. For example:

13.5 Corollary. *Let the $\mathscr{A}$-measurable numerical functions f and g on Ω satisfy $|f| \leq g$ μ-almost everywhere. Then along with g, the function f will also be μ-integrable.*

Proof. If we set $g' := g \vee |f|$, then g' is measurable, $g' = g$ almost everywhere and $|f| \leq g'$. From 13.4 part (b) we see first of all that g' is integrable, and then from 12.2 f is as well. □

Of special importance is the realization that integrability imposes limitations on how often a function can assume the values $\pm\infty$, or indeed any non-zero value. This is made precise in

13.6 Theorem. *Every μ-integrable numerical function f on Ω is μ-almost everywhere real-valued. Moreover, the set $\{f \neq 0\}$ is of σ-finite measure.*

Where a set $A \in \mathscr{A}$ is said to *possess σ-finite measure* if it is the union of a sequence of sets in $\mathscr{A}$ each of which has finite measure. This means nothing other than that the restriction of μ to $A \cap \mathscr{A}$ is a σ-finite measure.

Proof. The set $N := \{|f| = +\infty\}$ lies in $\mathscr{A}$ and for every real $\alpha \geq 0$ satisfies $\alpha 1_N \leq |f|$. Consequently, $\alpha\mu(N) \leq \int |f| \, d\mu < +\infty$, from which follows the first claim, $\mu(N) = 0$. To prove the second claim we pass over to $|f|$ and thereby assume that $f \geq 0$. Then

$$\{f \neq 0\} = \{f > 0\} = \bigcup_{n \in \mathbb{N}} \{f \geq n^{-1}\}.$$

Every set $A_n := \{f \geq n^{-1}\} = \{nf \geq 1\}$ satisfies $1_{A_n} \leq nf$ and therewith

$$\mu(A_n) \leq n \int f \, d\mu < +\infty.$$

This holds for all $n \in \mathbb{N}$, confirming the σ-finiteness claim. $\square$

Theorem 13.6 has yet another consequence: Let N be a μ-nullset and f a numerical function which is defined on $M := \complement N$ and is $M \cap \mathscr{A}$-measurable. Such a function is described as being a $(\mu\text{-})$*almost everywhere defined $(\mathscr{A}\text{-})$measurable function*. The function f_M introduced in 12.6 extends it to an $\mathscr{A}$-measurable function on Ω. Any other extension of f to Ω must agree with f_M almost everywhere. According to 13.4 therefore either every such extension is integrable or none is. In the first case moreover all extensions have the same μ-integral. These observations justify the following definition:

13.7 Definition. Let f be a μ-*almost everywhere defined*, $\mathscr{A}$-measurable numerical function on Ω. It will be called $(\mu\text{-})$*integrable* if it can be extended to a $(\mu\text{-})$integrable function f' defined on the whole of Ω. $\int f' \, d\mu$ will then be called the $(\mu\text{-})$integral of f and denoted $\int f \, d\mu$.

We will only occasionally be concerned with this extension of the integral concept, but its utility is already shown by the following

Remark. Suppose f and g are integrable numerical functions on Ω. According to 13.6 each is almost everywhere finite. Because the union of two nullsets is itself a nullset, there is a nullset N such that both $|f(\omega)| < +\infty$ and $|g(\omega)| < +\infty$ for all $\omega \in \complement N$. But then

$$\omega \mapsto f(\omega) + g(\omega) \qquad\qquad (\omega \in \complement N)$$

is an almost everywhere defined measurable function. This fact, in conjunction with what was shown above, shows that the explicit hypothesis made in 12.3 that $f + g$ be everywhere defined is of little significance. For two integrable numerical functions f and g on Ω the sum $f + g$ is almost everywhere defined, and in the sense of 13.7 integrable. The equality

$$\int (f + g) \, d\mu = \int f \, d\mu + \int g \, d\mu$$

prevails unrestrictedly.

Exercises.

1. The numerical functions f and g on the measure space $(\Omega, \mathscr{A}, \mu)$ satisfy $f = g$ μ-almost everywhere. Show via an example that in general the $\mathscr{A}$-measurability

of g does not follow from that of f. Show however that in case $(\Omega, \mathscr{A}, \mu)$ is complete, the $\mathscr{A}$-measurability of g is equivalent to that of f.

2. Let $(\Omega, \mathscr{A}, \mu)$ be a measure space, $(\Omega, \mathscr{A}_0, \mu_0)$ its completion. Prove that $f : \Omega \to \overline{\mathbb{R}}$ is $\mathscr{A}_0$-measurable just if $\mathscr{A}$-measurable numerical functions f_1, f_2 on Ω exist with the properties $f_1 \le f \le f_2$ everywhere in Ω and $f_1 = f_2$ μ-almost everywhere. If f is μ_0-integrable, then any functions f_1, f_2 with these properties are μ-integrable, and $\int f_1 \, d\mu = \int f_2 \, d\mu = \int f \, d\mu_0$. (This supplements Exercise 7 in §5 and generalizes Exercise 1 in §10.)

3. Even if the f in the preceding exercise is real-valued, the functions f_1, f_2 which were proved to exist there cannot always be chosen to be real-valued. Prove this for the case where Ω is any infinite set, $\mathscr{A} := \{\emptyset, \Omega\}$ and $\mu := 0$.

§14. The spaces $\mathscr{L}^p(\mu)$

According to 9.4 the product of two measurable functions is again measurable. By contrast however the product of two integrable functions is not generally integrable, as the next example shows:

Example. $(\Omega, \mathscr{A}, \mu)$ is the measurable space described in Example 2 of §12 and Example 2 of §11, with $\alpha_n := n^{-p-1}$ for each $n \in \mathbb{N}$, where $1 < p < +\infty$. The identity function, $f(n) := n$ for all $n \in \mathbb{N}$, is integrable, but its p-th power is not. Thus for $p = 2$, $f^2 = f \cdot f$ is not integrable.

This observation suggests the investigation of those measurable functions f on Ω for which $|f|^p$ is integrable.

In what follows p will designate a real number, $p \ge 1$. For every $\mathscr{A}$-measurable function f on Ω, $|f|$ and then also $|f|^p$ is measurable, because (adopting the usual convention that $(+\infty)^p := \infty$) for every real α

$$\{|f|^p \ge \alpha\} = \begin{cases} \Omega & \text{if } \alpha \le 0 \\ \{|f| \ge \alpha^{1/p}\} & \text{if } \alpha > 0. \end{cases}$$

For such an f

(14.1) $$N_p(f) := \left(\int |f|^p \, d\mu \right)^{1/p}$$

is therefore defined. It satisfies $0 \le N_p(f) \le +\infty$ and, clearly,

(14.2) $$N_p(\alpha f) = |\alpha| \, N_p(f) \qquad \text{for all } \alpha \in \mathbb{R}.$$

Two deeper properties will now be established:

14.1 Theorem. $p > 1$ *is a real number and* $q > 1$ *is defined by the equation*

$$\frac{1}{p} + \frac{1}{q} = 1.$$

Then for any measurable numerical functions f, g *on* Ω

(14.3) $$N_1(fg) \leq N_p(f)N_q(g) \qquad \text{(HÖLDER's inequality)}.$$

Proof. It is clear from definition (14.1) that we may assume $f \geq 0$ and $g \geq 0$. Setting

$$\sigma := N_p(f) \quad \text{and} \quad \tau := N_q(g),$$

we can also assume that both these numbers are positive. For if, say $\sigma = 0$, then by 13.2 f^p, whence also f, is almost everywhere equal to 0. The same is then true of $f \cdot g$ (remember that $0 \cdot (+\infty) = 0$), so that again by 13.2 we have $N_1(fg) = 0$, and (14.3) holds. Once σ, τ are each positive, no loss of generality is incurred by assuming that each is also finite, which we now do.

Applying the mean-value theorem of the differential calculus to the function $\eta \mapsto (1 + \eta)^{1/p}$, there follows at once the well-known *Bernoulli inequality*

$$(1 + \eta)^{1/p} \leq \frac{\eta}{p} + 1 \qquad \text{for all } \eta \in \mathbb{R}_+$$

or

$$\xi^{1/p} \leq \frac{\xi}{p} + \frac{1}{q} \qquad \text{for all } \xi \geq 1.$$

If now x and y are positive real numbers, then one of xy^{-1} and $x^{-1}y$ is such a ξ. Inserting this ξ into the last inequality (and reversing the roles of p and q if necessary), gives

$$x^{1/p}y^{1/q} \leq \frac{1}{p}x + \frac{1}{q}y.$$

This inequality – really equivalent to the concavity of the (natural) logarithm function – holds as well for all $x, y \in \mathbb{R}_+$. If finally we take $x := (\sigma^{-1}f(\omega))^p$ and $y := (\tau^{-1}g(\omega))^q$ for an $\omega \in \{f < +\infty\} \cap \{g < +\infty\}$, we get

$$\frac{1}{\sigma\tau}fg \leq \frac{1}{\sigma^p p}f^p + \frac{1}{\tau^q q}g^q,$$

valid throughout Ω, since it trivially prevails as well in the complementary set $\{f = +\infty\} \cup \{g = +\infty\}$. Integration of this inequality leads at once to (14.3). $\square$

14.2 Theorem. *For all measurable numerical functions* f *and* g *on* Ω *whose sum* $f + g$ *is defined throughout* Ω, *and for every* $p \in [1, +\infty[$

(14.4) $$N_p(f + g) \leq N_p(f) + N_q(g) \qquad \text{(MINKOWSKI's inequality)}.$$

Proof. Since $|f + g| \leq |f| + |g|$,

$$N_p(f + g) \leq N_p(|f| + |g|),$$

which shows that we may assume $f \geq 0$ and $g \geq 0$. In case $p = 1$ there is then even equality in (14.4), by (11.7). Therefore, for the rest we can assume that $1 < p < +\infty$, and then again define q by $p^{-1} + q^{-1} = 1$. We may further assume that both $N_p(f)$ and $N_p(g)$ are finite, that is, that f^p and g^p are integrable. 12.2(c) and the estimates

$$(f + g)^p \leq [2(f \vee g)]^p = 2^p[f^p \vee g^p] \leq 2^p(f^p + g^p)$$

then insure the integrability of $(f + g)^p$, that is, $N_p(f + g) < +\infty$. Now write

$$\int (f + g)^p \, d\mu = \int (f + g)^{p-1} f \, d\mu + \int (f + g)^{p-1} g \, d\mu$$

and apply Hölder's inequality to each integral on the right to get

$$\int (f + g)^p \, d\mu \leq N_q((f + g)^{p-1}) N_p(f) + N_q((f + g)^{p-1}) N_p(g)$$

which thanks to the fact $(p - 1)q = p$ reads

$$(N_p(f + g))^p \leq (N_p(f + g))^{p-1}[N_p(f) + N_p(g)] \, .$$

The desired inequality (14.4) follows from this and the finiteness of $N_p(f + g)$.   □

14.3 Definition. A numerical function f on Ω is called *p-fold (μ-)integrable* or *integrable of order p* or p^{th}-*power integrable*, for some $p \in [1, +\infty[$, if f is measurable and $|f|^p$ is μ-integrable; that is, f is measurable and $N_p(f) < +\infty$.

According to 12.2, 1-fold integrable functions are indeed just the integrable functions. In the case $p = 2$ we also speak of *square-integrability*.

It is immediate from the definition that a measurable function f is p-fold integrable if and only if $|f|$ is p-fold integrable; equivalently, if and only if there is a p-fold integrable function $g \geq 0$ with $|f| \leq g$. Further properties, already known to hold when $p = 1$, are codified in:

14.4 Theorem. *Consider $p \in [1, +\infty[$ and p-fold integrable functions f and g. Then for every $\alpha \in \mathbb{R}$*

$$\alpha f, \quad f \vee g \quad \text{and} \quad f \wedge g$$

are p-fold integrable, and in case it is defined throughout Ω, the function $f + g$ is p-fold integrable.

Proof. Because a function f is p-fold integrable just if it is measurable and $N_p(f)$ is finite, the claims about αf and $f + g$ follow from (14.2) and (14.4). The p-fold integrability of $f \vee g$ and $f \wedge g$ then follow as in the case $p = 1$ from the estimates

$$|f \vee g| \leq |f| + |g| \quad \text{and} \quad |f \wedge g| \leq |f| + |g| \, . \quad □$$

14.5 Corollary. *For $1 \leq p < +\infty$ a numerical function f on Ω is p-fold integrable just if its positive part f^+ and its negative part f^- are both p-fold integrable.*

Proof. Since $f = f^+ - f^-$, from the p-fold integrability of f^+ and f^- follows that of f, by 14.4. The converse also follows from 14.4, and the equalities

$$f^+ = f \vee 0 \quad \text{and} \quad f^- = (-f) \vee 0. \quad \square$$

Now for each $p \in [1, +\infty[$ we define

(14.5) $\mathscr{L}^p(\mu) :=$ the set of all *real p-fold μ-integrable functions on* Ω.

Then from 14.4 we get the property, already known for $p = 1$:

(14.6) $\mathscr{L}^p(\mu)$ *is a vector space over* $\mathbb{R}$.

In view of (14.5) real-valued p-fold integrable functions are also known as $\mathscr{L}^p$-*functions*.

From (14.3) we immediately get:

14.6 Theorem. *The product of a p-fold and a q-fold integrable numerical function is integrable (where $1 < p < +\infty$ and $\frac{1}{p} + \frac{1}{q} = 1$).*

In particular, the *product* of two *square-integrable functions is always integrable.*

14.7 Corollary. *If $1 < p < +\infty$ and the measure μ is finite, then every p-fold integrable function is integrable.*

Proof. Because $\mu(\Omega) < +\infty$, the constant function 1 is q-fold integrable on Ω, for each $q \in [1, +\infty[$. So the present claim follows from 14.6 upon writing any p-fold integrable f as $f \cdot 1$. $\square$

Remark. 1. Without the hypothesis $\mu(\Omega) < +\infty$ the conclusion of 14.7 may fail. For example, in Example 2 of §12 choose the measure μ by requiring $\alpha_n = n^{-1/2}$ for all n. Then the function f defined on $\Omega = \mathbb{N}$ by $f(n) := \alpha_n$ for all $n \in \mathbb{N}$ lies in $\mathscr{L}^2(\mu)$ but not in $\mathscr{L}^1(\mu)$.

More generally when μ is finite, from p-fold integrability follows p'-fold integrability for every $p' \in [1, p]$ – cf. Exercise 3 below.

Related to 14.6 we have:

14.8 Theorem. *Let $1 \leq p < +\infty$, $f : \Omega \to \overline{\mathbb{R}}$ p-fold integrable and $g : \Omega \to \overline{\mathbb{R}}$ a measurable almost-everywhere bounded function. Then the product fg is p-fold integrable.*

Proof. The boundedness hypothesis on g means that there is an $\alpha \in \mathbb{R}$ such that $|g| \leq \alpha$ almost everywhere. Then of course $|fg| \leq \alpha|f|$ almost everywhere. Because of the p-fold integrability of $\alpha|f|$, the claim follows from this inequality via 13.5. $\square$

In particular, along with every $f \in \mathscr{L}^p(\mu)$ and $A \in \mathscr{A}$ the function $1_A f$ lies in $\mathscr{L}^p(\mu)$.

It seems natural to formulate the analog of Theorem 14.6 in case $p = 1$. To this end we define

(14.7) $\mathscr{L}^\infty(\mu) :=$ the set of all *real*, $\mathscr{A}$-measurable, μ-almost everywhere bounded functions on Ω.

One immediately perceives that $\mathscr{L}^\infty(\mu)$ is also a *vector space* over $\mathbb{R}$. The union of Theorems 14.6 and 14.8 results in the assertion

(14.8) $f \in \mathscr{L}^p(\mu), g \in \mathscr{L}^q(\mu), 1 \leq p \leq +\infty, \ p^{-1} + q^{-1} = 1 \ \Rightarrow \ fg \in \mathscr{L}^1(\mu),$

where of course the convention $(+\infty)^{-1} = 0$ has to be recalled. Functions which are $(\mu\text{-})$almost everywhere bounded are also called $(\mu\text{-})$*essentially bounded*.

In closing it may be noted that for counting measure ζ on $\mathscr{P}(\Omega)$, with $\Omega := \{1,\ldots,n\}$, (14.3) and (14.4) go over into discrete versions of the Hölder and Minkowski inequalities. When $p = 2$ we get the Cauchy–Schwarz inequality and the triangle inequality for the euclidean norm, familiar from linear algebra and analytic geometry.

Remark. 2. Definition (14.1) of N_p obviously makes sense for every real $p > 0$, thus also for those $0 < p < 1$ heretofore excluded from consideration. For these p, however, the fundamental properties (14.3) and (14.4) are lost and the q determined by $p^{-1}+q^{-1} = 1$ is negative. (On this point, compare Exercise 5 below.) Remark 3. at the end of §15 will show that pathologies occur when $0 < p < 1$. All subsequent work will therefore be restricted to the case $p \geq 1$.

Exercises.

1. Let $(\Omega, \mathscr{A}, \mu)$ be a finite measure space, $1 \leq p \leq +\infty$. Show that every function f on Ω which is the uniform limit of a sequence (f_n) from $\mathscr{L}^p(\mu)$ itself lies in $\mathscr{L}^p(\mu)$.

2. For an arbitrary measure space $(\Omega, \mathscr{A}, \mu)$ and $1 \leq p < +\infty$, show that a real function f on Ω is p-fold integrable if and only if $f\,|f|^{p-1}$ is integrable. (In the "if" direction, measurability of f itself is not part of the hypothesis.)

3. Let $(\Omega, \mathscr{A}, \mu)$ be a finite measure space, $1 \leq p' \leq p < +\infty$, and f a measurable numerical function on Ω. Then

$$N_{p'}(f) \leq N_p(f) \cdot \mu(\Omega)^{1/p'-1/p} \quad \text{and} \quad \mathscr{L}^p(\mu) \subset \mathscr{L}^{p'}(\mu).$$

4. For any finite number of measurable numerical functions $f_1, \ldots, f_n$ on a measure space and real numbers $p_1, \ldots, p_n \in \,]1, +\infty[$ satisfying $\sum_{j=1}^{n} p_j^{-1} = 1$, prove the *generalized Hölder inequality*

$$N_1(f_1 \cdot \ldots \cdot f_n) \leq N_{p_1}(f_1) \cdot \ldots \cdot N_{p_n}(f_n).$$

5. Let $(\Omega, \mathscr{A}, \mu)$ be a measure space, $p \in \,]0,1[$ and $q < 0$ be defined by $p^{-1}+q^{-1} = 1$. Consider non-negative $f \in \mathscr{L}^p(\mu)$ and a measurable $g : \Omega \to \,]0, +\infty[$ satisfying $0 < N_q(g) := (\int g^q\,d\mu)^{1/q} < +\infty$. By an appropriate application of Hölder's

inequality show that

$$\int fg\, d\mu \geq N_p(f)N_q(g)\,.$$

Infer that

$$N_p(f+g) \geq N_p(f) + N_p(g)$$

and find an example to show that generally equality does not prevail here.

§15. Convergence theorems

Again consider $1 \leq p < +\infty$ and a measure space $(\Omega, \mathscr{A}, \mu)$. The function N_p is real-valued on the vector space $\mathscr{L}^p(\mu)$, and in fact a *semi-norm*, that is, a mapping

$$N_p : \mathscr{L}^p(\mu) \to \mathbb{R}_+$$

having properties (14.2) and (14.4). From the second of those properties, the Minkowski inequality, it follows that the function

$$d_p(f,g) := N_p(f-g) \qquad\qquad f,g \in \mathscr{L}^p(\mu),$$

satisfies the triangle inequality, that is,

$$d_p(f,g) \leq d_p(f,h) + d_p(h,g) \qquad \text{for all } f,g,h \in \mathscr{L}^p(\mu).$$

Evidently d_p thus has all the properties of a metric on $\mathscr{L}^p(\mu)$, with one exception: According to 13.2 and 13.3

$$d_p(f,g) = 0$$

is not equivalent to $f = g$, but only to

$$f = g \quad \mu\text{-almost everywhere.}$$

Distance-like functions without the property that "distance between two elements equal zero entails equality of the elements", are usually called *pseudometrics*. N_p and d_p are called the $\mathscr{L}^p$-*semi-norm* and the $\mathscr{L}^p$-*pseudometric*, also the *semi-norm* or the *pseudometric of convergence in the pth mean* or in $\mathscr{L}^p$-*convergence*.

To elaborate: If (f_n) is a sequence in $\mathscr{L}^p(\mu)$, then it is said to converge *in p^{th} mean* to $f \in \mathscr{L}^p(\mu)$, or to be $\mathscr{L}^p$-*convergent* if

(15.1) $$\lim_{n\to\infty} N_p(f_n - f) = \lim_{n\to\infty} d_p(f_n, f) = 0\,.$$

By virtue of what was noted above, the limit function f is only almost everywhere uniquely determined. (14.2) and (14.4) insure that linear computations with convergent sequences are like those we are accustomed to involving real numbers. In immediately apprehensible symbolic form these say:

$$f_n \xrightarrow{\mathscr{L}^p} f, \quad g_n \xrightarrow{\mathscr{L}^p} g \quad \Rightarrow \quad \alpha f_n + \beta g_n \xrightarrow{\mathscr{L}^p} \alpha f + \beta g$$

for any $\alpha, \beta \in \mathbb{R}$.

From (14.4) also follows a triangle inequality "from below"

(15.2) $$|N_p(f) - N_p(g)| \le N_p(f \pm g) \qquad \text{for all } f, g \in \mathscr{L}^p(\mu),$$

simply because

$$N_p(f) = N_p(f - g + g) \le N_p(f - g) + N_p(g),$$

$N_p(-g) = N_p(g)$, and the roles of f and g can be interchanged throughout.

In case $p = 1$ we speak of simply *convergence in mean*, and when $p = 2$ of *mean-square convergence*.

Taking note of the inequalities

(15.3) $$\left| \int_A f \, d\mu - \int_A g \, d\mu \right| \le \int_A |f - g| \, d\mu \le N_1(f - g),$$

valid for all $A \in \mathscr{A}$, $f, g \in \mathscr{L}^1(\mu)$, and

(15.4) $$|N_p(1_A f) - N_p(1_A g)| \le N_p((f - g)1_A) \le N_p(f - g)$$

valid for all $A \in \mathscr{A}$, $f, g \in \mathscr{L}^p(\mu)$, we immediately get:

15.1 Theorem. *Every sequence (f_n) in $\mathscr{L}^1(\mu)$ (resp., in $\mathscr{L}^p(\mu)$) which converges in mean (resp., in p^{th} mean) to a function f from $\mathscr{L}^1(\mu)$ (resp., from $\mathscr{L}^p(\mu)$) also satisfies*

(15.5) $$\lim_{n \to \infty} \int_A f_n \, d\mu = \int_A f \, d\mu \qquad \text{for every } A \in \mathscr{A}$$

(resp.,

(15.6) $$\lim_{n \to \infty} \int_A |f_n|^p \, d\mu = \int_A |f|^p \, d\mu \qquad \text{for every } A \in \mathscr{A}.)$$

Proof. (15.5) follows from (15.3). Correspondingly (15.6) follows from (15.4), which gives $\lim_{n \to \infty} N_p(1_A f_n) = N_p(1_A f)$, upon taking p^{th} powers in this last limit and using the continuity of the mapping $x \mapsto x^p$ on $\mathbb{R}_+$. $\square$

(15.5) and (15.6) say nothing other than that for each $A \in \mathscr{A}$ the mappings

$$f \mapsto \int_A f \, d\mu \quad \text{and} \quad f \mapsto \int_A |f|^p \, d\mu$$

on $\mathscr{L}^1(\mu)$ and $\mathscr{L}^p(\mu)$, respectively, are continuous with respect to $\mathscr{L}^1$-convergence and $\mathscr{L}^p$-convergence, respectively.

Further developments require a lemma which is fundamental for the whole of integration theory as well as its applications in probability theory and which goes back to P. FATOU (1878–1929):

15.2 Lemma (of Fatou). *Every sequence* $(f_n)_{n\in\mathbb{N}}$ *in* $E^*(\Omega,\mathscr{A})$, *that is consisting of* $\mathscr{A}$-*measurable numerical functions* $f_n \geq 0$, *satisfies*

$$\int \liminf_{n\to\infty} f_n \, d\mu \leq \liminf_{n\to\infty} \int f_n \, d\mu \, .$$

Proof. According to 9.5 and 11.6 the functions

$$f := \liminf_{m\to\infty} f_m \quad \text{and} \quad g_n := \inf_{m\geq n} f_m \qquad \text{for all } n \in \mathbb{N}$$

lie in $E^*(\Omega,\mathscr{A})$. By definition of limit inferior, $g_n \uparrow f$ and thus by 11.4

$$\int f \, d\mu = \sup_{n\in\mathbb{N}} \int g_n \, d\mu = \lim_{n\to\infty} \int g_n \, d\mu \, .$$

From this the claimed inequality follows, because by isotoneity

$$\int g_n \, d\mu \leq \inf_{m\geq n} \int f_m \, d\mu$$

holds for all $n \in \mathbb{N}$. □

If we choose for (f_n) a sequence (1_{A_n}) of indicator functions of measurable $A_n \subset \Omega$, then $\liminf_{n\to\infty} 1_{A_n}$ is the indicator function of the set

(15.7) $$\liminf_{n\to\infty} A_n := \bigcup_{n\in\mathbb{N}} \bigcap_{m\geq n} A_m \, .$$

This is the set of $\omega \in \Omega$ which lie in ultimately all of the sets A_n. Dual to it one defines

(15.8) $$\limsup_{n\to\infty} A_n := \bigcap_{n\in\mathbb{N}} \bigcup_{m\geq n} A_m \, ,$$

the set of $\omega \in \Omega$ which lie in infinitely many of the sets A_n, more correctly, the ω which lie in A_n for infinitely many n. Evidently

(15.9) $$\complement\left(\limsup_{n\to\infty} A_n\right) = \liminf_{n\to\infty} \complement A_n \, .$$

Hence we get the following corollary:

15.3 Corollary. *For every sequence* $(A_n)_{n\in\mathbb{N}}$ *of sets in the* σ-*algebra* $\mathscr{A}$

(15.10) $$\mu\left(\liminf_{n\to\infty} A_n\right) \leq \liminf_{n\to\infty} \mu(A_n) \, ,$$

and if the measure μ *is finite, the inequality*

(15.11) $$\limsup_{n\to\infty} \mu(A_n) \leq \mu\left(\limsup_{n\to\infty} A_n\right)$$

holds as well.

Proof. (15.10) is an immediate consequence of 15.2. In turn, if we apply (15.10) to the sequence $(\complement A_n)$ and use (15.9), we get

$$\mu(\Omega) - \mu\left(\limsup_{n\to\infty} A_n\right) = \mu\left(\complement \limsup_{n\to\infty} A_n\right) = \mu\left(\liminf_{n\to\infty} \complement A_n\right)$$
$$\leq \liminf_{n\to\infty} \mu(\complement A_n) = \mu(\Omega) - \limsup_{n\to\infty} \mu(A_n),$$

confirming (15.11). □

Fatou's lemma leads – in the hands of NOVINGER [1972] – surprisingly simply to the first *convergence theorem*, by which is meant a mechanism for inferring convergence in p^{th} mean from almost everywhere convergence. The result itself goes back to F. RIESZ (1880–1956); cf. RIESZ [1911].

15.4 Theorem (of F. Riesz). *Suppose* $1 \leq p < +\infty$ *and the sequence* $(f_n)_{n\in\mathbb{N}}$ *in* $\mathscr{L}^p(\Omega)$ *converges almost everywhere in* Ω *to a function* $f \in \mathscr{L}^p(\Omega)$. *Then the condition*

$$(15.12) \qquad \lim_{n\to\infty} \int |f_n|^p \, d\mu = \int |f|^p \, d\mu$$

is (*necessary and*) *sufficient for the convergence of* (f_n) *to* f *in* p^{th} *mean.*

Proof. The necessity of (15.12) follows (even without the hypothesis of almost-everywhere convergence) immediately from 15.1. The proof of sufficiency proceeds from the inequality

$$(\alpha + \beta)^p \leq 2^p(\alpha^p + \beta^p) \qquad\qquad (\alpha, \beta \in \mathbb{R}_+)$$

which has already been used in the proof of (14.4). Since $|\alpha - \beta| \leq \alpha + \beta$ this inequality yields

$$|\alpha - \beta|^p \leq 2^p(|\alpha|^p + |\beta|^p) \qquad\qquad (\alpha, \beta \in \mathbb{R}).$$

This inequality insures that

$$g_n := 2^p(|f_n|^p + |f|^p) - |f_n - f|^p, \qquad\qquad n \in \mathbb{N},$$

are non-negative functions. They lie in $\mathscr{L}^1(\mu)$ and by hypothesis they converge almost everywhere to $2^{p+1}|f|^p$. In particular, $2^{p+1}|f| = \liminf g_n$ almost everywhere. Therefore Fatou's lemma in conjunction with (15.12) delivers the relations

$$2^{p+1}\int |f|^p \, d\mu = \int \liminf_{n\to\infty} g_n \, d\mu \leq \liminf_{n\to\infty} \int g_n \, d\mu$$
$$= 2^{p+1}\int |f|^p \, d\mu - \limsup_{n\to\infty} \int |f_n - f|^p \, d\mu.$$

Since $2^{p+1}\int |f|^p \, d\mu < +\infty$, we infer by subtracting it that

$$\limsup_{n\to\infty} \int |f_n - f|^p \, d\mu \leq 0,$$

which asserts the claimed $\mathscr{L}^p$-convergence. □

In preparation for the proof of the next convergence theorem we extend Minkowski's inequality to series of non-negative functions.

15.5 Lemma. *Every sequence $(f_n)_{n\in\mathbb{N}}$ of functions from $E^*(\Omega,\mathscr{A})$ satisfies*

$$(15.13) \qquad N_p\Big(\sum_{n=1}^{\infty} f_n\Big) \leq \sum_{n=1}^{\infty} N_p(f_n) \qquad \text{for every } p \in [1,+\infty[.$$

Proof. If for each $n \in \mathbb{N}$ we set $s_n := f_1 + \ldots + f_n$, then by (14.4)

$$N_p(s_n) \leq \sum_{j=1}^{n} N_p(f_j) \leq \sum_{j=1}^{\infty} N_p(f_j).$$

The sequence (s_n) is isotone and $\sum_{n=1}^{\infty} f_n$ is its upper envelope; the same holds for the pth powers. Therefore from the monotone convergence theorem 11.4 follows

$$N_p\Big(\sum_{n=1}^{\infty} f_n\Big) = \sup_{n\in\mathbb{N}} N_p(s_n)$$

and together with the preceding inequalities this gives (15.13). $\square$

We come now to a second convergence theorem. It goes back to H. Lebesgue and is therefore frequently called *Lebesgue's convergence theorem*.

15.6 Theorem (on dominated convergence). *Let $1 \leq p < +\infty$ and $(f_n)_{n\in\mathbb{N}}$ be a sequence from $\mathscr{L}^p(\mu)$ which converges almost everywhere on Ω. Suppose there exists a p-integrable numerical function $g \geq 0$ on Ω such that*

$$(15.14) \qquad |f_n| \leq g \qquad \text{for all } n \in \mathbb{N}.$$

Then there is a real-valued measurable function f on Ω to which (f_n) converges almost everywhere. Every such f lies in $\mathscr{L}^p(\mu)$ and the sequence (f_n) converges to f in p^{th} mean.

Proof. By assumption there is a nullset M_1 such that $\lim f_n(\omega)$ exists (in $\overline{\mathbb{R}}$) for every $\omega \in \complement M_1$. Because of the integrability of g^p there is, according to 13.6, another nullset M_2 with $g(\omega) < +\infty$ for every $\omega \in \complement M_2$. If we set

$$f(\omega) := \begin{cases} \lim_{n\to\infty} f_n(\omega), & \omega \in \complement(M_1 \cup M_2) \\ 0, & \omega \in M_1 \cup M_2, \end{cases}$$

then f is real-valued and $\mathscr{A}$-measurable, and the sequence (f_n) converges almost everywhere to f. Consider now *any* function f with these properties. Then $|f| \leq g$ almost everywhere, so along with g^p the function $|f|^p$ is also integrable, that is, $f \in \mathscr{L}^p(\mu)$, by 13.5. We set, for each $n \in \mathbb{N}$

$$g_n := |f_n - f|^p$$

and then what has to be shown is that $\lim \int g_n \, d\mu = 0$. From the definition of g_n,

$$0 \le g_n \le (|f_n| + |f|)^p \le (g + |f|)^p \,.$$

Since the function $h := (g + |f|)^p$ is integrable, so is each g_n (by 14.4 and 12.2). Fatou's lemma applies to the sequence $(h - g_n)$ and says that

$$\int \liminf_{n\to\infty}(h - g_n) \, d\mu \le \liminf_{n\to\infty} \int (h - g_n) \, d\mu = \int h \, d\mu - \limsup_{n\to\infty} \int g_n \, d\mu \,.$$

Since (f_n) converges almost everywhere to f, $(h - g_n)$ converges almost everywhere to h. In particular,

$$\liminf_{n\to\infty}(h - g_n) = h \qquad\qquad \text{almost everywhere}$$

and so

$$\int \liminf_{n\to\infty}(h - g_n) \, d\mu = \int h \, d\mu \,.$$

The preceding inequality therefore yields $\limsup \int g_n \, d\mu \le 0$. Since all g_n are non-negative, this is equivalent to the desired $\lim \int g_n \, d\mu = 0$. $\square$

The concept of a Cauchy sequence makes sense in any pseudometric space, in particular therefore in $\mathscr{L}^p(\mu)$. A sequence (f_n) of functions from $\mathscr{L}^p(\mu)$ is said to be a *Cauchy sequence* in $\mathscr{L}^p(\mu)$ if for every $\varepsilon > 0$

$$d_p(f_m, f_n) = N_p(f_m - f_n) \le \epsilon$$

holds for ultimately all m, n. Every $\mathscr{L}^p(\mu)$-convergent sequence is a Cauchy sequence, as Minkowski's inequality shows. That the converse of this is also true, that, in other words, the space $\mathscr{L}^p(\mu)$ is (*metrically*) *complete*, is the content of the third convergence theorem. Its special case $p = 2$ goes back to F. RIESZ and E. FISCHER (1875–1956).

15.7 Theorem. *For each $1 \le p < +\infty$, every Cauchy sequence $(f_n)_{n\in\mathbb{N}}$ in $\mathscr{L}^p(\mu)$ converges in p^{th} mean to an $f \in \mathscr{L}^p(\mu)$. Some subsequence of (f_n) converges almost everywhere to f.*

Proof. Straight from the definition of Cauchy sequence we can construct $1 < n_1 < n_2 < \dots$ such that $N_p(f_{n_{k+1}} - f_{n_k}) \le 2^{-k}$ for all $k \in \mathbb{N}$. We define

$$g_k := f_{n_{k+1}} - f_{n_k} \quad \text{for each } k \in \mathbb{R}, \text{ and } g := \sum_{k=1}^{\infty} |g_k|.$$

Then from 15.5

$$N_p(g) \le \sum_{k=1}^{\infty} N_p(g_k) \le \sum_{k=1}^{\infty} 2^{-k} = 1 \,.$$

Consequently, the $\mathscr{A}$-measurable, non-negative numerical function g is p^{th}-power integrable and therefore (by 13.6) it is almost everywhere real-valued, that is, the series $\sum g_k$ is absolutely convergent almost everywhere. The kth partial sum of this series is $f_{n_{k+1}} - f_{n_1}$, so we see that the sequence $(f_{n_k})_{k \in \mathbb{N}}$ converges almost everywhere in Ω. Moreover,

$$|f_{n_{k+1}}| = |g_1 + \ldots + g_k + f_{n_1}| \leq g + |f_{n_1}|$$

and by 14.4 the sum $g + |f_{n_1}|$ is p^{th}-power integrable. Thus the sequence $(f_{n_k})_{k \in \mathbb{N}}$ satisfies all the hypotheses of the dominated convergence theorem, according to which it therefore converges in p^{th} mean to an $f \in \mathscr{L}^p(\mu)$ and

$$\lim_{k \to \infty} f_{n_k} = f \qquad\qquad \text{almost everywhere.}$$

Since (f_n) is a Cauchy sequence, this subsequence behavior entails the convergence in p^{th} mean of the whole sequence: Given $\varepsilon > 0$ there is an $m_\varepsilon \in \mathbb{N}$ such that

$$N_p(f_n - f_m) \leq \varepsilon \qquad\qquad \text{for all } m, n \geq m_\varepsilon.$$

Then there is a $k \in \mathbb{N}$ with $n_k > m_\varepsilon$ such that

$$N_p(f_{n_k} - f) \leq \varepsilon.$$

The triangle inequality then insures that

$$N_p(f_n - f) \leq N_p(f_n - f_{n_k}) + N_p(f_{n_k} - f) \leq 2\varepsilon$$

holds for any $n \geq m_\varepsilon$.   □

Passage to a subsequence cannot generally be circumvented if one wants almost everywhere convergence, as the next example illustrates.

Example. Consider $\Omega := [0, 1[$, $\mathscr{A} := \Omega \cap \mathscr{B}^1$ and $\mu := \lambda_\Omega^1$. Every natural number n is representable as $n = 2^h + k$ for a unique pair of integers $h \geq 0$, $k \geq 0$ with $k < 2^h$. Set $A_n := [k2^{-h}, (k+1)2^{-h}[$ and let f_n denote its indicator function. Then $\int f_n^p \, d\mu = \int f_n \, d\mu = \mu(A_n) = 2^{-h} < 2/n$, so (f_n) converges to 0 in p^{th} mean (for any $1 \leq p < +\infty$) and is therefore certainly a Cauchy sequence in $\mathscr{L}^p(\mu)$. But the sequence $(f_n(\omega))_{n \in \mathbb{N}}$ in $\{0, 1\}$ is *not* convergent for *any* $\omega \in \Omega$. Indeed, given $\omega \in \Omega$ and $h = 0, 1, \ldots$, there is exactly one $k = 0, \ldots, 2^h - 1$, such that $\omega \in [k2^{-h}, (k+1)2^{-h}[$, that is, $\omega \in A_{2^h+k}$. In case $k < 2^h - 1$, $\omega \notin A_{2^h+k+1}$. In case $k = 2^h - 1$ and $h \geq 1$, $\omega \notin A_{2^{h+1}}$.

We record the following simple corollary to 15.7:

15.8 Corollary. *If the Cauchy sequence (f_n) in $\mathscr{L}^p(\mu)$ converges almost everywhere to an $\mathscr{A}$-measurable real function f on Ω, then f lies in $\mathscr{L}^p(\mu)$ and the sequence converges to it in p^{th} mean.*

Proof. According to 15.7 there is an $f^* \in \mathscr{L}^p(\mu)$ to which (f_n) converges in p^{th} mean and to which a subsequence of (f_n) converges almost everywhere. Outside the union of this exceptional nullset and that in the hypothesis the two limits f and f^* must agree. Hence $f = f^*$ almost everywhere.   □

Corresponding to Theorem 14.6 and its corollary we have finally the following two convergence assertions:

15.9 Theorem. *The sequence (f_n) in $\mathscr{L}^p(\mu)$ converges in p^{th} mean to a function $f \in \mathscr{L}^p(\mu)$ and the sequence (g_n) in $\mathscr{L}^q(\mu)$ converges in q^{th} mean to $g \in \mathscr{L}^q(\mu)$. If $1 < p < +\infty$ and $p^{-1} + q^{-1} = 1$, then the sequence $(f_n g_n)$ of products converges in mean to fg.*

Proof. The triangle inequality in $\mathbb{R}$ yields

$$|f_n g_n - fg| \leq |f_n - f|\,|g_n| + |f|\,|g_n - g| \qquad (n \in \mathbb{N})$$

which the Hölder inequality (14.3) transforms into

$$N_1(f_n g_n - fg) \leq N_p(f_n - f)N_q(g_n) + N_p(f)N_q(g_n - g) \qquad (n \in \mathbb{N}).$$

Our claim follows from this when we recall from (15.2) or (15.6) that the sequence $(N_q(g_n))_{n \in \mathbb{N}}$ is convergent, hence bounded. □

15.10 Corollary. *If the measure μ is finite, then every sequence $(f_n)_{n \in \mathbb{N}}$ in $\mathscr{L}^p(\mu)$ which converges in p^{th} mean to an $f \in \mathscr{L}^p(\mu)$ for some $1 \leq p < +\infty$, also converges to f in mean.*

Proof. For $p = 1$ there is nothing to prove. For $1 < p < +\infty$ the claim follows from the theorem upon taking every function g_n there to be the constant function 1; because of the finiteness of μ the constant functions lie in $\mathscr{L}^q(\mu)$ for every $q \in [1, +\infty[$. □

The reader should convince himself via an example like that in the remark after 14.7 that the converse of the assertion in this corollary is not true. However, the conclusion of the corollary can be refined somewhat; namely, under its hypotheses there is $\mathscr{L}^{p'}$-convergence of (f_n) to f for every $p' \in [1, p]$. Cf. Exercise 2 below.

Remarks. 1. Because

$$N_p : \mathscr{L}^p(\mu) \to \mathbb{R}_+$$

is a semi-norm, the set

$$\mathscr{N} := N_p^{-1}(0)$$

is a linear subspace of $\mathscr{L}^p(\mu)$. It is independent of p because it consists of all measurable real functions on Ω which are almost everywhere equal to 0. The quotient vector space

$$L^p(\mu) := \mathscr{L}^p(\mu)/\mathscr{N}$$

becomes a normed space in a natural way: Letting $f \mapsto \tilde{f}$ denote the canonical mapping of $\mathscr{L}^p(\mu)$ onto $L^p(\mu)$, we define

$$\|\tilde{f}\|_p := N_p(f) \qquad\qquad \text{for all } \tilde{f} \in L^p(\mu).$$

One checks effortlessly that $\tilde{f} \mapsto \|\tilde{f}\|_p$ is thereby well defined and provides a norm on $L^p(\mu)$. Theorem 15.7 says that $L^p(\mu)$ is complete with respect to this norm, that is, it is a *Banach space* (for $1 \leq p < +\infty$).

$L^2(\mu)$ is even a *Hilbert space*. For the product fg of two functions $f, g \in \mathscr{L}^2(\mu)$ is integrable, by 14.6, and it is clear that the integral $\int fg \, d\mu$ depends only on the canonical images $\tilde{f}, \tilde{g}$ of these functions, which means that

$$(\tilde{f}, \tilde{g}) \mapsto \int fg \, d\mu$$

is a well-defined mapping. A short calculation suffices to confirm that it provides a scalar product in $L^2(\mu)$.

2. $f \in \mathscr{L}^\infty(\mu)$ means that the set W_f of all $\alpha \in \mathbb{R}_+$ such that $|f| \leq \alpha$ almost everywhere is not empty. We can set

$$N_\infty(f) := \inf W_f$$

and show easily that $N_\infty : \mathscr{L}^\infty(\mu) \to \mathbb{R}_+$ is a semi-norm on $\mathscr{L}^\infty(\mu)$. Also in this case $N_\infty^{-1}(0)$ coincides with the space $\mathscr{N}$ described in 1. In the quotient space

$$L^\infty(\mu) := \mathscr{L}^\infty(\mu)/\mathscr{N}$$

a norm $\tilde{f} \mapsto \|\tilde{f}\|_\infty$ can be defined via N_∞ just as before. One checks that $L^\infty(\mu)$ thus also becomes a *Banach space*.

3. For every measure space $(\Omega, \mathscr{A}, \mu)$ and every $p \in \,]0, 1[$ the set $\mathscr{L}^p(\mu)$ (cf. Remark 2 in §14) turns out to be a vector space. N_p is generally not a semi-norm (cf. Exercise 5, §14), but $d_p(f, g) := N_p^p(f - g)$ is a complete pseudometric – with, however, strange properties: The unit "ball" centered at 0 is generally not convex. For L-B measure on $[0, 1]$, *every* $f \in \mathscr{L}^p$ is actually a convex combination of functions in this ball. See BOURBAKI [1965], chap. 4, §6, exer. 13.

Exercises.

1. Let (f_n) be a sequence of numerical measurable functions on a measure space $(\Omega, \mathscr{A}, \mu)$. Under the hypothesis that a μ-integrable function g satisfying $|f_n| \leq g$ for every $n \in \mathbb{N}$ exists, show that $\liminf f_n$ and $\limsup f_n$ are μ-integrable functions and satisfy

$$\int \liminf_{n \to \infty} f_n \, d\mu \leq \liminf_{n \to \infty} \int f_n \, d\mu \leq \limsup_{n \to \infty} \int f_n \, d\mu \leq \int \limsup_{n \to \infty} f_n \, d\mu \,.$$

Show by an explicit example that this chain of inequalities can fail if there is no such majorizing function g. (To this end, cf. Exercise 6 in §21.)

2. Let μ be a finite measure, $1 \leq p' \leq p < +\infty$. Show that if a sequence in $\mathscr{L}^p(\mu)$ converges in p^{th} mean to a function $f \in \mathscr{L}^p(\mu)$, then it also converges in p'^{th} mean to f. (Cf. Exercise 3 in §14.)

3. Let $(\Omega, \mathscr{A}, \mu)$ be a finite measure space and on $\mathscr{A}$ consider the pseudometric $d_\mu(A, B) := \mu(A \triangle B) = \int |1_A - 1_B| \, d\mu$ introduced in Exercise 7 of §3. Show that the pseudometric space $(\mathscr{A}, d_\mu)$ is complete.

4. Show that if $1 \leq p < +\infty$ and $f, f_n \in \mathscr{L}^p(\mu)$ satisfy

$$\sum_{n=1}^{\infty} N_p(f_n - f) < +\infty \,,$$

then the sequence (f_n) converges almost everywhere to f.

5. Show that the conclusion of Theorem 15.9 remains valid for $p = 1$ and $q = +\infty$.

§16. Applications of the convergence theorems

We will now demonstrate the applicability of the convergence theorems by means of three examples which will be important in the sequel. The first concerns the behavior of parameter-dependent integrals, the second the connection between the Riemann and the Lebesgue integral, and the third the calculation of the (Gaussian) integral

(16.1) $$G := \int e^{-x^2} \lambda^1(dx) \,.$$

I. Parameter-dependent integrals. The question of the continuity and differentiability of functions which are defined by integrals will be answered in the following lemmas and corollary. Throughout, $(\Omega, \mathscr{A}, \mu)$ is an arbitrary measure space.

16.1 Lemma (Continuity lemma). *Let E be a metric space and $f : E \times \Omega \to \mathbb{R}$ a function with the properties*
(a) *$\omega \mapsto f(x, \omega)$ is μ-integrable for every $x \in E$;*
(b) *$x \mapsto f(x, \omega)$ is continuous at $x_0 \in E$ for every $\omega \in \Omega$;*
(c) *there is a μ-integrable function $h \geq 0$ on Ω such that*
$$|f(x, \omega)| \leq h(\omega) \qquad \text{for all } (x, \omega) \in E \times \Omega.$$
Then the function defined on E by
$$\varphi(x) := \int f(x, \omega) \mu(d\omega)$$
is continuous at x_0.

Proof. The continuity of φ at x_0 is proved if we show that for every sequence (x_n) in E with $\lim x_n = x_0$,
$$\lim_{n \to \infty} \varphi(x_n) = \varphi(x_0)$$

holds. To accomplish this, we introduce the sequence (f_n) by
$$f_n(\omega) := f(x_n, \omega) \qquad\qquad (n \in \mathbb{Z}_+, \, \omega \in \Omega).$$

By hypothesis these are integrable functions, each satisfies $|f_n| \leq h$, and for every fixed $\omega \in \Omega$, $\lim_{n \to \infty} f_n(\omega) = f_0(\omega)$. From the theorem on dominated convergence therefore follows that

$$\lim_{n\to\infty} \int f_n \, d\mu = \int f_0 \, d\mu = \int f(x_0, \omega) \mu(d\omega) \,,$$

that is, indeed $\lim \varphi(x_n) = \varphi(x_0)$. □

In the following applications of this lemma the space E will frequently be an interval in $\mathbb{R}$ or, more generally, a subset of $\mathbb{R}^d$.

16.2 Lemma (Differentiation lemma). *Let I be a non-degenerate (meaning, containing more than one point) interval in $\mathbb{R}$, and $f : I \times \Omega \to \mathbb{R}$ be a function with the properties*
(a) *$\omega \mapsto f(x, \omega)$ is μ-integrable for each $x \in I$;*
(b) *$x \mapsto f(x, \omega)$ is differentiable on I for each $\omega \in \Omega$, the derivative at x being denoted by $f'(x, \omega)$;*
(c) *there is a μ-integrable function $h \geq 0$ on Ω such that*

$$|f'(x, \omega)| \leq h(\omega) \qquad\qquad \text{for all } (x, \omega) \in I \times \Omega.$$

Then the function defined on I by

(16.2) $$\varphi(x) := \int f(x, \omega) \mu(d\omega)$$

is differentiable, for each $x \in I$ the function $\omega \mapsto f'(x, \omega)$ is μ-integrable, and

(16.3) $$\varphi'(x) = \int f'(x, \omega) \mu(d\omega) \qquad\qquad \text{for every } x \in I.$$

In short, under the stated conditions (16.2) can be differentiated under the integral sign.

Proof. Fix $x_0 \in I$ and consider any sequence $(x_n)_{n\in\mathbb{N}} \subset I \setminus \{x_0\}$ which converges to x_0. Then the function defined on Ω by

$$g_n(\omega) := \frac{f(x_n, \omega) - f(x_0, \omega)}{x_n - x_0}$$

is μ-integrable, for each $n \in \mathbb{N}$, and

$$\lim_{n\to\infty} g_n(\omega) = f'(x_0, \omega) \qquad\qquad \text{for all } \omega \in \Omega.$$

It is a consequence of hypothesis (c) that $|g_n| \leq h$ for $n \in \mathbb{N}$, as we now confirm. It suffices to apply the mean-value theorem of differential calculus. According to it, for each $x \in I \setminus \{x_0\}$ and each fixed $\omega \in \Omega$ there is a point t, in the open interval whose endpoints are x and x_0, such that

$$\frac{f(x, \omega) - f(x_0, \omega)}{x - x_0} = f'(t, \omega)$$

and therefore by (c) this quotient is majorized by $h(\omega)$. In particular,

$$|g_n(\omega)| \le h(\omega) \qquad \text{for all } \omega \in \Omega, \text{ and every } n \in \mathbb{N}.$$

Now the dominated convergence theorem comes into play to insure that the function $\omega \mapsto f'(x_0, \omega)$ to which the g_n converge is μ-integrable and

$$\lim_{n \to \infty} \int g_n \, d\mu = \int f'(x_0, \omega) \mu(d\omega).$$

Claim (16.3) follows from this because

$$\int g_n \, d\mu = \frac{\varphi(x_n) - \varphi(x_0)}{x_n - x_0} \qquad \text{for all } n \in \mathbb{N}. \quad \square$$

Passage to the multi-dimensional analog is painless:

16.3 Corollary. *Let U be an open subset of $\mathbb{R}^d$, $i \in \{1, \ldots, d\}$, and $f : U \times \Omega \to \mathbb{R}$ a function with the properties*
(a) *$\omega \mapsto f(x, \omega)$ is μ-integrable for each $x \in U$;*
(b) *$x \mapsto f(x, \omega)$ has an ith partial derivative at each point of U, for every $\omega \in \Omega$;*
(c) *there is a μ-integrable function $h \ge 0$ on Ω such that*

$$\left| \frac{\partial f}{\partial x_i}(x, \omega) \right| \le h(\omega) \qquad \text{for all } (x, \omega) \in U \times \Omega.$$

Then the function defined on U by

$$\varphi(x) := \int f(x, \omega) \mu(d\omega)$$

has an ith partial derivative at every $x \in U$, the function $\omega \mapsto \dfrac{\partial f}{\partial x_i}(x, \omega)$ is μ-integrable, and

$$\frac{\partial \varphi}{\partial x_i}(x) = \int \frac{\partial f}{\partial x_i}(x, \omega) \mu(d\omega) \qquad \text{for every } x \in U.$$

This follows at once from the differentiation lemma: Given $\bar{x} = (\bar{x}_1, \ldots, \bar{x}_d) \in U$, there is an open interval $I \subset \mathbb{R}$ containing $\bar{x}_i$ such that for each $t \in I$ the point $(\bar{x}_1, \ldots, \bar{x}_{i-1}, t, \bar{x}_{i+1}, \ldots \bar{x}_d)$ lies in U, and we can apply 16.2 to the function $(t, \omega) \mapsto f(\bar{x}_1, \ldots, \bar{x}_{i-1}, t, \bar{x}_{i+1}, \ldots \bar{x}_d, \omega)$.

II. Comparison of the Riemann and Lebesgue integrals. For every d-dimensional Borel set $B \in \mathscr{B}^d$ and suitable Borel measurable numerical functions f on B the integral $\int_B f \, d\lambda^d$ was defined in §12 and identified with $\int f \, d\lambda^d_B$. This integral is called for short the *Lebesgue integral* of f over B. A frequently encountered alternative way of writing it is

(16.4) $$\int_B f(x) \, dx := \int_B f \, d\lambda^d = \int f \, d\lambda^d_B.$$

In case $d = 1$ and $B = [\alpha, \beta]$, or $] -\infty, \alpha]$, or $\mathbb{R}$, etc. the notations $\int_\alpha^\beta f(x)\,dx$, or $\int_{-\infty}^\alpha f(x)\,dx$, or $\int_{-\infty}^{+\infty} f(x)\,dx$, etc., are also common.

Since in basic analysis courses it is frequently only the Riemann integral that is dealt with, the following remarks relating it to what has been done here may be useful.

16.4 Theorem. *Consider a Borel measurable real function f defined on a compact interval $I := [\alpha, \beta]$ in $\mathbb{R}$. If f is Riemann integrable (which in particular means it is bounded), then it is also Lebesgue integrable, and the values of the two integrals of f coincide.*

Proof. To every finite subdivision

$$\jmath := \{\alpha = \alpha_0 \le \alpha_1 \le \ldots \le \alpha_n = \beta\}$$

of I the Riemann theory associates the lower and upper sums

$$L_\jmath := \sum_{i=1}^n \gamma_i(\alpha_i - \alpha_{i-1}) \quad \text{and} \quad U_\jmath := \sum_{i=1}^n \Gamma_i(\alpha_i - \alpha_{i-1})$$

in which

$$\gamma_i := \inf f([\alpha_{i-1}, \alpha_i]) \quad \text{and} \quad \Gamma_i := \sup f([\alpha_{i-1}, \alpha_i]), \quad i = 1, \ldots, n.$$

If we set $\mu := \lambda_I^1$ and $A_i := [\alpha_{i-1}, \alpha_i]$, $i = 1, \ldots, n$, then

$$l_\jmath := \sum_{i=1}^n \gamma_i 1_{A_i} \quad \text{and} \quad u_\jmath := \sum_{i=1}^n \Gamma_i 1_{A_i}$$

are μ-integrable functions on I for which

$$L_\jmath = \int l_\jmath \, d\mu \quad \text{and} \quad U_\jmath = \int u_\jmath \, d\mu.$$

Riemann integrability of f means, by definition, that there is a sequence $(\jmath_n)$ of subdivisions of I such that each $\jmath_{n+1}$ refines its predecessor $\jmath_n$ and the sequences $(L_{\jmath_n})$ and $(U_{\jmath_n})$ tend to the same real limit value, the Riemann integral ϱ of f over I.

Because of the refinement feature of the sequence $(\jmath_n)$, $(l_{\jmath_n})$ is an isotone and $(u_{\jmath_n})$ is an antitone sequence. Hence

$$q := \lim_{n\to\infty} (u_{\jmath_n} - l_{\jmath_n})$$

exists (in $\mathbb{R}$) on I. If therefore we apply Fatou's lemma 15.2 to the sequence of functions $u_{\jmath_n} - l_{\jmath_n} \ge 0$, there follows

$$0 \le \int q \, d\mu \le \lim_{n\to\infty} (U_{\jmath_n} - L_{\jmath_n}) = 0$$

and so by 13.2, $q = 0$ μ-almost everywhere. Since in addition for every n, $l_{\jmath_n} \le f \le u_{\jmath_n}$ holds μ-almost everywhere (everywhere except possibly at the points of $\jmath_n$),

$q = 0$ almost everywhere entails that

$$\lim_{n \to \infty} l_{\mathfrak{z}_n} = f \qquad \qquad \mu\text{-almost everywhere on } I.$$

As has been noted, f is bounded, say $|f| \le M \in \mathbb{R}$. The sequence $(|l_{\mathfrak{z}_n}|)$ is therefore majorized by the constant M, a μ-integrable function, and so Theorem 15.6 on dominated convergence delivers the μ-integrability of f as well as the convergence of $(l_{\mathfrak{z}_n})$ to f in mean. From 15.1 finally follows

$$\int f \, d\mu = \lim \int l_{\mathfrak{z}_n} \, d\mu = \lim L_{\mathfrak{z}_n} = \varrho \,,$$

which finishes the proof. □

Remarks. 1. Consider once again Dirichlet's jump function f on the unit interval (cf. Exercise 2 of §10). Being the indicator function of $\mathbb{Q} \cap [0, 1]$, it is Borel measurable and almost everywhere 0 with respect to L-B measure $\lambda^1_{[0,1]}$. Consequently it is Lebesgue integrable and $\int_0^1 f(x) \, dx = 0$. But f is not Riemann integrable. So the roles of Riemann and Lebesgue integration cannot be reversed in 16.4.

2. Borel measurability of f need not be hypothesized: the above proof shows, even without it, that $\lim l_{\mathfrak{z}_n} = f$ μ-almost everywhere and so f is μ-almost everywhere equal to the Borel function $\overline{\lim} \, l_{\mathfrak{z}_n}$. However, in this case it can well happen that f itself is not Borel measurable.

3. The ideas in the proof of Theorem 16.4 can be amplified into a non-trivial criterion for Riemann integrability. Namely, $f : [\alpha, \beta] \to \mathbb{R}$ is Riemann integrable if and only if it is bounded and is continuous at λ^1-almost every point of $[\alpha, \beta]$. See Theorem 2.5.1 of COHN [1980] or the multi-part Exercise 12.51 of HEWITT and STROMBERG [1965].

16.5 Corollary. *The non-negative, real-valued, Borel measurable function f is Riemann integrable over every compact interval. Then f is Lebesgue integrable over $\mathbb{R}$ if and only if the improper Riemann integral*

$$\varrho := \lim_{n \to \infty} \int_{-n}^{+n} f(x) \, dx$$

exists. In this case $\varrho = \int f \, d\lambda^1$.

Proof. Denote by ϱ_n the Riemann integral of f over $A_n := [-n, +n]$ for each $n \in \mathbb{N}$. According to the theorem just proved

$$\varrho_n = \int_{A_n} f \, d\lambda^1 = \int 1_{A_n} f \, d\lambda^1 \,.$$

From 11.4 and the fact that $1_{A_n} f \uparrow f$ we get

$$\sup_n \varrho_n = \int f \, d\lambda^1 \,.$$

The improper Riemann integral exists, by definition, just if this supremum is finite and in that case its value ϱ is that supremum. From these observations and the monotone convergence theorem our present result follows. $\square$

Utilizing the decomposition $f = f^+ - f^-$ into positive and negative parts, it follows from 12.2 and 16.5 that every Borel measurable real function f on $\mathbb{R}$ with absolutely convergent improper Riemann integral is also Lebesgue integrable and $\int f \, d\lambda^1$ coincides with the improper Riemann integral of f. Obviously too, any open or half-open interval $I \subset \mathbb{R}$ can take over the role of $\mathbb{R}$ in 16.5.

By contrast, from the existence of the improper Riemann integral of f does not follow the Lebesgue integrability of f, even for continuous functions. Consider, for example, the function $f : \mathbb{R} \to \mathbb{R}$ defined by $f(x) := (\sin x)/x$ when $x \neq 0$ and $f(0) := \lim_{x \to 0} (\sin x)/x = \sin'(0) = 1$. Of course, it is continuous. If for each $k \in \mathbb{N}$ we set

$$a_k := \int_{k\pi}^{(k+1)\pi} \frac{\sin x}{x} \, dx = (-1)^k \int_0^\pi \frac{\sin t}{t + k\pi} \, dt \, ,$$

we see that the signs of the a_k alternate, their moduli decrease as k increases, and

$$|a_k| \leq \int_{k\pi}^{(k+1)\pi} \frac{1}{x} \, dx = \log\left(\frac{(k+1)\pi}{k\pi}\right) = \log\left(1 + \frac{1}{k}\right) \to 0 \text{ as } k \to \infty.$$

Therefore the series $\sum_{k=1}^\infty a_k$ converges. Using this it is very easy to confirm that the improper Riemann integral

$$\lim_{R \to +\infty} \int_0^R \frac{\sin x}{x} \, dx$$

exists. On the other hand,

$$\int_{k\pi}^{(k+1)\pi} \frac{|\sin x|}{x} \, dx = \int_0^\pi \frac{\sin t}{t + k\pi} \, dt \geq \int_0^\pi \frac{\sin t}{\pi + k\pi} \, dt = \frac{2}{(k+1)\pi}$$

and so for every $n \in \mathbb{N}$

$$\int_{\mathbb{R}_+} |f| \, d\lambda^1 \geq \int_{[\pi,(n+1)\pi]} |f| \, d\lambda^1 = \sum_{k=1}^n \int_{k\pi}^{(k+1)\pi} \frac{|\sin x|}{x} \, dx \geq \frac{2}{\pi} \sum_{k=1}^n \frac{1}{k+1} \, .$$

Since the harmonic series diverges, these inequalities show that $\int_{\mathbb{R}_+} |f| \, d\lambda^1 = +\infty$, and so by 12.2 f is not Lebesgue integrable over $\mathbb{R}_+$.

III. Calculation of the integral G. The preceding considerations show that integrals which the reader may already have encountered as Riemann integrals can, in the stated circumstances, be immediately interpreted as Lebesgue integrals. Known formulas and computational rules for the Riemann integral thereby become available to the Lebesgue theory as well.

As an illustration, consider the non-negative function

(16.5)
$$f(x,\omega) := \frac{e^{-x(1+\omega^2)}}{1+\omega^2} \qquad (x,\omega) \in \mathbb{R}_+ \times \mathbb{R}.$$

Both f and the function $(x,\omega) \mapsto f'(x,\omega) := -e^{-x(1+\omega^2)}$ are continuous. For fixed $x_0 > 0$ form the auxiliary functions

$$h_0(\omega) := e^{-2x_0|\omega|} \quad \text{and} \quad h(\omega) := (1+\omega^2)^{-1}, \qquad \omega \in \mathbb{R}.$$

Their λ^1-integrability (over $\mathbb{R}$) follows from Corollary 16.5 and the fundamental theorem of calculus. For example,

$$\int_{-\infty}^{+\infty} (1+\omega^2)^{-1}\, d\omega = \lim_{n\to\infty} [\arctan(\omega)]_{-n}^{n} = \pi.$$

Obviously $f(x,\omega) \le h(\omega)$ for all $(x,\omega) \in \mathbb{R}_+ \times \mathbb{R}$. It follows from 12.2 that for each $x \in \mathbb{R}_+$ the function $\omega \mapsto f(x,\omega)$ is λ^1-integrable. And the real function defined by

(16.6)
$$\varphi(x) := \int f(x,\omega)\, d\omega \qquad x \in \mathbb{R}_+$$

is continuous by the continuity lemma 16.1. Note that $\varphi(0) = \pi$. Since $2|\omega| \le 1+\omega^2$ for all $\omega \in \mathbb{R}$, we have $|f'(x,\omega)| \le h_0(\omega)$ for all $(x,\omega) \in [x_0, +\infty[\times\mathbb{R}$. Consequently the differentiation lemma 16.2 insures that φ is differentiable in $]x_0, +\infty[$, for every $x_0 > 0$, that is, differentiable in $]0, +\infty[$, and

(16.7)
$$\varphi'(x) = -\int e^{-x(1+\omega^2)}\lambda^1(d\omega) \qquad \text{for } x > 0$$

and via the substitution $t = \omega\sqrt{x}$ this reads

(16.8)
$$\varphi'(x) = -Gx^{-1/2}e^{-x} \qquad \text{for } x > 0$$

where G designates the integral (16.1) that we are trying to explicitly compute. Its existence is already part of the preceding analysis, but can also be inferred from the majorization $e^{-t^2} \le e^{-t}$, which holds for $t \ge 1$. From (16.6), (16.8) and the fundamental theorem of calculus

$$\varphi(x) - \varphi(\alpha) = G \int_x^\alpha t^{-1/2} e^{-t} \, dt = 2G \int_{\sqrt{x}}^{\sqrt{\alpha}} e^{-\omega^2} \, d\omega \, ,$$

for $x > 0$ and $\alpha > 0$. Upon letting α run to $+\infty$, we will get

(16.9)
$$\varphi(x) = 2G \int_{\sqrt{x}}^{+\infty} e^{-\omega^2} \, d\omega$$

if we notice that $\varphi(\alpha) \to 0$ as $\alpha \to +\infty$, which in turn is a consequence of the inequalities

$$\varphi(\alpha) \leq e^{-\alpha} \int (1 + \omega^2)^{-1} \lambda^1(d\omega) = \varphi(0) e^{-\alpha} \qquad \text{for all } \alpha > 0.$$

Because φ is continuous on $\mathbb{R}_+$ we can pass to the limit $x \to 0^+$ in (16.9) and get

$$\pi = \varphi(0) = 2G \int_0^{+\infty} e^{-\omega^2} \, d\omega = G^2 \, ,$$

using the obvious (on grounds of symmetry) fact that $\int_{-\infty}^0 e^{-\omega^2} \, d\omega = \int_0^{+\infty} e^{-\omega^2} \, d\omega$. Since $G > 0$, it follows finally that $G = \sqrt{\pi}$. That is,

(16.10)
$$\int e^{-x^2} \, dx = \sqrt{\pi}$$

or equivalently, in the form seen in probability theory,

(16.10′)
$$\int e^{-x^2/2} \, dx = \sqrt{2\pi} \, .$$

This derivation goes back to ANONYME [1889] and VAN YZEREN [1979]. A particularly short alternative one is made possible by Tonelli's theorem (cf. Exercise 4 in §23).

Exercises.

1. Which of the two functions below are integrable, which are square-integrable with respect to Lebesgue–Borel measure on the indicated intervals?

(a) $\qquad\qquad f(x) := x^{-1}, \qquad x \in I := [1, +\infty[\, ;$

(b) $\qquad\qquad f(x) := x^{-1/2}, \qquad x \in I := \,]0, 1] \, .$

2. Show that for every real number $\alpha > 0$ the function $x \mapsto e^{-x^\alpha}$ is λ^1-integrable over $\mathbb{R}_+$.

3. Show that for every real number $\alpha > 0$ the function

$$x \mapsto e^{-\alpha x} \left[\frac{\sin x}{x} \right]^3$$

is λ^1-integrable over $]0, +\infty[$ and that

$$\alpha \mapsto \int_0^{+\infty} e^{-\alpha x} \left[\frac{\sin x}{x} \right]^3 \lambda^1(dx)$$

is continuous on $]0, +\infty[$.

§17. Measures with densities: the Radon–Nikodym theorem

Again let $(\Omega, \mathscr{A}, \mu)$ be an arbitrary measure space and $E^* = E^*(\Omega, \mathscr{A})$ the set of all $\mathscr{A}$-measurable, non-negative numerical functions on Ω. In 12.4 we defined the integral of every function $f \in E^*$ over every set $A \in \mathscr{A}$. We are interested here in how this integral behaves with respect to A.

17.1 Theorem. *For each function $f \in E^*$ the equation*

(17.1) $$\nu(A) := \int_A f \, d\mu$$

defines a measure ν on $\mathscr{A}$.

Proof. $\nu(\emptyset) = 0$ and $\nu \geq 0$. For every sequence $(A_n)_{n \in \mathbb{N}}$ of pairwise disjoint sets from $\mathscr{A}$ with $A := \bigcup_{n \in \mathbb{N}} A_n$

$$1_A f = \sum_{n=1}^{\infty} 1_{A_n} f$$

and so by 11.5

$$\nu(A) = \sum_{n=1}^{\infty} \nu(A_n),$$

the final property needing to be checked in confirming that ν is a measure on $\mathscr{A}$. □

17.2 Definition. *If f is a non-negative $\mathscr{A}$-measurable, numerical function on Ω, then the measure ν defined on $\mathscr{A}$ by (17.1) is called the measure having density f with respect to μ.* It will be denoted by

(17.2) $$\nu = f\mu.$$

Concerning the relationship between ν- and μ-integrals we will show

17.3 Theorem. *Let $f, \varphi \in E^*$, $\nu := f\mu$. Then*

(17.3) $$\int \varphi \, d\nu = \int \varphi f \, d\mu$$

or, written out,

$$(17.3') \qquad\qquad \int \varphi \, d(f\mu) = \int \varphi f \, d\mu \, .$$

An $\mathscr{A}$-measurable function $\varphi : \Omega \to \overline{\mathbb{R}}$ is ν-integrable if and only if φf is μ-integrable. In this case (17.3) is again valid.

Proof. First suppose $\varphi = \sum\limits_{i=1}^{n} \alpha_i 1_{A_i}$ is an $\mathscr{A}$-elementary function. In this case (17.3) holds because

$$\int \varphi \, d\nu = \sum_{i=1}^{n} \alpha_i \nu(A_i) = \sum_{i=1}^{n} \alpha_i \int 1_{A_i} f \, d\mu = \int \varphi f \, d\mu \, .$$

For an arbitrary $\varphi \in E^*$ there is a sequence (u_n) in E such that $u_n \uparrow \varphi$. Since then $u_n f \uparrow \varphi f$ as well, (17.3) follows from 11.4. Finally, consider any $\mathscr{A}$-measurable numerical function φ on Ω. By now we know that

$$\int \varphi^+ \, d\nu = \int \varphi^+ f \, d\mu = \int (\varphi f)^+ \, d\mu \quad \text{and} \quad \int \varphi^- \, d\nu = \int \varphi^- f \, d\mu = \int (\varphi f)^- \, d\mu \, .$$

From these equations and the definition of integrability follows the second part of the theorem. $\square$

It now follows that the formation of measures with densities is *transitive*:

17.4 Corollary. *Let $f, g \in E^*$, $\nu := f\mu$ and $\varrho := g\nu$. Then $\varrho = (gf)\mu$, that is,*

$$(17.4) \qquad\qquad g(f\mu) = (gf)\mu \, .$$

Proof. For every $A \in \mathscr{A}$

$$\varrho(A) = \int_A g \, d\nu = \int 1_A g \, d\nu$$

and furthermore, according to 17.3

$$\int 1_A g \, d\nu = \int 1_A g f \, d\mu = \int_A (gf) \, d\mu \, .$$

We thus obtain $\varrho(A) = \int_A gf \, d\mu$, for all $A \in \mathscr{A}$; which is what had to be proved. $\square$

On the question of uniqueness of density functions we have

17.5 Theorem. *For functions $f, g \in E^*$*

$$(17.5) \qquad f = g \quad \mu\text{-almost everywhere} \quad \Rightarrow \quad f\mu = g\mu \, .$$

If either f or g is μ-integrable, the converse implication holds as well.

Proof. If f and g coincide μ-almost everywhere, then so do $1_A f$ and $1_A g$ for each $A \in \mathscr{A}$, whence

$$\int_A f \, d\mu = \int_A g \, d\mu \qquad \text{for all } A \in \mathscr{A},$$

which just says that $f\mu = g\mu$.

Now suppose that f is μ-integrable and that $f\mu = g\mu$. Since $g \geq 0$ and $\int g \, d\mu = \int f \, d\mu < +\infty$, g is also μ-integrable. Let us show that the set

$$N := \{f > g\},$$

which lies in $\mathscr{A}$ by 9.3, is a μ-nullset. For every $\omega \in N$, $f(\omega) - g(\omega)$ is defined and is positive, which means that the definition

$$h := 1_N f - 1_N g$$

makes sense. The functions $1_N f$, $1_N g$, being majorized by the μ-integrable functions f, g, are themselves integrable. Because $f\mu = g\mu$, they have the same μ-integral. From this we get that

$$\int h \, d\mu = \int_N f \, d\mu - \int_N g \, d\mu = 0.$$

Since $N = \{h > 0\}$, this equality and 13.2 tell us that $\mu(N) = 0$. With the roles of f and g reversed, this conclusion reads $\mu(N') = 0$, where $N' := \{g > f\}$. Since $\{f \neq g\} = N \cup N'$, the desired conclusion, namely that $\{f \neq g\}$ is a μ-nullset, is obtained. $\square$

The converse of implication (17.5) is not valid without some additional hypothesis on the densities f and g. The next example illustrates this.

Example. 1. As in Example 2 of §3 let Ω be an uncountable set, $\mathscr{A}$ the σ-algebra of countable and co-countable subsets of Ω (see Example 2 in §1). But the measure μ will be defined on $\mathscr{A}$ by $\mu(A) := 0$ or $+\infty$, according as A or $\complement A$ is countable. If f and g are the constant functions on Ω with the respective values 1 and 2, then indeed $f\mu = g\mu$, yet $f(\omega) = g(\omega)$ holds for *no* $\omega \in \Omega$. Of course, it then follows from 17.5 that neither f nor g is μ-integrable.

Before turning to the principal problem of this section, we will examine another characterization of σ-finite measures which is important for what follows and is of interest in its own right.

17.6 Lemma. *Let $(\Omega, \mathscr{A}, \mu)$ be a measure space. The measure is σ-finite if and only if there exists a μ-integrable function h on Ω which satisfies*

$$(17.6) \qquad\qquad 0 < h(\omega) < +\infty \qquad\qquad \text{for every } \omega \in \Omega.$$

Proof. If μ is σ-finite, there is a sequence $(A_n)_{n\in\mathbb{N}}$ in $\mathscr{A}$ such that $\mu(A_n) < +\infty$ for each $n \in \mathbb{N}$ and $A_n \uparrow \Omega$. Choose positive real numbers η_n satisfying both $\eta_n \leq 2^{-n}$

and $\eta_n \mu(A_n) \leq 2^{-n}$, for each $n \in \mathbb{N}$. Then the function

$$h := \sum_{n=1}^{\infty} \eta_n 1_{A_n}$$

does what is wanted. It is measurable, $0 < h(\omega) \leq 1$ for each $\omega \in \Omega$, and $\int h \, d\mu \leq 1$. The converse implication is already known: it is contained in the second part of 13.6. □

In the light of 13.2 this lemma has another formulation: For each σ-finite measure μ there exists a real, measurable function $h \geq 0$ such that the measure $h\mu$ is finite and has the same nullsets as μ.

We come now to the main problem, already alluded to: On the σ-algebra $\mathscr{A}$ of the measurable space $(\Omega, \mathscr{A})$ two measures ν and μ are given. We pose the question of how to decide whether ν has a density with respect to μ, that is, whether there is an $\mathscr{A}$-measurable, non-negative, numerical function f on Ω satisfying $\nu = f\mu$, satisfying in other words

$$\nu(A) = \int_A f \, d\mu \qquad\qquad \text{for all } A \in \mathscr{A}.$$

For an affirmative answer it is necessary, as 13.3 shows, that every μ-null set in $\mathscr{A}$ be a ν-null set as well.

17.7 Definition. A measure ν on $\mathscr{A}$ is called *continuous with respect to a measure* μ on $\mathscr{A}$, for short, μ-*continuous*, if every μ-nullset from $\mathscr{A}$ is also a ν-nullset.

In the case of a finite measure ν there is a condition equivalent to μ-continuity which clarifies and justifies the terminology:

17.8 Theorem. *A finite measure ν on $\mathscr{A}$ is μ-continuous if and only if for every $\varepsilon > 0$ there exists $\delta > 0$ such that*

$$(17.7) \qquad A \in \mathscr{A} \quad \text{and} \quad \mu(A) \leq \delta \quad \Rightarrow \quad \nu(A) \leq \varepsilon.$$

Proof. From (17.7) it follows that $\nu(A) \leq \varepsilon$ holds for every $\varepsilon > 0$ if A is a μ-nullset. Hence $\nu(A) = 0$ and ν is thus a μ-continuous measure, even without the finiteness hypothesis. For the converse we will show that if (17.7) fails, then ν is not μ-continuous. Thus, for some $\varepsilon > 0$ there is no δ, which means there is a sequence $(A_n)_{n \in \mathbb{N}}$ in $\mathscr{A}$ with the properties

$$\mu(A_n) \leq 2^{-n} \quad \text{and} \quad \nu(A_n) > \varepsilon \qquad \text{for each } n \in \mathbb{N}.$$

We set

$$A := \limsup_{n \to \infty} A_n := \bigcap_{n \in \mathbb{N}} \bigcup_{m \geq n} A_m$$

and have a set in $\mathscr{A}$ which on the one hand satisfies

$$\mu(A) \leq \mu\left(\bigcup_{m \geq n} A_m\right) \leq \sum_{m=n}^{\infty} \mu(A_m) \leq \sum_{m=n}^{\infty} 2^{-m} = 2^{-n+1} \qquad \text{for every } n \in \mathbb{N},$$

whence $\mu(A) = 0$, and on the other hand, due to the finiteness of ν and 15.3, satisfies

$$\nu(A) \geq \limsup_{n \to \infty} \nu(A_n) \geq \varepsilon > 0\,,$$

which proves that ν is not μ-continuous. $\square$

Examples. 2. Let Ω be an uncountable set, $\mathscr{A}$ the σ-algebra of countable and co-countable subsets of $\mathscr{A}$ (Example 2 in §1). As in the preceding Example, consider the measure ν on $\mathscr{A}$ which assigns to a set the value 0 or $+\infty$ according as the set or its complement is countable. Let μ denote the counting measure ζ on $\mathscr{A}$ (from Example 6, §3). Since $\emptyset$ is the only μ-nullset, ν is trivially μ-continuous. However, ν cannot have a density with respect to μ. For from $\nu = f\mu$ with $f \in E^*$ it would follow that

$$0 = \nu(\{\omega\}) = \int_{\{\omega\}} f\, d\mu = f(\omega)\mu(\{\omega\}) = f(\omega)$$

for every $\omega \in \Omega$, making $f = 0$ and therefore $\nu = f\mu = 0$, which is not the case because Ω is uncountable.

3.   Let $(\mathbb{R}, \mathscr{B}^1, \mu)$ be the 1-dimensional Lebesgue–Borel measure space (so $\mu = \lambda^1$) and denote by $\mathscr{N}$ the system of all μ-nullsets. Then $\mathscr{N}$ is an example of a σ-*ideal* in $\mathscr{B}^1$: The union of any sequence of its sets is another, as are the intersections of its sets with those of $\mathscr{B}^1$ (cf. Exercise 5, §3). These properties insure that

$$\nu(A) := \begin{cases} 0 & \text{if } A \in \mathscr{N} \\ +\infty & \text{if } A \in \mathscr{B}^1 \setminus \mathscr{N} \end{cases}$$

defines a measure on $\mathscr{B}^1$ (cf. Exercise 6, §3). From its definition it is clear that ν is μ-continuous. Here however (17.7) fails, since for every $\delta > 0$

$$\mu([0, \delta[) = \delta \quad \text{and} \quad \nu([0, \delta[) = +\infty\,.$$

Thus the finiteness hypothesis on ν in 17.8 is not superfluous. Example 2 shows that for the existence of a density $f \in E^*$ with $\nu = f\mu$, the μ-continuity of ν, while necessary, is not sufficient. All the more noteworthy is the theorem of Radon and Nikodym which we will prove, after a preparatory lemma.

17.9 Lemma. *Let σ and τ be finite measures on a σ-algebra $\mathscr{A}$ of subsets of Ω and let $\varrho := \tau - \sigma$ denote their difference. Then there is a set $\Omega_0 \in \mathscr{A}$ with the properties*

(17.8) $\varrho(\Omega_0) \geq \varrho(\Omega)\,;$

(17.9) $\varrho(A) \geq 0$ *for all $A \in \Omega_0 \cap \mathscr{A}$.*

Proof. Let us first proof the weaker claim:

(*) For every $\varepsilon > 0$ there exists $\Omega_\varepsilon \in \mathscr{A}$ with the properties

(17.8') $\varrho(\Omega_\varepsilon) \geq \varrho(\Omega)\,;$

(17.9') $\varrho(A) \geq -\varepsilon$ for all $A \in \Omega_\varepsilon \cap \mathscr{A}$.

We may obviously suppose that $\varrho(\Omega) \geq 0$, since otherwise $\Omega_\varepsilon := \emptyset$ does what is wanted. If then $\varrho(A) > -\varepsilon$ for all $A \in \mathscr{A}$, it suffices to choose $\Omega_\varepsilon := \Omega$. So we consider the case that some $A_1 \in \mathscr{A}$ satisfies $\varrho(A_1) \leq -\varepsilon$. From the definition of ϱ and the subtractivity of the finite measures σ and τ,

$$\varrho(\complement A_1) = \varrho(\Omega) - \varrho(A_1) \geq \varrho(\Omega) + \varepsilon > \varrho(\Omega).$$

Therefore, if $\varrho(A) > -\varepsilon$ for all $A \in (\complement A_1) \cap \mathscr{A}$, we can set $\Omega_\varepsilon := \complement A_1$ and be done. In the contrary case there is a set $A_2 \in (\complement A_1) \cap \mathscr{A}$ with $\varrho(A_2) \leq -\varepsilon$. Then because A_1, A_2 are disjoint

$$\varrho(\complement(A_1 \cup A_2)) = \varrho(\Omega) - \varrho(A_1) - \varrho(A_2) \geq \varrho(\Omega) + 2\varepsilon > \varrho(\Omega)$$

and the preceding dichotomy presents itself anew. If after finitely many repetitions of this procedure we have not reached our goal, then we will have generated a sequence $(A_n)_{n \in \mathbb{N}}$ of pairwise disjoint sets in $\mathscr{A}$ with

$$\varrho(\Omega \setminus (A_1 \cup \ldots \cup A_n)) > \varrho(\Omega) \quad \text{and} \quad \varrho(A_n) \leq -\varepsilon \qquad \text{for every } n \in \mathbb{N}.$$

Because of the finite additivity of σ and τ, this would have the consequence that

$$\varrho(A_1 \cup \ldots \cup A_n) = \sum_{i=1}^{n} \varrho(A_i) \leq -n\varepsilon \qquad \text{for every } n \in \mathbb{N}$$

and entail the divergence of the series $\sum\limits_{n=1}^{\infty} \varrho(A_n)$. But the latter is untenable, because when the σ-additivity of σ and τ is applied to the disjoint union $A := \bigcup\limits_{n \in \mathbb{N}} A_n$ it shows this series to be convergent:

$$\sum_{n=1}^{\infty} \varrho(A_n) = \sum_{n=1}^{\infty} (\tau(A_n) - \sigma(A_n)) = \tau(A) - \sigma(A) \in \mathbb{R}.$$

This contradiction proves that the construction procedure must terminate after some finite number n of steps, with the set $\Omega_\varepsilon := \complement(A_1 \cup \ldots \cup A_n)$ then satisfying (17.8′) and (17.9′).

We now take $\varepsilon = 1/n$ in (*) for successive $n \in \mathbb{N}$. The sets Ω_ε can be chosen with the additional property of isotoneity. For if $\Omega_1 \supset \Omega_{1/2} \supset \ldots \supset \Omega_{1/n}$ has already been realized, we simply apply (*) to $\Omega_{1/n}$ as a new base space in the role of Ω, that is, we consider the restriction of the measures σ and τ to $\Omega_{1/n} \cap \mathscr{A}$. Finally, the set $\Omega_0 := \bigcap\limits_{n \in \mathbb{N}} \Omega_{1/n}$ will be seen to do the desired job. For since $\Omega_{1/n} \downarrow \Omega_0$, (17.8) follows from (17.8′), and (17.9) follows from (17.9′), which insures that $\varrho(A) \geq -1/n$ for all $n \in \mathbb{N}$ and every $A \in \Omega_0 \cap \mathscr{A}$. $\square$

As indicated, this puts us in a position to answer the important question we posed earlier.

17.10 Theorem (Radon–Nikodym). *Let μ and ν be measures on a σ-algebra $\mathscr{A}$ in a set Ω. If μ is σ-finite, the following two assertions are equivalent:*

(i) ν *has a density with respect to* μ.
(ii) ν *is* μ-*continuous*.

Proof. Only the implication (ii)$\Rightarrow$(i) is still in need of proof. To that end we distinguish three cases.

First Case: The measures μ and ν are each finite. Form the set $\mathscr{G}$ of all $\mathscr{A}$ measurable numerical functions $g \geq 0$ on Ω which satisfy $g\mu \leq \nu$, that is, which satisfy

$$\int_A g \, d\mu \leq \nu(A) \qquad\qquad \text{for all } A \in \mathscr{A}.$$

The constant function $g = 0$ lies in $\mathscr{G}$, so $\mathscr{G}$ is not empty. $\mathscr{G}$ is moreover sup-stable, that is, $g \vee h \in \mathscr{G}$ whenever $g, h \in \mathscr{G}$. Indeed, setting $A_1 := \{g \geq h\}$, $A_2 := \complement A_1$, every $A \in \mathscr{A}$ satisfies

$$\int_A g \vee h \, d\mu = \int_{A \cap A_1} g \, d\mu + \int_{A \cap A_2} h \, d\mu \leq \nu(A \cap A_1) + \nu(A \cap A_2) = \nu(A).$$

Since $\int g \, d\mu \leq \nu(\Omega) < +\infty$ for every $g \in \mathscr{G}$, the number

$$\gamma := \sup\left\{ \int g \, d\mu : g \in \mathscr{G} \right\}$$

is finite and there is a sequence (g_n') in $\mathscr{G}$ such that $\lim \int g_n' \, d\mu = \gamma$. Due to sup-stability the functions $g_n := g_1' \vee \ldots \vee g_n'$ lie in $\mathscr{G}$, and consequently $\gamma \geq \int g_n \, d\mu \geq \int g_n' \, d\mu$ (since $g_n \geq g_n'$) for all $n \in \mathbb{N}$. Which shows that $\lim \int g_n \, d\mu = \gamma$. As the sequence (g_n) is isotone, the monotone convergence theorem can be applied, assuring that $f := \sup g_n$ is a function in $\mathscr{G}$ and that $\int f \, d\mu = \gamma$. All this proves that the function $g \mapsto \int g \, d\mu$ on $\mathscr{G}$ assumes its maximum value at f.

Now we prove that $\nu = f\mu$. In any case we have $f\mu \leq \nu$, since $f \in \mathscr{G}$, and so

$$\tau := \nu - f\mu$$

is a finite measure on $\mathscr{A}$, evidently μ-continuous since ν is by hypothesis. We have to show that $\tau = 0$. So let us assume contrariwise that $\tau(\Omega) > 0$. Due to the μ-continuity of τ, this entails that $\mu(\Omega) > 0$ as well, and we may form the real number

$$\beta := \frac{1}{2} \cdot \frac{\tau(\Omega)}{\mu(\Omega)} > 0,$$

which satisfies $\tau(\Omega) = 2\beta\mu(\Omega) > \beta\mu(\Omega)$. The preceding lemma applied to τ and $\sigma := \beta\mu$ supplies a set $\Omega_0 \in \mathscr{A}$ which satisfies

$$\tau(\Omega_0) - \beta\mu(\Omega_0) \geq \tau(\Omega) - \beta\mu(\Omega) > 0 \quad \text{and} \quad \tau(A) \geq \beta\mu(A) \text{ for all } A \in \Omega_0 \cap \mathscr{A}.$$

The $\mathscr{A}$-measurable, non-negative function $f_0 := f + \beta 1_{\Omega_0}$ therefore has the property

$$\int_A f_0 \, d\mu = \int_A f \, d\mu + \beta\mu(\Omega_0 \cap A) \leq \int_A f \, d\mu + \tau(\Omega_0 \cap A) \leq \int_A f \, d\mu + \tau(A) = \nu(A)$$

for every $A \in \mathscr{A}$. These inequalities put f_0 in $\mathscr{G}$. Since τ is μ-continuous and $\tau(\Omega_0) > \beta\mu(\Omega_0)$, we must have $\mu(\Omega_0) > 0$, leading to

$$\int f_0 \, d\mu = \int f \, d\mu + \beta\mu(\Omega_0) = \gamma + \beta\mu(\Omega_0) > \gamma \,,$$

an inequality which is incompatible with the definition of γ and the fact that $f_0 \in \mathscr{G}$. The assumption $\tau(\Omega) > 0$ is therefore untenable, and $\tau = 0$, as desired.

Second Case: The measure μ is finite and the measure ν is infinite. We will produce a decomposition $\Omega = \bigcup\limits_{n=0}^{\infty} \Omega_n$ of Ω into pairwise disjoint sets from $\mathscr{A}$ with the following properties

(a) $A \in \Omega_0 \cap \mathscr{A}$ $\Rightarrow$ either $\mu(A) = \nu(A) = 0$ or $0 < \mu(A) < \nu(A) = +\infty$.

(b) $\nu(\Omega_n) < +\infty$ for all $n \in \mathbb{N}$.

To this end let $\mathscr{Q}$ denote the system of all $Q \in \mathscr{A}$ with $\nu(Q) < +\infty$ and define

$$\alpha := \sup\{\mu(Q) : Q \in \mathscr{Q}\} \,.$$

This is a real number because the measure μ is finite. There is a sequence $(Q_m)_{m \in \mathbb{N}}$ in $\mathscr{Q}$ with $\lim \mu(Q_m) = \alpha$. Since $\mathscr{Q}$ is evidently closed under finite unions, (Q_m) may be assumed to be isotone. $Q_0 := \bigcup\limits_{m \in \mathbb{N}} Q_m$ is then a set from $\mathscr{A}$ satisfying $\mu(Q_0) = \alpha$. We will show that $\Omega_0 := \complement Q_0$ satisfies (a). So consider $A \in \Omega_0 \cap \mathscr{A}$ with $\nu(A) < +\infty$. We need to see that $\mu(A) = \nu(A) = 0$, and since ν is μ-continuous we really only need to confirm that $\mu(A) = 0$. Since $\nu(A) < +\infty$ and, as noted already, $\mathscr{Q}$ is closed under union, each $Q_m \cup A$ lies in $\mathscr{Q}$, so that $\mu(Q_m \cup A) \leq \alpha$, and consequently

$$\mu(Q_0 \cup A) = \lim_{m \to \infty} \mu(Q_m \cup A) \leq \alpha \,.$$

Since A is disjoint from Ω_0, $\mu(Q_0 \cup A) = \alpha + \mu(A)$. Conjoined with the preceding inequality and the finiteness of α this says that indeed $\mu(A)$ must be 0. Finally, to take care of (b) we merely define $\Omega_1 := Q_1$, and $\Omega_m := Q_m \setminus Q_{m-1}$ for all integers $m \geq 2$ in order to get a decomposition of Ω with the desired properties.

Now let μ_n, ν_n denote the restrictions of μ, ν to the trace σ-algebra $\Omega_n \cap \mathscr{A}$, for $n = 0, 1, \ldots$ and note that each ν_n is a μ_n-continuous measure. Moreover, for all $n \geq 1$ both μ_n and ν_n are finite. Case 1 therefore supplies $\Omega_n \cap \mathscr{A}$-measurable functions $f_n \geq 0$ on Ω_n with $\nu_n = f_n\mu_n$. Taking f_0 to be the constant function $+\infty$ on Ω_0, $\nu_0 = f_0\mu_0$ also holds, thanks to (a). Finally, "putting all the pieces together" gives our result in this second case. Namely, the function f on Ω defined to coincide on each Ω_n with f_n $(n = 0, 1, \ldots)$ is non-negative, $\mathscr{A}$-measurable and satisfies $\nu = f\mu$.

Third Case: This is the general case: only the σ-finiteness of μ is demanded. There is according to 17.6 a strictly positive function $h \in \mathscr{L}^1(\mu)$. The measure $h\mu$ is therefore finite and possesses exactly the same nullsets as does μ. Consequently ν is also $(h\mu)$-continuous. By what has already been proved there is then an

$\mathscr{A}$-measurable function $f \geq 0$ on Ω with $\nu = f(h\mu)$. According to 17.4 ν then has the density fh with respect to μ. $\square$

The question arises whether, in the situation of Theorem 17.10 the density f of ν is μ-almost everywhere uniquely determined. From 17.5 we at least get a positive answer when f is μ-integrable, that is, when ν is a finite measure. But more is true:

17.11 Theorem. *Let $\nu = f\mu$ be a measure having a density f with respect to a σ-finite measure μ on $\mathscr{A}$. Then f is μ-almost everywhere uniquely determined. The measure ν is σ-finite exactly when f is μ-almost everywhere real-valued.*

Proof. First we show that f is μ-almost everywhere uniquely determined if the measure μ is *finite*. In proving this we may assume that $\nu(\Omega) = +\infty$, since its truth is otherwise a consequence of the second part of 17.5. Furthermore, as we now find ourselves in case 2 of the preceding proof, the decomposition of Ω into $\Omega_0, \Omega_1, \ldots$ employed there lets us confine our attention to Ω_0, as 17.5 takes care of the remaining Ω_n $(n \in \mathbb{N})$. So it suffices to treat the case $\Omega = \Omega_0$, that is, to assume that μ and ν are linked by the alternative:

$$A \in \mathscr{A} \quad \Rightarrow \quad \text{either } \mu(A) = \nu(A) = 0 \quad \text{or} \quad 0 < \mu(A) < \nu(A) = +\infty.$$

The constant function $+\infty$ is then a density for ν with respect to μ and what has to be shown for uniqueness is that $f = +\infty$ holds μ-almost everywhere. And for that it suffices to show that

$$\mu(\{f \leq n\}) = 0 \qquad\qquad \text{for each } n \in \mathbb{N},$$

which in turn is a consequence of the above alternative and the inequalities

$$\nu(\{f \leq n\}) = \int_{\{f \leq n\}} f \, d\mu \leq n\mu(\{f \leq n\}) < +\infty$$

coming from the finiteness of μ.

We will use 17.6 to reduce the general case of σ-finite μ to the case just treated. That lemma supplies a strictly positive function $h \in \mathscr{L}^1(\mu)$. The measure $h\nu = h(f\mu) = f(h\mu)$ has the density f with respect to the finite measure $h\mu$, so f is $(h\mu)$-almost everywhere uniquely determined. Since the measures μ and $h\mu$ have the same nullsets, f is therefore also uniquely determined μ-almost everywhere.

Next, suppose ν is σ-finite. From 17.6 once again we get a strictly positive function $k \in \mathscr{L}^1(\nu)$. Then $k\nu = (fk)\mu$ is a finite measure, that is, fk is μ-integrable, consequently also μ-almost everywhere real-valued. Because k takes only non-zero real values, this means that f itself is real μ-almost everywhere.

Conversely, suppose that f is μ-almost everywhere real-valued. We want to see that ν is σ-finite. First of all, there is a decomposition $\Omega = \bigcup_{n=0}^{\infty} \Omega_n$ of Ω into a sequence of pairwise disjoint sets from $\mathscr{A}$ each of finite μ-measure. Set $A_n := \{n - 1 \leq f < n\}$ for each $n \in \mathbb{N}$ and $A_0 := \{f = +\infty\}$, the present

hypothesis being just that $\mu(A_0) = 0$. $\Omega = \bigcup\limits_{i,j=0}^{\infty} (\Omega_i \cap A_j)$ is a decomposition of Ω
into a (doubly-indexed) sequence of pairwise disjoint sets from $\mathscr{A}$. If each has finite
ν-measure, this proves that ν is σ-finite. Consider any $i \in \mathbb{Z}_+$. Because $\mu(A_0) = 0$
and $\nu = f\mu$, we have $\nu(\Omega_i \cap A_0) \leq \nu(A_0) = 0$. Because $\nu = f\mu$ and $f \leq j$ in A_j,
we have $\nu(\Omega_i \cap A_j) \leq j\mu(\Omega_i) < +\infty$ for all $j \in \mathbb{N}$ as well. Thus all is proven. $\square$

In the generality presented here Theorem 17.10 was proved in 1930 by O.M. NI-
KODYM (1888–1974). H. Lebesgue proved the theorem in 1910 for the case where
μ is the L-B measure λ^1. J. RADON (1887-1956) pushed things further in a fundamental work which appeared in 1913. So 17.10 is often also called the *theorem of*
Lebesgue–Radon–Nikodym. The uniquely determined density f in 17.11 is called
the *Radon–Nikodym density* or the *Radon–Nikodym integrand* (of ν with respect
to μ). A beautiful proof of 17.10 by elementary Hilbert-space methods was discovered in 1940 by J. VON NEUMANN (1903–1957) and appears in many textbooks,
e.g., in RUDIN [1987], p. 130–131.

The history of the result to be presented next, the Lebesgue decomposition
theorem, runs somewhat parallel, Radon and Nikodym having also made significant contributions. We need a concept complementary to μ-continuity, namely
μ-singularity:

17.12 Definition. Let $(\Omega, \mathscr{A})$ be a measurable space, μ and ν measures defined
on $\mathscr{A}$. Let us write $\nu \ll \mu$ if ν is μ-continuous. ν is said to be *singular with respect
to μ* (or *μ-singular*), written $\nu \perp \mu$, if a set $N \in \mathscr{A}$ exists with $\mu(N) = 0 = \nu(\complement N)$.

It is obvious that the relation $\nu \perp \mu$ is *symmetric* in μ and ν, so it is also expressed as μ and ν are *singular to each other* (or *mutually singular*). The definition
of $\nu \perp \mu$ expresses the fact that for a suitable μ-nullset $N \in \mathscr{A}$

$$(17.10) \qquad\qquad \nu(A) = \nu(A \cap N) \qquad\qquad \text{for all } A \in \mathscr{A},$$

as follows from $\nu(A) = \nu(A \cap N) + \nu(A \cap \complement N)$ and $\nu(\complement N) = 0$. The condition
that $\nu \perp \mu$ thus says that the measure ν is "*carried* by a μ-nullset". From $\nu \ll \mu$
and $\nu \perp \mu$ together follows that $\nu(N) = 0$, and so $\nu = 0$. In this sense the
concepts μ-continuity and μ-singularity are diametral or antipodal. Relative to
L-B measure λ^d every Dirac measure ε_x on $\mathscr{B}^d$ obviously satisfies $\lambda^d \perp \varepsilon_x$.

17.13 Theorem (Lebesgue's decomposition theorem). *If μ and ν are σ-finite
measures on a σ-algebra $\mathscr{A}$ in a set Ω, then ν can be decomposed in just one way
as $\nu = \nu_c + \nu_s$ with measures ν_c, ν_s on $\mathscr{A}$ that satisfy $\nu_c \ll \mu$ and $\nu_s \perp \mu$.*

ν_c is called the *continuous part* of ν with respect to μ, ν_s the *singular part*. The
Radon–Nikodym theorem is applicable to the part ν_c.

Proof. We will carry out the proof in detail only for *finite* μ and ν and indicate in
Exercise 4 how the reader can then handle the general case himself.

Existence of a decomposition: Let $\mathcal{N}_\mu$ designate the system of all μ-nullsets in $\mathcal{A}$. Since $\nu(A) \leq \nu(\Omega) < +\infty$ for every $A \in \mathcal{A}$,

$$(17.11) \qquad \alpha := \sup\{\nu(A) : A \in \mathcal{N}_\mu\}$$

is a real number. Since $\mathcal{N}_\mu$ is closed under countable unions, there exists an isotone sequence (A_n) in $\mathcal{N}_\mu$ with $\nu(A_n) \uparrow \alpha$. Since ν is continuous from below, it follows that

$$\nu(N) = \alpha$$

for the set $N := \bigcup_{n \in \mathbb{N}} A_n \in \mathcal{N}_\mu$. We will show that via

$$\nu_c(A) := \nu(A \cap \complement N) \quad \text{and} \quad \nu_s(A) := \nu(A \cap N)$$

two measures are defined on $\mathcal{A}$ that do what is wanted. Evidently $\nu = \nu_c + \nu_s$, and $\nu_s \perp \mu$ since $N \in \mathcal{N}_\mu$. To prove that $\nu_c \ll \mu$, it must be shown that $\nu(A') = 0$ whenever $A \in \mathcal{N}_\mu$ and $A' := A \cap \complement N$. As a subset of $A \in \mathcal{N}_\mu$, the set A' and then also the set $A' \cup N$, is μ-null. Therefore $\nu(A' \cup N) \leq \alpha$ by definition of α. But $A' \cap N = \emptyset$ and $\nu(N) = \alpha$. Hence

$$\alpha + \nu(A') = \nu(N) + \nu(A') = \nu(N \cup A') \leq \alpha,$$

from which follows $\nu(A') = 0$ as desired, since α is finite.

Uniqueness of the decomposition: Suppose

$$(17.12) \qquad \nu = \nu_c + \nu_s = \nu_c' + \nu_s'$$

are two decompositions of the kind described in the theorem. The measures ν_s, ν_s' are carried by μ-nullsets N, N' in the sense of (17.10); which means that

$$(17.13) \qquad \nu_s(A) = \nu_s(A \cap N) \quad \text{and} \quad \nu_s'(A) = \nu_s'(A \cap N') \qquad \text{for all } A \in \mathcal{A}.$$

Setting $N_0 := N \cup N'$ gives a set in $\mathcal{N}_\mu$, so that from $\nu_c, \nu_c' \ll \mu$ follows

$$\nu_c(A \cap N_0) = \nu_c'(A \cap N_0) = 0 \qquad \text{for every } A \in \mathcal{A}.$$

Therefore (17.12) and (17.13) give

$$\nu(A \cap N_0) = \nu_c(A \cap N_0) + \nu_s(A \cap N_0) = \nu_s(A \cap N_0) = \nu_s(A \cap N_0 \cap N)$$
$$= \nu_s(A \cap N) = \nu_s(A), \qquad \text{for every } A \in \mathcal{A}.$$

Analogously of course, $\nu(A \cap N_0) = \nu_s'(A)$ for every $A \in \mathcal{A}$. Thus we have $\nu_s = \nu_s'$. A return to (17.12), recalling that all measures are finite, gives $\nu_c = \nu_c'$ as well. $\quad\square$

There is a short, elementary proof of 17.13 that does not make use of the Radon–Nikodym theorem; see Woo [1971].

Exercises.

1. Show that the Dirac measure ε_x on $\mathcal{B}^d$ has no density with respect to λ^d, for any $x \in \mathbb{R}^d$. (Physicists occasionally work with such a "symbolic" density d_x, calling it the Dirac *function* at the point x. The correct mathematical object is nevertheless the Dirac *measure* ε_x.)

2. Show that the relation $\ll$ on the set of measures on a σ-algebra $\mathscr{A}$ is reflexive and transitive. The relation $\mu \sim \nu$ defined as $\mu \ll \nu$ and $\nu \ll \mu$ is then an equivalence relation. Two measures μ and ν stand in this relation just when they have the same nullsets. For σ-finite measures μ and ν on $\mathscr{A}$ show that $\mu \sim \nu$ is equivalent to $\nu = f\mu$ for a density f which satisfies $0 < f(\omega) < +\infty$ for μ-almost all (or even for all) $\omega \in \Omega$.

3. On a σ-algebra $\mathscr{A}$ in a set Ω two measures μ and ν are related by $\nu \leq \mu$. Show that if further μ is σ-finite, then there is an $\mathscr{A}$-measurable function f satisfying $0 \leq f \leq 1$ such that $\nu = f\mu$.

4. Lebesgue's decomposition theorem was proved for finite measures μ and ν. Show how to infer its validity for σ-finite measures from this. [*Hint*: For the existence proof use 17.6. For the uniqueness proof choose a sequence (A_n) in $\mathscr{A}$ with $A_n \uparrow \Omega$ and $\mu(A_n), \nu(A_n)$ finite for each n, and consider the measures $\nu_n(A) := \nu(A \cap A_n)$, $A \in \mathscr{A}$, $n \in \mathbb{N}$.]

5. Let $\nu = \nu_c + \nu_s$ be the Lebesgue decomposition of a σ-finite measure ν on $\mathscr{A}$ with respect to a σ-finite measure μ. The singular part ν_s has the form $\nu_s(A) = \nu(A \cap N)$ for all $A \in \mathscr{A}$ and a suitable μ-nullset $N \in \mathscr{A}$. Show that if N' is any other μ-nullset with this property, then $\mu(N \triangle N') = \nu(N \triangle N') = 0$.

6. Let $(\Omega, \mathscr{A}, \mu)$ be a measure space, $\nu = f\mu$ a σ-finite measure on $\mathscr{A}$ having density f with respect to μ. Show that this density function is μ-almost everywhere uniquely determined and is μ-almost everywhere real-valued. Show that if f is strictly positive, then μ itself is σ-finite.

7. Let $(\Omega, \mathscr{A})$ be a measurable space. For every measure μ on $\mathscr{A}$ let $\mathscr{N}_\mu$ denote the σ-ideal of its nullsets. Show that for any sequence $(\mu_n)_{n \in \mathbb{N}}$ of σ-finite measures on $\mathscr{A}$ there is a finite measure μ on $\mathscr{A}$ for which $\mathscr{N}_\mu = \bigcap_{n \in \mathbb{N}} \mathscr{N}_{\mu_n}$.

8. The set $\Omega :=]0, +\infty[$ is a group with respect to multiplication. Show that the measure on $\Omega \cap \mathscr{B}^1$ defined by $\mu := h\lambda_\Omega^1$ with density function $h(x) := 1/x$ is invariant under each self-mapping $x \mapsto \alpha x$ of Ω ($\alpha \in \Omega$). μ is thus the Haar measure of the group Ω in the sense of the remark immediately following 8.2.

§18*. Signed measures

It is worthwhile turning our attention back to Lemma 17.9. The measure concept in this book is that formulated in Definition 3.3: Measures are premeasures μ on a σ-algebra $\mathscr{A}$, and so are *non-negative* σ-additive functions on $\mathscr{A}$ satisfying the additional condition $\mu(\emptyset) = 0$. In Lemma 17.9 we encountered a real-valued, σ-additive function ϱ which is the difference of two finite measures. Similarly for any $f \in \mathscr{L}^1(\mu)$ the function $A \mapsto \int_A f \, d\mu$ on $\mathscr{A}$ is the difference of two finite measures, for example $f^+\mu$ and $f^-\mu$.

We will call a real function $\varrho : \mathscr{A} \to \mathbb{R}$ on a σ-algebra a *finite signed measure* if it is σ-additive in the sense of (3.2), non-negativity not being required. From

σ-additivity applied to the constant sequence $\emptyset, \emptyset, \ldots$ follows immediately that (3.1) is also satisfied, that is, $\varrho(\emptyset) = 0$, because ϱ is only allowed to take real values. A second pass through the proof of Lemma 17.9 will convince the reader that this lemma is in fact valid for every finite signed measure. As a corollary we immediately get the following theorem on the existence a *Hahn decomposition* of ϱ, a theorem that goes back to H. HAHN (1879–1934).

18.1 Theorem. *Let ϱ be a finite signed measure on a σ-algebra $\mathscr{A}$ in a set Ω. Then there are sets $\Omega^+, \Omega^- \in \mathscr{A}$ with $\Omega = \Omega^+ \cup \Omega^-$, $\Omega^+ \cap \Omega^- = \emptyset$, and $\varrho(A) \geq 0$ for all A in the trace σ-algebra $\Omega^+ \cap \mathscr{A}$, and $\varrho(A) \leq 0$ for all $A \in \Omega^- \cap \mathscr{A}$.*

Proof. Set

$$\gamma := \sup\{\varrho(A) : A \in \mathscr{A}\}$$

and choose a sequence (A_n) in $\mathscr{A}$ with $\lim \varrho(A_n) = \gamma$. By applying 17.9 to the restriction of ϱ to $A_n \cap \mathscr{A}$, we may replace A_n by a set $P_n \in \mathscr{A}$ satisfying $\varrho(P_n) \geq \varrho(A_n)$ and $\varrho(A) \geq 0$ for all $A \in P_n \cap \mathscr{A}$. We will then have

(18.1) $$\gamma = \sup\{\varrho(P_n) : n \in \mathbb{N}\}.$$

The decomposition of Ω that is sought can be realized by

$$\Omega^+ := \bigcup_{n \in \mathbb{N}} P_n, \qquad \Omega^- := \Omega \setminus \Omega^+.$$

Indeed, all $A \in \Omega^+ \cap \mathscr{A}$ satisfy $\varrho(A) \geq 0$ because such an A has the form

$$A = \bigcup_{n \in \mathbb{N}} B_n$$

with pairwise disjoint sets $B_n \in P_n \cap \mathscr{A}$ (by the disjointification procedure used in the verification of (3.10)). From this representation of A and the σ-additivity follows $\varrho(A) = \sum_{n=1}^{\infty} \varrho(B_n) \geq 0$. Thus ϱ assumes only non-negative real values on $\Omega^+ \cap \mathscr{A}$, that is, the restriction of ϱ to $\Omega^+ \cap \mathscr{A}$ is a finite measure. Moreover, because $\varrho(P_n) \leq \varrho(\Omega^+) \leq \gamma$ and (18.1) this measure satisfies

$$\gamma = \varrho(\Omega^+).$$

In particular, $\gamma < +\infty$ since ϱ assumes only real values. $\varrho(A) > 0$ cannot hold for any $A \in \Omega^- \cap \mathscr{A}$, for otherwise $\varrho(\Omega^+ \cup A) = \varrho(\Omega^+) + \varrho(A) > \gamma$. Thus, $\varrho(A) \leq 0$ for all $A \in \Omega^- \cap \mathscr{A}$. $\square$

Measures (in the sense of Definition 3.3) have occasionally been interpreted as *mass distributions* on the underlying set Ω. A finite signed measure can be analogously interpreted as an (electric) *charge distribution* smeared over Ω. The foregoing theorem justifies this metaphor by showing that as with charge in electrostatics, there are two disjoint sets, one carrying all the positive charge, the other all the negative charge.

From this theorem another important feature of signed measures becomes evident: The difference ϱ in Lemma 17.9 is more than an illustrative example of a signed measure – it is the typical signed measure:

18.2 Corollary. *Every finite signed measure ϱ on a σ-algebra $\mathscr{A}$ in Ω is the difference of two finite measures on $\mathscr{A}$.*

Proof. Let $\Omega = \Omega^+ \cup \Omega^-$ be a Hahn decomposition in the sense of 18.1. Then evidently

$$\varrho^+(A) := \varrho(A \cap \Omega^+) \quad \text{and} \quad \varrho^-(A) := -\varrho(A \cap \Omega^-), \qquad A \in \mathscr{A}$$

define measures on $\mathscr{A}$, which satisfy $\varrho = \varrho^+ - \varrho^-$, since each $A \in \mathscr{A}$ is the disjoint union $(A \cap \Omega^+) \cup (A \cap \Omega^-)$. $\quad \square$

With this result the circle closes: finite signed measures are nothing more than the differences of finite measures. It is however possible to dispense with the finiteness hypothesis if σ-additivity is handled with sufficient care, but we will not go into this further.

In the final analysis it is because of the preceding corollary that we only consider measures with non-negative values in this book. Often to emphasize the distinction with signed measures, what we call simply measures are called *positive measures*.

Exercises.

1. Show that every finite signed measure on a σ-algebra is bounded and assumes a largest and a smallest value.

2. Let ϱ be a finite signed measure on σ-algebra $\mathscr{A}$ in Ω, and $\Omega = \Omega_1^+ \cup \Omega_1^-$, $\Omega = \Omega_2^+ \cup \Omega_2^-$ be two Hahn decompositions for it. Show that $\Omega_1^+ \triangle \Omega_2^+$ and $\Omega_1^- \triangle \Omega_2^-$ are *totally ϱ-nullsets*, meaning that $\varrho(N) = 0$ for every $N \in \mathscr{A}$ which is subset of either of them. Conclude that to within such totally ϱ-nullsets there is only one Hahn decomposition for ϱ.

3. Let ϱ be a finite signed measure on a σ-algebra $\mathscr{A}$ in Ω. Show that the specific representation $\varrho = \varrho^+ - \varrho^-$ of ϱ as the difference of the two measures on $\mathscr{A}$ which was produced in the proof of 18.2 is characterized by the following minimality property: In every representation $\varrho = \varrho_1 - \varrho_2$ as the difference of measures ϱ_1, ϱ_2 on $\mathscr{A}$, $\varrho_1 = \varrho^+ + \delta$ and $\varrho_2 = \varrho^- + \delta$ for an appropriate finite measure δ on $\mathscr{A}$, and indeed if $\Omega = \Omega^+ \cup \Omega^-$ is any Hahn decomposition of Ω corresponding to ϱ, $\delta = (1_{\Omega^+})\varrho_2 + (1_{\Omega^-})\varrho_1$. (Conversely, of course, every finite non-zero measure δ on $\mathscr{A}$ generates in this way a different representation of ϱ.) Infer that the only measure ν on $\mathscr{A}$ which satisfies $\nu(A) \leq \min\{\varrho^+(A), \varrho^-(A)\}$ for every $A \in \mathscr{A}$ is the identically 0 measure. [Remark: The representation $\varrho = \varrho^+ - \varrho^-$ uniquely determined by this minimality condition is called the *Jordan decomposition* of the finite signed measure ϱ. As with functions, ϱ^+ and ϱ^- are called the *positive part* and the *negative part* of ϱ.]

§19. Integration with respect to an image measure

Along with the measure space $(\Omega, \mathscr{A}, \mu)$ a measurable space $(\Omega', \mathscr{A}')$ and an $\mathscr{A}$-$\mathscr{A}'$-measurable mapping

$$T : (\Omega, \mathscr{A}) \to (\Omega', \mathscr{A}')$$

are given. Then the image measure

$$\mu' := T(\mu)$$

is defined in (7.5). The connection between μ-integrals and μ'-integrals is elucidated by:

19.1 Theorem. *For every $\mathscr{A}'$-measurable numerical function $f' \geq 0$ on Ω'*

$$(19.1) \qquad \int f' \, dT(\mu) = \int f' \circ T \, d\mu \, .$$

Proof. The non-negative function $f' \circ T$ is $\mathscr{A}$-measurable, by 7.3. The integral on the right-hand side of (19.1) is therefore defined. To prove the equality there we first consider only $\mathscr{A}'$-elementary f':

$$f' := \sum_{i=1}^{n} \alpha_i 1_{A'_i}$$

(with coefficients $\alpha_i \in \mathbb{R}^+$ and sets $A'_i \in \mathscr{A}'$). For such f'

$$f' \circ T = \sum_{i=1}^{n} \alpha_i 1_{A_i}$$

with $A_i := T^{-1}(A'_i)$, so this composite is an $\mathscr{A}$-elementary function. Since

$$T(\mu)(A'_i) = \mu(A_i) \qquad\qquad (i = 1, \ldots, n)$$

holds by definition of image measures, (19.1) follows in this case. For an arbitrary $\mathscr{A}'$-measurable $f' \geq 0$ there is an isotone sequence (u'_n) of $\mathscr{A}'$-elementary functions for which $u'_n \uparrow f'$. Then $(u'_n \circ T)$ is a sequence of $\mathscr{A}$-elementary functions for which $u'_n \circ T \uparrow f' \circ T$. From the validity of (19.1) for the u'_n and Definition 11.3 of the integral in general, we get (19.1) for f'. $\square$

19.2 Corollary 1. *Let f' be an $\mathscr{A}'$-measurable numerical function on Ω'. Then the $T(\mu)$-integrability of f' entails the μ-integrability of $f' \circ T$, and conversely. In case of integrability*

$$(19.2) \qquad \int f' \, dT(\mu) = \int f' \circ T \, d\mu \, .$$

Proof. From 19.1

$$\int (f')^+ \, dT(\mu) = \int (f')^+ \circ T \, d\mu \quad \text{and} \quad \int (f')^- \, dT(\mu) = \int (f')^- \circ T \, d\mu \,,$$

and of course

$$(f' \circ T)^+ = (f')^+ \circ T \quad \text{and} \quad (f' \circ T)^- = (f')^- \circ T \,.$$

Both claims therefore follow from the definition of the integral 12.1. □

19.3 Corollary 2. *The mapping $T : \Omega \to \Omega'$ is bijective and $\mathscr{A}$-$\mathscr{A}'$-measurable, with $\mathscr{A}'$-$\mathscr{A}$-measurable inverse T^{-1}. Further f' is a numerical function on Ω'. Then the $T(\mu)$-integrability of f' is equivalent to the μ-integrability of $f' \circ T$, and in its presence equality (19.2) prevails.*

One has only to note that the integrability of $f' \circ T$ entails the measurability of $f' \circ T$ and therewith that of $f' = f' \circ T \circ T^{-1}$. □

The content of 19.1–19.3 constitutes what is called the "*general transformation theorem for integrals*".

As the behavior of the L-B measure with respect to C_1-diffeomorphisms is known from (8.16′), the *transformation theorem for Lebesgue integrals* follows at once:

19.4 Theorem. *Let G, G' be open subsets of $\mathbb{R}^d$, $\varphi : G \to G'$ a C_1-diffeomorphism of G onto G'. A numerical function f' on G' is λ^d-integrable if and only if the function $f' \circ \varphi \, |\det D\varphi|$ is λ^d-integrable over G, and in this case*

$$(19.3) \qquad \int_{G'} f' \, d\lambda^d = \int_G f' \circ \varphi \, |\det D_\varphi| \, d\lambda^d \,.$$

Proof. The λ^d-integrability of f' over G' and that of $f' \circ \varphi \, |\det D\varphi|$ over G means the $\lambda^d_{G'}$-integrability and the λ^d_G-integrability of those functions, respectively. According to (8.16′)

$$\varphi^{-1}(\lambda^d_{G'}) = |\det D\varphi| \, \lambda^d_G \,;$$

furthermore, the Borel measurability of f' is equivalent to that of $f' \circ \varphi$. According to 17.3 therefore $f' \circ \varphi$ is integrable with respect to the measure $\mu := |\det D\varphi| \, \lambda^d_G$ if and only if $f' \circ \varphi \, |\det D\varphi|$ is integrable with respect to λ^d_G. Consequently the present claim follows from Corollary 19.3 applied to $T := \varphi^{-1}$, because $f' = f' \circ \varphi \circ \varphi^{-1}$ and

$$\int f' \, d\lambda^d_{G'} = \int f' \circ \varphi \circ \varphi^{-1} \, d\lambda^d_{G'} = \int f' \circ \varphi \, d\varphi^{-1}(\lambda^d_{G'})$$

$$= \int f' \circ \varphi \, d\mu = \int f' \circ \varphi \, |\det D\varphi| \, d\lambda^d_G \,. \quad □$$

Because of Theorem 19.1, equality (19.3) holds as well for all non-negative, Borel measurable, numerical functions on G'.

Exercises.

1. Let $(\Omega, \mathscr{A}, \mu)$ be a measure space, $T : \Omega \to \Omega$ a mapping which together with its inverse is an $\mathscr{A}$-$\mathscr{A}$-measurable bijection. Show that for every $f \in E^*(\Omega, \mathscr{A})$ the image measure $T(f\mu)$ has a density with respect to $T(\mu)$, namely $f \circ T^{-1}$.

2. Let $(\Omega, \mathscr{A}, \mu)$ be a σ-finite measure space, $T : \Omega \to \Omega$ an $\mathscr{A}$-$\mathscr{A}$-measurable mapping such that $T^{-1}(A)$ is a μ-nullset whenever A is. Prove the existence of a measurable function $q \geq 0$ such that

$$\int_A fq\,d\mu = \int_{T^{-1}(A)} f \circ T\,d\mu$$

for all $\mathscr{A}$-measurable numerical functions $f \geq 0$ on Ω, and all $A \in \mathscr{A}$.

§20. Stochastic convergence

Let us return to the study of p-fold integrable functions begun in §14. Our goal will be to replace the almost-everywhere convergence concept that underlies the theorems proved there with a weaker convergence concept. It is suggested by a simple but very useful inequality.

The setting is once again an arbitrary measure space $(\Omega, \mathscr{A}, \mu)$.

20.1 Lemma. *For every measurable numerical function f on Ω and every pair of real numbers $p > 0$ and $\alpha > 0$ the* Chebyshev–Markov inequality

$$(20.1) \qquad \mu(\{|f| \geq \alpha\}) \leq \frac{1}{\alpha^p} \int |f|^p\,d\mu$$

holds.

For $p = 2$ this is also known simply as *Chebyshev's inequality*.

Proof. The set $A_\alpha := \{|f| \geq \alpha\}$ lies in $\mathscr{A}$ and

$$\int |f|^p\,d\mu \geq \int_{A_\alpha} |f|^p\,d\mu \geq \int_{A_\alpha} \alpha^p\,d\mu = \alpha^p\mu(A_\alpha),$$

which is what (20.1) claims. $\square$

Therefore if $\int |f|^p\,d\mu$ is finite, which when $p \geq 1$ means just that f is p-fold integrable, it follows from (20.1) that

$$(20.2) \qquad \lim_{\alpha \to +\infty} \mu(\{|f| \geq \alpha\}) = 0.$$

One can also study the dependence on $n \in \mathbb{N}$ of the measures of the sets $\{|f_n - f| \geq \alpha\}$ when $f, f_1, f_2, \ldots$ are measurable real functions. That leads to the aforementioned new convergence concept.

20.2 Definition. A sequence $(f_n)_{n\in\mathbb{N}}$ of measurable real functions on Ω is said to be $(\mu\text{-})$*stochastically convergent* (or to be *convergent in μ-measure*) to a measurable real function f on Ω, if for each real number $\alpha > 0$ and each $A \in \mathscr{A}$ of *finite* measure

$$(20.3) \qquad\qquad \lim_{n\to\infty} \mu(\{|f_n - f| \geq \alpha\} \cap A) = 0\,.$$

In this case we also write

$$(20.4) \qquad\qquad \mu\text{-}\lim f_n = f$$

and call f a $(\mu\text{-})$*stochastic limit* of the sequence (f_n).

Remarks. 1. For a finite measure μ we may take $A = \Omega$ in (20.3) and in this case stochastic convergence of (f_n) to f is equivalent to the requirement

$$(20.5) \qquad\qquad \lim_{n\to\infty} \mu(\{|f_n - f| \geq \alpha\}) = 0 \qquad\qquad \text{for every } \alpha > 0.$$

The more complicated condition (20.3) is dictated by the desire to treat infinite, and especially σ-finite, measures as well as finite ones.

2. For σ-finite measures μ the stochastic convergence of a sequence (f_n) to f is generally not equivalent to (20.5), as the next example illustrates.

Example. 1. Let $\Omega := \mathbb{N}$, $\mathscr{A} := \mathscr{P}(\mathbb{N})$, μ the measure (obviously σ-finite) defined on $\mathscr{A}$ by the equations

$$\mu(\{n\}) = \frac{1}{n} \qquad\qquad \text{for every } n \in \mathbb{N}$$

and the requirement of σ-additivity. With $A_n := \{n, n+1, \ldots\}$ and $f_n := 1_{A_n}$ for each $n \in \mathbb{N}$, the sequence (f_n) converges stochastically to 0: For every $\alpha \in \,]0,1[$, $\{f_n \geq \alpha\} = A_n$, and since $A_n \downarrow \emptyset$, it follows from 3.2 that $\lim \mu(A_n \cap A) = 0$ for every $A \in \mathscr{A}$ having finite measure. On the other hand, $\mu(A_n) = +\infty$ for every $n \in \mathbb{N}$.

Remark. 3. Let f be a stochastic limit of a sequence (f_n) and consider any measurable real function f^* on Ω. If $f^* = f$ μ-almost everywhere in every $A \in \mathscr{A}$ which has finite measure, then f^* is also a stochastic limit of the sequence (f_n). This is because the sets

$$\{|f_n - f^*| \geq \alpha\} \cap A \quad \text{and} \quad \{|f_n - f| \geq \alpha\} \cap A$$

differ from each other only in an (n-independent) nullset.

The converse of this is important:

20.3 Theorem. *For every σ-finite measure μ, any two stochastic limits of a sequence of measurable real functions are μ-almost everywhere equal to each other.*

Proof. If f and f^* are stochastic limits of the sequence (f_n), then from the triangle inequality in $\mathbb{R}$

$$\{|f - f^*| \geq \alpha\} \subset \{|f_n - f| \geq \alpha/2\} \cup \{|f_n - f^*| \geq \alpha/2\},$$

whence

$$\mu(\{|f - f^*| \geq \alpha\} \cap A) \leq \mu(\{|f_n - f| \geq \alpha/2\} \cap A) + \mu(\{|f_n - f^*| \geq \alpha/2\} \cap A)$$

for every $n \in \mathbb{N}$ and every $A \in \mathscr{A}$. Letting $n \to \infty$ shows that

$$\mu(\{|f - f^*| \geq \alpha\} \cap A) = 0$$

for every $\alpha > 0$ and every $A \in \mathscr{A}$ of finite measure. Then however, $f = f^*$ μ-almost everywhere in every such set A, since

$$\{f \neq f^*\} \cap A = \bigcup_{k \in \mathbb{N}} \{|f - f^*| \geq 1/k\} \cap A$$

is a μ-nullset. Upon taking for A the sets in a sequence (A_n) in $\mathscr{A}$ which satisfies $\mu(A_n) < +\infty$ for all n and $A_n \uparrow \Omega$, the μ-almost everywhere equality of f and f^* follows. $\square$

To supplement this fact we mention:

Remark. 4. Stochastic limits f and f^* of the same sequence (f_n) are almost everywhere equal without any hypotheses on the measure itself if both functions are p-fold integrable for some $p \in [1, +\infty[$. This is because for every real $\alpha > 0$ the set $\{|f - f^*| \geq \alpha\}$ has finite measure, by (20.1), and so $f = f^*$ μ-almost everywhere in this set, whence $\{|f - f^*| > 0\} = \bigcup_{n \in \mathbb{N}} \{|f - f^*| \geq 1/n\}$ is a countable union of μ-nullsets. This just says that $f = f^*$ μ-almost everywhere in Ω. But the next example shows that it may fail if one of the functions is not in any $\mathscr{L}^p$-space.

Example. 2. Consider the measure space $(\Omega, \mathscr{P}(\Omega), \mu)$, where Ω consists of exactly two elements ω_0, ω_1 and $\mu(\{\omega_0\}) = 0$, $\mu(\{\omega_1\}) = +\infty$, $f_n = f = 0$ for every $n \in \mathbb{N}$. These functions lie in every $\mathscr{L}^p(\mu)$ and the sequence (f_n) converges stochastically to f, as well as to *every* real-valued function f^* on Ω. Every such f^* which is non-zero at ω_1, however, lies in no $\mathscr{L}^p(\mu)$ with $1 \leq p < +\infty$ and fails to coincide μ-almost everywhere in Ω with f.

The considerations with which we began this section lead to an important class of stochastically convergent sequences:

20.4 Theorem. *If the sequence (f_n) in $\mathscr{L}^p(\mu)$ converges in p^{th} mean to a function $f \in \mathscr{L}^p(\mu)$ for some $1 \leq p < +\infty$, then it also converges to f μ-stochastically.*

Proof. The Chebyshev–Markov inequality tells us that

$$\mu(\{|f_n - f| \geq \alpha\} \cap A) \leq \mu(\{|f_n - f| \geq \alpha\}) \leq \alpha^{-p} \int |f_n - f|^p \, d\mu$$

holds for every $n \in \mathbb{N}$, every $\alpha > 0$ and every $A \in \mathscr{A}$. The claimed stochastic convergence, that is, the convergence to 0 of the left end of this chain as $n \to \infty$, follows because $\int |f_n - f|^p \, d\mu \to 0$ as $n \to \infty$ is the definition of convergence in p^{th} mean. $\square$

The proof shows that convergence in p^{th} mean actually entails the stronger form of stochastic convergence in (20.5). The situation is different when the given sequence is almost everywhere convergent. (On this point cf. also Remark 5.)

20.5 Theorem. *If a sequence $(f_n)_{n \in \mathbb{N}}$ of measurable real functions on Ω converges μ-almost everywhere in Ω – or even just μ-almost everywhere in each set $A \in \mathscr{A}$ of finite measure – to a measurable real function f on Ω, then this sequence also converges μ-stochastically to f.*

Proof. For every $\alpha > 0$,

$$\{|f_n - f| \geq \alpha\} \subset \left\{ \sup_{m \geq n} |f_m - f| \geq \alpha \right\}$$

and so

$$\mu(\{|f_n - f| \geq \alpha\} \cap A) \leq \mu\left(\left\{ \sup_{m \geq n} |f_m - f| \geq \alpha \right\} \cap A \right)$$

for every $A \in \mathscr{A}$. The present claim therefore follows from our next lemma, applied to the restriction of μ to $A \cap \mathscr{A}$ for each A of finite measure. $\square$

20.6 Lemma. *If the measure μ is finite, then each of the following three conditions on a sequence $(f_n)_{n \in \mathbb{N}}$ of measurable real functions is equivalent to (f_n) converging μ-almost everywhere to 0:*

$$(20.6) \qquad \lim_{n \to \infty} \mu\left(\left\{ \sup_{m \geq n} |f_m| \geq \alpha \right\} \right) = 0 \qquad \text{for every } \alpha > 0,$$

$$(20.6') \qquad \lim_{n \to \infty} \mu\left(\left\{ \sup_{m \geq n} |f_m| > \alpha \right\} \right) = 0 \qquad \text{for every } \alpha > 0,$$

$$(20.7) \qquad \mu\left(\limsup_{n \to \infty} \{|f_n| > \alpha\} \right) = 0 \qquad \text{for every } \alpha > 0.$$

Proof. To prove the equivalence of (20.6) with the almost everywhere convergence of (f_n) to 0, we set, for each $\alpha > 0$ and each $n \in \mathbb{N}$

$$A_n^\alpha := \left\{ \sup_{m \geq n} |f_m| \geq \alpha \right\}.$$

Obviously both $n \mapsto A_n^\alpha$ and $\alpha \mapsto A_n^\alpha$ are antitone mappings; then $k \mapsto A_n^{1/k}$ is isotone on $\mathbb{N}$. If we also set

$$A := \{\omega \in \Omega : \lim_{n \to \infty} f_n(\omega) = 0\} = \{\omega \in \Omega : \limsup_{n \to \infty} |f_n|(\omega) = 0\},$$

then these lie in $\mathscr{A}$, either by appeal to 9.5 or by noticing that each $A_n^\alpha \in \mathscr{A}$ and

$$A = \bigcap_{k \in \mathbb{N}} \bigcup_{n \in \mathbb{N}} \complement A_n^{1/k} \,.$$

Passing to complements,

$$\complement A = \bigcup_{k \in \mathbb{N}} \bigcap_{n \in \mathbb{N}} A_n^{1/k}$$

and so

$$\bigcap_{n \in \mathbb{N}} A_n^{1/k} \uparrow \complement A \quad \text{as } k \to \infty, \quad \text{and} \quad A_n^{1/k} \downarrow \bigcap_{m \in \mathbb{N}} A_m^{1/k} \quad \text{as } n \to \infty.$$

Consequently,

$$(20.8) \qquad \mu(\complement A) = \sup_{k \in \mathbb{N}} \mu\left(\bigcap_{n \in \mathbb{N}} A_n^{1/k} \right) = \sup_{k \in \mathbb{N}} \inf_{n \in \mathbb{N}} \mu(A_n^{1/k}),$$

because the finite measure μ is both continuous from above and continuous from below, by 3.2. Thus (f_n) converges almost everywhere to 0 just when the number defined by (20.8) is 0. In turn, the latter occurs exactly in case

$$\inf_{n \in \mathbb{N}} \mu(A_n^{1/k}) = \lim_{n \to \infty} \mu(A_n^{1/k}) = 0$$

for every $k \in \mathbb{N}$. The first equivalence follows from this. The equivalence of (20.6) with (20.6$'$) follows from the observation that for any numerical function g on Ω

$$\{g > \alpha\} \subset \{g \geq \alpha\} \subset \{g > \alpha'\}$$

whenever $0 < \alpha' < \alpha$.

Finally, the equivalence of (20.6$'$) with (20.7) follows from the validity, for every $\alpha > 0$, of the equality

$$(20.9) \qquad \lim_{n \to \infty} \mu\left(\left\{ \sup_{m \geq n} |f_m| > \alpha \right\} \right) = \mu\left(\limsup_{n \to \infty} \{|f_n| > \alpha\} \right).$$

For the proof of which we introduce

$$B_n := \bigcup_{m \geq n} \{|f_m| > \alpha\} \quad \text{and} \quad B := \limsup_{n \to \infty} \{|f_n| > \alpha\}.$$

On the one hand, $B_n \downarrow B$ and consequently $\lim \mu(B_n) = \mu(B)$. On the other hand, however,

$$B_n = \bigcup_{m \geq n} \{|f_m| > \alpha\} = \left\{ \sup_{m \geq n} |f_m| > \alpha \right\}.$$

From this finally we get the needed (20.9). □

The conditions involved in Theorems 20.4 and 20.5 are indeed sufficient to insure stochastic convergence, but they are not necessary for it, as the following examples show.

Examples. 3. Let $\Omega := [0, 1[$, $\mathscr{A} := \Omega \cap \mathscr{B}^1$ and $\mu := \lambda_\Omega^1$, a finite measure. With $A_n :=]0, 1/n[\in \mathscr{A}$, the sequence $(n 1_{A_n})_{n \in \mathbb{N}}$ converges to 0 at every point of Ω and so, either by appeal to 20.4 or by virtue of

$$\mu(\{n 1_{A_n} \geq \alpha\}) = \mu(A_n) = \frac{1}{n} \qquad \text{whenever } 0 < \alpha \leq n \in \mathbb{N},$$

this sequence also converges stochastically to 0. By contrast

$$\int (n 1_{A_n})^p \, d\mu = n^p \mu(A_n) = n^{p-1}$$

shows that the sequence does not converge to 0 in p^{th} mean for any $p \geq 1$.

4. Let $(\Omega, \mathscr{A}, \mu)$ be the measure space of the preceding example. Write each $n \in \mathbb{N}$ as $n = 2^h + k$ with non-negative integers h and k satisfying $0 \leq k < 2^n$ (which uniquely determines them) and set

$$A_n := [k 2^{-h}, (k+1) 2^{-h}[, \quad f_n := 1_{A_n}, \qquad\qquad n \in \mathbb{N}.$$

It was shown in the example in §15 that the sequence $(f_n(\omega))_{n \in \mathbb{N}}$ converges for no $\omega \in \Omega$. Nevertheless the sequence (f_n) does converge stochastically to 0, since for every $\alpha > 0$ and $n \in \mathbb{N}$

$$\mu(\{|f_n| \geq \alpha\}) \leq 2^{-h} < \frac{2}{n}.$$

In this example stochastic convergence can also be inferred from 20.4, since the example in §15 showed that (f_n) converges to 0 in p^{th} mean for every $p \in [1, +\infty[$.

The connection between stochastic convergence and almost-everywhere convergence is nevertheless closer than one would be led to suspect on the basis of the last example.

20.7 Theorem. *If a sequence $(f_n)_{n \in \mathbb{N}}$ of measurable real functions converges μ-stochastically to a measurable real function f, then for every $A \in \mathscr{A}$ of finite μ-measure some subsequence of (f_n) converges to f μ-almost everywhere in A.*

Proof. For $A \in \mathscr{A}$ with $\mu(A) < +\infty$, the measure μ_A, which is the restriction of μ to $A \cap \mathscr{A}$, is finite. It therefore suffices to deal with the case of a finite measure μ; moreover, in that case we can simply take A to be Ω itself.

For $\alpha > 0$ and $m, n \in \mathbb{N}$ the triangle inequality shows that

$$\{|f_m - f_n| \geq \alpha\} \subset \{|f_m - f| \geq \alpha/2\} \cup \{|f_n - f| \geq \alpha/2\};$$

thus by hypothesis $\mu(\{|f_m - f_n| \geq \alpha\})$ can be made arbitrarily small by taking m and n sufficiently large. If therefore $(\eta_k)_{k \in \mathbb{N}}$ is a sequence of positive real numbers with

$$\sum_{k=1}^{\infty} \eta_k < +\infty,$$

then for each $k \in \mathbb{N}$ there is an $n_k \in \mathbb{N}$ such that

$$\mu(\{|f_m - f_{n_k}| \geq \eta_k\}) \leq \eta_k \qquad \text{for all } m \geq n_k.$$

Clearly the sequence $(n_k)_{k \in \mathbb{N}}$ can be chosen strictly isotone: $n_k < n_{k+1}$ for every $k \in \mathbb{N}$. If now we set

$$A_k := \{|f_{n_{k+1}} - f_{n_k}| \geq \eta_k\}, \qquad k \in \mathbb{N},$$

then

$$\sum_{k=1}^{\infty} \mu(A_k) \leq \sum_{k=1}^{\infty} \eta_k < +\infty,$$

and consequently,

$$\lim_{n \to \infty} \sum_{k=n}^{\infty} \mu(A_k) = 0.$$

From this it follows that the set $A := \limsup\limits_{n \to \infty} A_n$ satisfies

$$\mu(A) = 0,$$

because $A \subset \bigcup\limits_{k \geq n} A_k$ for every $n \in \mathbb{N}$, entailing that $\mu(A) \leq \sum\limits_{k=n}^{\infty} \mu(A_k)$ for every n. The definition of A shows that if $\omega \in \complement A$, then the inequality

$$\left|f_{n_{k+1}}(\omega) - f_{n_k}(\omega)\right| \geq \eta_k$$

prevails for at most finitely many $k \in \mathbb{N}$. Therefore, along with the series $\sum \eta_k$, the series

$$\sum_{k=1}^{\infty} [f_{n_{k+1}}(\omega) - f_{n_k}(\omega)]$$

converges (absolutely); that is, the sequence $(f_{n_k}(\omega))_{k \in \mathbb{N}}$ converges in $\mathbb{R}$. In summary, the sequence (f_{n_k}) converges almost everywhere to a measurable real function f^* on Ω. By 20.5 f^* is also a stochastic limit of $(f_{n_k})_{k \in \mathbb{N}}$. But, as a subsequence of $(f_n)_{n \in \mathbb{N}}$, that sequence converges stochastically to f as well. Hence by 20.3, $f = f^*$ almost everywhere. We have shown therefore that $(f_{n_k})_{k \in \mathbb{N}}$ converges almost everywhere to f. $\square$

In terms of almost-everywhere convergence we can now even characterize stochastic convergence by a *subsequence principle*.

20.8 Corollary. *A sequence (f_n) of measurable real functions on Ω converges μ-stochastically to a measurable real function f on Ω if and only if for each $A \in \mathscr{A}$ of finite measure, each subsequence $(f_{n_k})_{k \in \mathbb{N}}$ of (f_n) contains a further subsequence which converges to f μ-almost everywhere in A.*

Proof. The preceding theorem establishes that the subsequence condition is necessary for the stochastic convergence of (f_n) to f, since every subsequence of (f_n)

likewise converges stochastically to f. Let us now assume that the subsequence condition is fulfilled, and fix an $A \in \mathscr{A}$ of finite measure. Since every subsequence (f_{n_k}) contains another which converges almost everywhere in A to f and by 20.5 this latter subsequence must also converge (in A) stochastically to f, we see that in the sequence of numbers

$$\mu(\{|f_{n_k} - f| \geq \alpha\} \cap A) \qquad\qquad (k \in \mathbb{N}),$$

in which $\alpha > 0$ is fixed, a subsequence exists which converges to 0. But, as an easy argument confirms, a sequence of real numbers whose subsequences have this property must itself converge to 0. That is, the sequence of real numbers

$$\mu(\{|f_n - f| \geq \alpha\} \cap A) \qquad\qquad (n \in \mathbb{N})$$

converges to 0. As this is true of every $A \in \mathscr{A}$ having finite measure and every $\alpha > 0$, the stochastic convergence of (f_n) to f is thereby confirmed. □

Remarks. 5. It is not to be expected that in 20.7 and 20.8 the reference to the finite-measure set $A \in \mathscr{A}$ can be stricken. This is already illustrated by Example 2 if one replaces the sequence (f_n) there with the sequence (f_n') defined by $f_n' := n1_{\{\omega_1\}}$, $n \in \mathbb{N}$. This new sequence also converges stochastically to $f := 0$. See however Exercise 5.

6. The second part of the proof of 20.7 shows that for *finite measures* μ there is a *Cauchy criterion* for the stochastic convergence of a sequence (f_n): Necessary and sufficient for the stochastic convergence of a sequence (f_n) to a measurable real function on Ω is the condition

$$\lim_{m,n\to\infty} \mu(\{|f_m - f_n| \geq \alpha\}) = 0 \qquad\qquad \text{for every } \alpha > 0.$$

7.   The sequence formed by alternately taking terms from each of two stochastically convergent sequences whose limit functions do not coincide almost everywhere shows that in Corollary 20.8 it does not suffice to demand that in each A some subsequence of the full sequence (f_n) converge almost everywhere.

A particularly useful consequence of 20.8 is:

20.9 Theorem. *If the sequence $(f_n)_{n \in \mathbb{N}}$ of measurable real functions on Ω converges stochastically to a measurable real function f on Ω, and $\varphi : \mathbb{R} \to \mathbb{R}$ is continuous, then the sequence $(\varphi \circ f_n)_{n \in \mathbb{N}}$ converges stochastically to $\varphi \circ f$.*

Proof. One exploits both directions of 20.8, noting that from the almost everywhere convergence of a subsequence (f_{n_k}) to f on an $A \in \mathscr{A}$ follows the almost everywhere convergence of $(\varphi \circ f_{n_k})$ to $\varphi \circ f$ on A. □

The general question of functions $\varphi : \mathbb{R} \to \mathbb{R}$ which preserve convergence, in the sense that $(\varphi \circ f_n)_{n \in \mathbb{N}}$ inherits the kind of convergence $(f_n)_{n \in \mathbb{N}}$ has, is investigated by BARTLE and JOICHI [1961]. They show how Theorem 20.9 can fail if the more restrictive definition (20.5) is adopted for stochastic convergence.

Exercises.

1. (f_n) and (g_n) are stochastically convergent sequences of measurable real functions, having limit functions f and g, respectively. Show that for all $\alpha, \beta \in \mathbb{R}$ the sequence $(\alpha f_n + \beta g_n)$ converges stochastically to $\alpha f + \beta g$, and the sequences $(f_n \wedge g_n)$, $(f_n \vee g_n)$ converge stochastically to $f \wedge g$, $f \vee g$, respectively.

2. For a measure space $(\Omega, \mathscr{A}, \mu)$ with finite measure μ let d_μ be the pseudometric on $\mathscr{A}$ constructed in Exercise 7 of §3. Show that a sequence (A_n) in $\mathscr{A}$ is d_μ-convergent to $A \in \mathscr{A}$ if and only if the sequence (1_{A_n}) of indicator functions converges stochastically to the indicator function 1_A.

3. For every pair of measurable real functions f and g on a measure space $(\Omega, \mathscr{A}, \mu)$ with finite measure μ define

$$D_\mu(f, g) := \inf\{\varepsilon > 0 : \mu(\{|f - g| \geq \varepsilon\}) \leq \varepsilon\}$$

and then prove that
(a) D_μ is a pseudometric on the set $M(\mathscr{A})$ of all measurable real functions.
(b) A sequence (f_n) in $M(\mathscr{A})$ converges stochastically to $f \in M(\mathscr{A})$ if and only if $\lim_{n \to \infty} D_\mu(f_n, f) = 0$.
(c) $M(\mathscr{A})$ is D_μ-complete, that is, every D_μ-Cauchy sequence in $M(\mathscr{A})$ converges with respect to D_μ to some function in $M(\mathscr{A})$.
What is the relation of D_μ to the d_μ of Exercise 2?

4. In the context of Exercise 3 define

$$D'_\mu(f, g) := \int \frac{|f - g|}{1 + |f - g|} \, d\mu,$$

for every pair of functions $f, g \in M(\mathscr{A})$. Show that D'_μ also enjoys the properties (a)–(c) proved for D_μ in the preceding exercise.

5. Let $(\Omega, \mathscr{A}, \mu)$ be a σ-finite measure space. Show that a sequence (f_n) of measurable real functions on Ω converges stochastically to a measurable real function f on Ω if and only if from every subsequence (f_{n_k}) of (f_n) a further subsequence can be extracted which converges almost everywhere in Ω to f. [*Hints:* Suppose (f_n) is stochastically convergent. Choose a sequence (A_k) from $\mathscr{A}$ with $\mu(A_k) < +\infty$ for each k and $A_k \uparrow \Omega$, and consider the finite measures $\mu_k(A) := \mu(A \cap A_k)$ on $\mathscr{A}$. The claim is true of each measure μ_k. Given a subsequence Φ of (f_n), there is for each $k \in \mathbb{N}$ a subsequence of $(g_n^{(k)})_{n \in \mathbb{N}}$ of Φ which converges μ_k-almost everywhere to f. It can be arranged that $(g_n^{(k+1)})$ is a subsequence of $(g_n^{(k)})$ for each k. Then the diagonal subsequence $(g_n^{(n)})_{n \in \mathbb{N}}$ does what is wanted.]

6. Give an "elementary" proof of 20.9 based directly on the relevant definition 20.2. To this end, show that for each $\varepsilon \in \,]0, 1[$ there exists $\delta > 0$ such that $\{|f| \leq 1/\varepsilon\} \cap \{|f_n - f| \leq \delta\} \subset \{|\varphi \circ f_n - \varphi \circ f| \leq \varepsilon\}$ for all $n \in \mathbb{N}$.

7. (Theorem of D.F. EGOROV (1869–1931)) Let $(\Omega, \mathscr{A}, \mu)$ be a measure space with *finite* measure μ. Show that: For every sequence $(f_n)_{n \in \mathbb{N}}$ of measurable real functions on Ω its convergence almost everywhere to a measurable real function f is equivalent to its so-called *almost-uniform* convergence to f. The latter means

that for every $\delta > 0$ there exists an $A_\delta \in \mathscr{A}$ such that $\mu(A_\delta) < \delta$ and (f_n) converges to f uniformly on $\complement A_\delta$. [*Hint*: Exercise 2 of §11.]

§21. Equi-integrability

The sufficient condition for convergence in p^{th} mean which is set out in Lebesgue's dominated convergence theorem can be transformed into a necessary as well as sufficient condition with the help of stochastic convergence. But we need the concept of equi-integrability, which is of fundamental significance.

In the following $(\Omega, \mathscr{A}, \mu)$ will again be an arbitrary measure space, and p is always a real number satisfying $1 \le p < +\infty$.

The point of departure is a simple observation. A measurable numerical function f on Ω is integrable if and only if for every $\varepsilon > 0$ there is a non-negative integrable function $g = g_\varepsilon$ such that

$$(21.1) \qquad \int_{\{|f| \ge g\}} |f| \, d\mu \le \varepsilon .$$

For if f is integrable and we take, as we then may, g to be $2|f|$, then $\{|f| \ge g\} = \{f = 0\} \cup \{|f| = +\infty\}$ and thanks to 13.6 the integral in (21.1) is actually equal to 0. Conversely, if we have (21.1) even for just one real $\varepsilon > 0$, then

$$\int |f| \, d\mu = \int_{\{|f| \ge g\}} |f| \, d\mu + \int_{\{|f| < g\}} |f| \, d\mu \le \varepsilon + \int g \, d\mu < +\infty$$

and hence f is integrable.

This observation induces us to make

21.1 Definition. A set M of $\mathscr{A}$-measurable numerical functions on Ω is called $(\mu\text{-})equi\text{-}integrable$ if for every $\varepsilon > 0$ there exists a μ-integrable function $g = g_\varepsilon \ge 0$ on Ω such that every $f \in M$ satisfies

$$(21.2) \qquad \int_{\{|f| \ge g\}} |f| \, d\mu \le \varepsilon .$$

Correspondingly a *family* $(f_i)_{i \in I}$ of measurable numerical functions on Ω is called *equi-integrable* if the *set* $\{f_i : i \in I\}$ is equi-integrable. Equi-integrable sets and families are sometimes also called "uniformly integrable".

From now on, any function g_ε as described in Definition 21.1 will be called an $\varepsilon\text{-}bound$ for the given set of functions. Obviously, along with an ε-bound g for a set of functions, any integrable $g' \ge g$ is also an ε-bound.

Examples. 1. If $M_1, \ldots, M_n$ are finitely many μ-equi-integrable sets of measurable functions on Ω, then their union is also μ-equi-integrable, because whenever g_j is an ε-bound for M_j $(j = 1, \ldots, n)$, then $g_1 \vee \ldots \vee g_n$ is an ε-bound for $M_1 \cup \ldots \cup M_n$.

2.   Every finite set of μ-integrable functions is μ-equi-integrable. This follows from Example 1 and the fact, demonstrated in the course of proving (21.1), that any set consisting of just one integrable function f is equi-integrable, the function $2\,|f|$ being an ε-bound for every $\varepsilon > 0$.

3.   Suppose M is a set of measurable numerical functions on Ω, $1 \leq p < +\infty$, and there is a p-fold μ-integrable majorant g for M, that is, every $f \in M$ satisfies

$$|f| \leq g \qquad\qquad \mu\text{-almost everywhere.}$$

Then the set
$$M^p := \{|f|^p : f \in M\}$$

is equi-integrable. Indeed, as in Example 2, the single integrable function $h := 2g^p$ is an ε-bound for every $\varepsilon > 0$, since by 13.6

$$\int_{\{|f|^p \geq h\}} |f|^p \, d\mu \leq \int_{\{g^p \geq h\}} g^p \, d\mu = \int_{\{g=+\infty\}} g^p \, d\mu = 0 \qquad \text{for every } f \in M.$$

This example shows that Theorem 15.6 on dominated convergence is really about an equi-integrable set of functions. Of course, one cannot expect that conversely from the equi-integrability of a subset of $\mathscr{L}^1(\mu)$ there should follow the existence of a single integrable majorant for the set. The following example confirms this.

4.   Consider the probability space $(\mathbb{N}, \mathscr{P}(\mathbb{N}), \mu)$, the finite measure μ being specified by $\mu(\{n\}) := 2^{-n}$ for each $n \in \mathbb{N}$. The sequence of functions $f_n := 2^n n^{-1} 1_{\{n\}}$ $(n \in \mathbb{N})$ is equi-integrable: For the constant function $1 \in \mathscr{L}^1(\mu)$ the inequality

$$\int_{\{f_n \geq 1\}} f_n \, d\mu \leq \frac{1}{n} \qquad\qquad \text{holds for all } n \in \mathbb{N}.$$

However, the smallest function g which majorizes every f_n is the non-μ-integrable function $n \mapsto 2^n n^{-1}$ on $\mathbb{N}$.

5. Let $(\Omega, \mathscr{A}, \mu)$ be the measure space of Example 3, §20, and $(f_n)_{n\in\mathbb{N}}$ the sequence of functions considered there: $A_n := [0, \frac{1}{n}[$ and $f_n := n 1_{A_n}$ for each $n \in \mathbb{N}$. This sequence is *not* equi-integrable, which we see as follows: for every integrable $g \geq 0$ and every $n \in \mathbb{N}$

$$\int_{\{|f_n| \geq g\}} |f_n| \, d\mu = \int_{A_n \cap \{n \geq g\}} n \, d\mu = \int_{A_n} n \, d\mu - \int_{A_n \cap \{n < g\}} n \, d\mu \geq 1 - \int_{A_n} g \, d\mu.$$

From the finiteness of the measure $g\mu$ and the fact that $A_n \downarrow \{0\}$, it follows that

$$\liminf_{n \to \infty} \int_{\{|f_n| \geq g\}} |f_n| \, d\mu \geq 1,$$

showing that g cannot be an ε-bound for any $\varepsilon \in \,]0, 1[$.

Here is a useful characterization of equi-integrability, which, for σ-finite measures, will be improved upon in 21.8.

21.2 Theorem. *A set M of measurable numerical functions on Ω is equi-integrable if and only if the following two conditions are satisfied:*

$$(21.3) \qquad\qquad \sup_{f \in M} \int |f| \, d\mu < \infty .$$

(21.4) *For every $\varepsilon > 0$ there exists a μ-integrable function $h \geq 0$ and a number $\delta > 0$ such that*

$$\int_A h \, d\mu \leq \delta \Rightarrow \int_A |f| \, d\mu \leq \varepsilon \qquad \text{for all } f \in M \text{ and all } A \in \mathscr{A}.$$

Proof. For every $A \in \mathscr{A}$, every measurable numerical function f on Ω, and every integrable function $g \geq 0$

$$\int_A |f| \, d\mu = \int_{A \cap \{|f| \geq g\}} |f| \, d\mu + \int_{A \cap \{|f| < g\}} |f| \, d\mu \leq \int_{\{|f| \geq g\}} |f| \, d\mu + \int_A g \, d\mu$$

and in particular for $A := \Omega$

$$\int |f| \, d\mu \leq \int_{\{|f| \geq g\}} |f| \, d\mu + \int g \, d\mu .$$

Assuming that the set M is equi-integrable, let us choose for g an $\frac{\varepsilon}{2}$-bound for it and then set $h := g$, $\delta := \frac{\varepsilon}{2}$. Then conditions (21.3) and (21.4) follow from the preceding inequalities.

Conversely, assume the two conditions are fulfilled and let $\varepsilon > 0$ be given. Let h and $\delta > 0$ be as furnished by (21.4). For each $f \in M$ and real $\alpha > 0$, consider the obviously valid inequality

$$\int |f| \, d\mu \geq \int_{\{|f| \geq \alpha h\}} |f| \, d\mu \geq \int_{\{|f| \geq \alpha h\}} \alpha h \, d\mu$$

or its equivalent

$$\int_{\{|f| \geq \alpha h\}} h \, d\mu \leq \frac{1}{\alpha} \int |f| \, d\mu .$$

The integrals $\int |f| \, d\mu$ here are bounded as f ranges over M, by (21.3). Therefore $\alpha > 0$ can be chosen so large that

$$\int_{\{|f| \geq \alpha h\}} h \, d\mu \leq \delta \qquad\qquad \text{for all } f \in M.$$

(21.4) then insures that $g := \alpha h$ is an ε-bound for M, which proves that this set is equi-integrable. $\quad\square$

21.3 Corollary. *Let $M \subset \mathscr{L}^p$ and the set $M^p := \{|f|^p : f \in M\}$ be equi-integrable, where $1 \leq p < +\infty$. Then the set*

$$M_*^p := \{|\alpha f + \beta g|^p : f, g \in M, \alpha, \beta \in \mathbb{R}, |\alpha| \leq 1, |\beta| \leq 1\}$$

is equi-integrable.

Proof. For every $f \in \mathscr{L}^p(\mu)$ and every $A \in \mathscr{A}$, $|1_A f| \leq |f|$ shows that $1_A f \in \mathscr{L}^p(\mu)$ too, and so for all $f_1, f_2 \in \mathscr{L}^p(\mu)$ Minkowski's inequality (14.4) gives

$$N_p(1_A f_1 + 1_A f_2) \leq N_p(1_A f_1) + N_p(1_A f_2),$$

whence

$$\int_A |f_1 + f_2|^p \, d\mu \leq \left[\left(\int_A |f_1|^p \, d\mu\right)^{1/p} + \left(\int_A |f_2|^p \, d\mu\right)^{1/p}\right]^p.$$

Applying this inequality to $f_1 = \alpha f_1$, $f_2 = \beta g$ with $f, g \in M$ $\alpha, \beta \in \mathbb{R}$ and $|\alpha| \leq 1$, $|\beta| \leq 1$, and bearing in mind that 21.2 is (by hypothesis) valid for the set M^p, one realizes that conditions (21.3) and (21.4) are fulfilled by M_*^p as well as by M^p, with the same function h in both cases. $\square$

We are now in a position to deliver the sharpened version of the dominated convergence theorem mentioned in the introduction to this section. That we really have to do with a sharpening here is attested to on the one hand by Example 3 and Theorem 20.4, according to which stochastic convergence follows from almost-everywhere convergence, and on the other by Example 4 of §20, which shows that there are situations in which the dominated convergence theorem is not applicable but the following theorem is.

21.4 Theorem. *For every sequence $(f_n)_{n \in \mathbb{N}}$ of p-fold, μ-integrable real functions on a measure space $(\Omega, \mathscr{A}, \mu)$ the following two assertions are equivalent:*
(i)   *The sequence (f_n) converges in p^{th} mean.*
(ii) *The sequence (f_n) converges μ-stochastically, and the sequence $(|f_n|^p)$ is μ-equi-integrable.*

Proof. (i)$\Rightarrow$(ii): Suppose (f_n) converges in p^{th} mean, to $f \in \mathscr{L}^p(\mu)$; thus

$$\lim_{n \to \infty} N_p(f_n - f) = 0.$$

In the light of 20.4 only the equi-integrability of the sequence $(|f_n|^p)$ has to be proved. By (15.2) the sequence $(N_p(f_n))_{n \in \mathbb{N}}$ converges to $N_p(f)$ and is therefore bounded, so the set $M := \{|f_n|^p : n \in \mathbb{N}\}$ satisfies (21.3).

For every $A \in \mathscr{A}$ and every $n \in \mathbb{N}$ we have by (15.4)

$$\left(\int_A |f_n| \, d\mu\right)^{1/p} \leq N_p(f_n - f) + \left(\int_A |f|^p \, d\mu\right)^{1/p}.$$

To every $\varepsilon > 0$ corresponds an $n_\varepsilon \in \mathbb{N}$ such that $N_p(f_n - f) < 2^{-1}\varepsilon^{1/p}$ for all $n > n_\varepsilon$. Therefore, if we set $\delta := 2^{-p}\varepsilon$ and

$$h := |f_1|^p \vee \ldots \vee |f_{n_\varepsilon}|^p \vee |f|^p,$$

condition (21.4) is also satisfied by M.

(ii)$\Rightarrow$(i): From the stochastic convergence of the sequence (f_n) and Remark 6 in §20 it follows that

(21.5) $$\lim_{n,m \to \infty} \mu(\{|f_m - f_n| \geq \alpha\} \cap A) = 0$$

for every $A \in \mathscr{A}$ of finite measure and every real $\alpha > 0$. We have to show that (f_n) is a Cauchy sequence in $\mathscr{L}^p(\mu)$, that is, that the doubly-indexed sequence of functions $f_{mn} := f_m - f_n$ satisfies

$$\lim_{m,n\to\infty} \int |f_{mn}|^p \, d\mu = 0 \,.$$

According to 21.3, along with the set $\{|f_n|^p : n \in \mathbb{N}\}$ the set $M_0 := \{|f_{mn}|^p : m, n \in \mathbb{N}\}$ is also equi-integrable. Hence to every $\varepsilon > 0$ corresponds an integrable function $g_\varepsilon \geq 0$ such that $\int_{\{f \geq g_\varepsilon\}} f \, d\mu \leq \varepsilon$ holds for all $f \in M_0$. If we set $g := g_\varepsilon^{1/p}$, then g is p-fold integrable and the preceding inequality can be written

$$\int_{\{|f_{mn}| \geq g\}} |f_{mn}|^p \, d\mu \leq \varepsilon \qquad\qquad \text{for all } m, n \in \mathbb{N}.$$

Because

$$\int |f_{mn}|^p \, d\mu = \int_{\{|f_{mn}| \geq g\}} |f_{mn}|^p \, d\mu + \int_{\{|f_{mn}| < g\}} |f_{mn}|^p \, d\mu$$

it suffices to show that

$$(21.6) \qquad\qquad \int_{\{|f_{mn}| < g\}} |f_{mn}|^p \, d\mu \leq 3\varepsilon$$

holds for all sufficiently large $m, n \in \mathbb{N}$. Now $g^p \mu$, being a finite measure on $\mathscr{A}$, is continuous from above. Since $\bigcap_{k\in\mathbb{N}} \{g < k^{-1}\} = \{g = 0\}$, $\eta > 0$ can therefore be chosen small enough that

$$\int_{\{g < \eta\}} g^p \, d\mu \leq \varepsilon \,.$$

Consequently we also have

$$(21.7) \qquad \int_{\{|f_{mn}| < g\} \cap \{g < \eta\}} |f_{mn}|^p \, d\mu \leq \int_{\{g < \eta\}} g^p \, d\mu \leq \varepsilon \qquad \text{for all } m, n \in \mathbb{N}.$$

The Chebyshev–Markov inequality insures that the set $\{g \geq \eta\}$ has finite μ-measure. According to (21.5) therefore the doubly-indexed sequence of sets

$$A_{mn} := \{|f_{mn}| \geq \alpha\} \cap \{g \geq \eta\} \qquad\qquad m, n \in \mathbb{N}$$

satisfies, whatever $\alpha > 0$ is involved,

$$\lim_{m,n\to\infty} \mu(A_{mn}) = 0 \,.$$

We choose the positive number α so as to have

$$\left(\frac{\alpha}{\eta}\right)^p \int g^p \, d\mu \leq \varepsilon \,.$$

The μ-continuity of the finite measure $g^p\mu$ and 17.8 provide for an $n_0 \in \mathbb{N}$ such that

$$\int_{A_{mn}} g^p \, d\mu \leq \varepsilon \qquad\qquad \text{for all } m, n \geq n_0.$$

Hence

(21.8) $$\int_{\{|f_{mn}|<g\}\cap A_{mn}} |f_{mn}|^p \, d\mu \leq \int_{A_{mn}} g^p \, d\mu \leq \varepsilon \qquad \text{for all } m, n \geq n_0.$$

A second application of the Chebyshev–Markov inequality furnishes the estimate

(21.9) $$\int_{\{|f_{mn}|<g\}\cap A'_{mn}} |f_{mn}|^p \, d\mu \leq \alpha^p \mu(\{g \geq \eta\}) \leq \left(\frac{\alpha}{\eta}\right)^p \int g^p \, d\mu \leq \varepsilon \qquad \text{for all } m, n \in \mathbb{N},$$

in which

$$A'_{mn} := \{|f_{mn}| < \alpha\} \cap \{g \leq \eta\}.$$

By adding the inequalities (21.7)–(21.9) we get finally inequality (21.6), whose confirmation was the last outstanding claim in the proof that (ii) implies (i). $\square$

Remark. 1. Theorem 21.4 does not claim that from the stochastic convergence of a sequence (f_n) to a measurable real function f, the p-fold integrability of f and the convergence of (f_n) to f in p^{th} mean follow as soon as the sequence $(|f_n|^p)$ is equi-integrable. Rather the theorem guarantees the existence of a p-fold integrable function among the possible stochastic limits of the sequence (f_n). The sequence (f_n) does converge in p^{th} mean to every such stochastic limit, as follows from the proof of the theorem in the light of Remark 4 of §20, according to which any two p-fold integrable stochastic limits must in fact coincide almost everywhere.

But stochastic limits that are not p-fold integrable do exist, a fact that can be demonstrated with the aid of the Example in §20: For the sequence (f_n) there, $(|f_n|^p)$ is equi-integrable. But among the stochastic limits f^* that occur there, $f^* \in \mathscr{L}^p(\mu)$ for some $p \in [1, +\infty[$ if and only if $f^*(\omega_1) = 0$.

However, the phenomenon discussed above does not occur for σ-finite measures. By 20.3 in that case any two stochastic limits are almost everywhere equal. Therefore we have

21.5 Corollary. *Suppose the measure μ is σ-finite. If a sequence (f_n) from $\mathscr{L}^p(\mu)$ converges stochastically to a (measurable, real) function f, and if the sequence $(|f_n|^p)$ is equi-integrable, then $f \in \mathscr{L}^p(\mu)$ and (f_n) converges in p^{th} mean to f.*

Theorem 21.4 can be sharpened by bringing in a further condition equivalent to (i) and (ii) which is suggested by F. Riesz' Theorem 15.3. En route to this sharpening the following lemma plays a key role. On the other hand, from the sharpening that we are aiming for, the lemma can in turn be deduced, as can the theorem of F. Riesz, even with its almost-everywhere convergence hypothesis weakened to stochastic convergence.

21.6 Lemma. *Suppose the sequence of functions $f_n \geq 0$ from $\mathcal{L}^1(\mu)$ converges stochastically to a function $f \geq 0$ from $\mathcal{L}^1(\mu)$. If in addition*

$$\lim_{n \to \infty} \int f_n \, d\mu = \int f \, d\mu \,,$$

then the sequence (f_n) converges to f in mean.

Proof. We consider the sequence $(f \wedge f_n)_{n \in \mathbb{N}}$. The inequalities

$$0 \leq f \wedge f_n \leq f$$

and Example 3 show that it is equi-integrable. Since

$$0 \leq f - f \wedge f_n \leq |f_n - f| \qquad\qquad \text{(for all } n \in \mathbb{N}),$$

stochastic convergence of (f_n) to f entails that of $(f \wedge f_n)$ to f. From Theorem 21.4 this new sequence then converges to f in mean. We therefore also have

(21.10) $$\lim_{n \to \infty} \int f \wedge f_n \, d\mu = \int f \, d\mu \,.$$

From this, the decomposition $f + f_n = f \vee f_n + f \wedge f_n$, and the convergence hypothesis follows the companion result

(21.10′) $$\lim_{n \to \infty} \int f \vee f_n \, d\mu = \int f \, d\mu \,.$$

But then the decomposition

$$|f_n - f| = f \vee f_n - f \wedge f_n$$

shows that the claimed mean convergence ensues upon subtracting (21.10) from (21.10′). □

Now we can get the sharpening of Theorems 21.4 and 15.4 mentioned earlier:

21.7 Theorem. *For every sequence (f_n) in $\mathcal{L}^p(\mu)$ which converges μ-stochastically to a function $f \in \mathcal{L}^p(\mu)$ the following three assertions are equivalent:*
(i)   *The sequence (f_n) converges in p^{th} mean to f.*
(ii) *The sequence $(|f_n|^p)$ is equi-integrable.*
(iii) $\lim_{n \to \infty} \int |f_n|^p \, d\mu = \int |f|^p \, d\mu.$

Proof. The equivalence of (i) and (ii) is contained in Theorem 21.4. We need therefore establish only two implications:
    (i)⇒(iii): Assertion (15.6) in Theorem 15.1 affirms this.
    (iii)⇒(ii): From the hypothesized stochastic convergence of the sequence (f_n) to f follows that of $(|f_n|^p)$ to $|f|^p$, via 20.9. And then from the preceding lemma it further follows that the sequence $(|f_n|^p)$ converges to $|f|^p$ in mean. Finally, Theorem 21.4 – with the p there chosen to be 1 – shows that the convergence in mean of this sequence entails its equi-integrability. □

For σ-finite measures μ, equi-integrability can be characterized in a way that is particularly convenient for applications. The σ-finiteness will be exploited in the form expressed by 17.6, that there is a strictly positive function h in $\mathscr{L}^1(\mu)$.

21.8 Theorem. *Let $(\Omega, \mathscr{A}, \mu)$ be a σ-finite measure space and h a strictly positive function from $\mathscr{L}^1(\mu)$. Then for any set M of $\mathscr{A}$-measurable numerical functions on Ω the following three assertions are equivalent:*
(i)   *M is equi-integrable.*
(ii) *For every $\varepsilon > 0$ some scalar multiple of h is an ε-bound for M.*
(iii) *M satisfies*

$$(21.11) \qquad \sup_{f \in M} \int |f|\, d\mu < +\infty$$

as well as the following: Given $\varepsilon > 0$ there exists $\delta > 0$ such that

$$(21.12) \qquad \int_A h\, d\mu \leq \delta \Rightarrow \int_A |f|\, d\mu \leq \varepsilon \qquad \text{for all } A \in \mathscr{A},\ f \in M.$$

Statement (ii) simply says that

$$(21.13) \qquad \lim_{\alpha \to +\infty} \int_{\{|f| \geq \alpha h\}} |f|\, d\mu = 0$$

holds *uniformly* for $f \in M$. Condition (21.12) is for obvious reasons (cf. 17.8) called the *equi-$(h\mu)$-continuity* of the measures $|f|\,\mu$, $f \in M$.

Proof. (i)$\Rightarrow$(ii): Let g be an $\frac{\varepsilon}{2}$-bound for M. Then for all $f \in M$ and all $\alpha > 0$

$$\int_{\{|f| \geq \alpha h\}} |f|\, d\mu = \int_{\{|f| \geq \alpha h\} \cap \{|f| \geq g\}} |f|\, d\mu + \int_{\{|f| \geq \alpha h\} \cap \{|f| < g\}} |f|\, d\mu$$

$$\leq \int_{\{|f| \geq g\}} |f|\, d\mu + \int_{\{g > \alpha h\}} g\, d\mu \leq \frac{\varepsilon}{2} + \int_{\{g > \alpha h\}} g\, d\mu.$$

According to 13.6, $\mu(\{g = +\infty\}) = 0$. Since $g\mu$ is a finite measure on $\mathscr{A}$, it is continuous from above. Hence the fact that

$$\bigcap_{\alpha > 0} \{g > \alpha h\} = \bigcap_{n \in \mathbb{N}} \{g > nh\} = \{g = +\infty\}$$

is a set of $(g\mu)$-measure 0 means that

$$\int_{\{g > \alpha h\}} g\, d\mu < \frac{\varepsilon}{2}$$

for all sufficiently large α. Coupled with the preceding inequality this shows that indeed αh is an ε-bound for all sufficiently large α, that is, (ii) holds.

(ii)⇒(iii): This can be gleaned from the inequality derived at the beginning of the proof of 21.2, αh being now eligible for the function g there:

$$\int_A |f|\, d\mu \le \int_{\{|f|\ge \alpha h\}} |f|\, d\mu + \alpha \int_A h\, d\mu \qquad \text{for all } f \in M.$$

(iii)⇒(i): 21.2 affirms this.   □

Theorem 21.8 is of special significance for *finite* measures μ. Then it is often expedient to choose for h the constant function 1. When one does, (21.13) assumes the equivalent form

$$(21.13') \qquad \lim_{\alpha \to +\infty} \int_{\{|f|\ge \alpha\}} |f|\, d\mu = 0 \qquad \text{uniformly for } f \in M.$$

This condition is thus – just as (21.13) for σ-finite measures – necessary and sufficient for equi-integrability of M.

Remark. 2. In part (iii) of Theorem 21.8 the $\mathscr{L}^1$-boundedness of M expressed by (21.11) cannot in general be dropped from the hypotheses. It suffices to consider the measure space $(\{a\}, \mathscr{P}(\{a\}), \varepsilon_a)$ consisting of a single point and the sequence of functions $f_n := n \cdot 1$. This sequence is not equi-integrable, although for every $\varepsilon > 0$ and every strictly positive h, (21.12) holds whenever $0 < \delta < h(a)$.

Let us close by deriving a sufficient condition for equi-integrability in the finite-measure case which generalizes the introductory Example 3.

21.9 Lemma. Let μ be a finite measure and $M \subset \mathscr{L}^1(\mu)$. Suppose that there is a μ-integrable function $g \ge 0$ such that

$$(21.14) \qquad \int_{\{|f|\ge \alpha\}} |f|\, d\mu \le \int_{\{|f|\ge \alpha\}} g\, d\mu$$

for all $f \in M$ and all $\alpha \in \mathbb{R}_+$. Then M is equi-integrable.

Proof. The case $\alpha := 0$ of (21.14) says that $\int |f|\, d\mu \le \int g\, d\mu < +\infty$ for all $f \in M$. Then Chebyshev's inequality tells us that

$$\mu(\{|f|\ge \alpha\}) \le \frac{1}{\alpha}\int |f|\, d\mu \le \frac{1}{\alpha}\int g\, d\mu \qquad \text{for all } \alpha > 0,\, f \in M.$$

It follows from this that

$$(21.15) \qquad \lim_{\alpha \to +\infty} \mu(\{|f|\ge \alpha\}) = 0 \qquad \text{uniformly in } f \in M.$$

For each $\varepsilon > 0$, 17.8 supplies a $\delta > 0$ such that

$$A \in \mathscr{A} \quad \text{and} \quad \mu(A) < \delta \qquad \Rightarrow \qquad \int_A g\, d\mu \le \varepsilon.$$

Putting this together with (21.14) and (21.15) gives us

$$\lim_{\alpha \to +\infty} \int_{\{|f| \geq \alpha\}} |f|\, d\mu = 0 \qquad \text{uniformly for } f \in M,$$

that is, (21.13′), which we have seen entails equi-integrability of M. $\square$

Exercises.

1. Show that for any measure space $(\Omega, \mathscr{A}, \mu)$ a set M of measurable numerical functions is equi-integrable if and only if for every $\varepsilon > 0$ there is an integrable function $h = h_\varepsilon \geq 0$ such that $\int (|f| - h)^+ \leq \varepsilon$ for all $f \in M$. [*Hint:* For sufficiently large $\eta > 0$, $g := \eta h$ will be a 2ε-bound for M.]

2. Let $(\Omega, \mathscr{A}, \mu)$ be an arbitrary measure space, $1 \leq p < +\infty$. Suppose the sequence (f_n) in $\mathscr{L}^p(\mu)$ converges almost everywhere on Ω to a measurable real function f. Show that f lies in $\mathscr{L}^p(\mu)$ and (f_n) converges to f in p^{th} mean if the sequence $(|f_n|^p)$ is equi-integrable.

3. Show that from the $\mathscr{L}^p$-convergence of a sequence (f_n) to a function $f \in \mathscr{L}^p(\mu)$ follows the $\mathscr{L}^1$-convergence of the sequence $(|f_n|^p)$ to $|f|$, for any $1 \leq p < +\infty$.

4. Consider a finite measure μ and an $M \subset \mathscr{L}^1(\mu)$. For each $n \in \mathbb{N}$, $f \in M$ set

$$a_n(f) := n\mu(\{n \leq |f| < n + 1\}).$$

Show that M is equi-integrable if and only the series $\sum\limits_{n=1}^{\infty} a_n(f)$ converges uniformly in $f \in M$. [Cf. Theorem 3.4 and its proof in BAUER [1996].]

5. Consider a finite measure μ and an $M \subset \mathscr{L}^1(\mu)$. Show that M is equi-integrable if there is a function $q : \mathbb{R}_+ \to \mathbb{R}_+$ with the properties

$$\lim_{t \to +\infty} \frac{q(t)}{t} = +\infty \quad \text{and} \quad \sup_{f \in M} \int q \circ |f|\, d\mu < +\infty.$$

(In fact we have to do here with a necessary as well as a sufficient condition, which goes back to CH. DE LA VALLÉE POUSSIN (1866–1962). Moreover, q can always be chosen to be convex and isotone. Cf. MEYER [1976], p. 19 or DELLACHERIE and MEYER [1975], p. 38.)

6. Let $(\Omega, \mathscr{A}, \mu)$ be a measure space with $\mu(\Omega) < +\infty$, $(f_n)_{n \in \mathbb{N}}$ a sequence of measurable numerical functions $f_n \geq 0$, and set $f^* := \limsup\limits_{n \to \infty} f_n$. Show that:

(a) If the sequence (f_n) is equi-integrable (or at least satisfies condition (21.12)), then the following "dual version" of Fatou's lemma is valid:

$$(*) \qquad \limsup_{n \to \infty} \int_A f_n\, d\mu \leq \int_A f^*\, d\mu \qquad \text{for all } A \in \mathscr{A}.$$

How does the corresponding result in Exercise 1 of §15 fit in? [*Hint:* Exercise 2 of §11.]

(b) Under the hypothesis $\int f^*\, d\mu < +\infty$, the sequence (f_n) is equi-integrable if and only if $(*)$ holds. [In proving the "if" direction, argue indirectly.]

(c) Result (b) can fail in case $\int f^* \, d\mu = +\infty$. Try to corroborate this with a sequence $(\alpha_n f_n)$ derived by appropriate choice of (sufficiently large) numbers $\alpha_n > 0$ from the sequence (f_n) in the Example from §15.

7. Let $(\Omega, \mathscr{A}, \mu)$ be a measurable space with $\mu(\Omega) < +\infty$, and let $(\nu_i)_{i \in I}$ be a family of finite and μ-continuous measures on $\mathscr{A}$. Suppose this family is *equi-continuous at $\emptyset$*, meaning that to every sequence $(A_n)_{n \in \mathbb{N}}$ in $\mathscr{A}$ with $A_n \downarrow \emptyset$ and to every $\varepsilon > 0$ there is an $n_\varepsilon \in \mathbb{N}$ such that $\nu_i(A_n) \leq \varepsilon$ for all $n \geq n_\varepsilon$, and all $i \in I$. Show that then this family is *equi-μ-continuous* in the following sense (cf. (21.12)): To every $\varepsilon > 0$ there corresponds a $\delta = \delta_\varepsilon > 0$ such that

$$A \in \mathscr{A} \quad \text{and} \quad \mu(A) \leq \delta \qquad \Rightarrow \qquad \nu_i(A) \leq \varepsilon \text{ for all } i \in I.$$

What does this result say in view of Theorem 21.8? [*Hint*: Review the proof of Theorem 17.8.]

Chapter III

Product Measures

In this short chapter we will investigate whether and how one can associate a product with finitely many measure spaces. And for the product measures thus gotten we will want to see about how to integrate with respect to them in terms of their factors. We will recognize the L-B measure λ^d as being a special product measure when $d \geq 2$. One important application of product measures is the introduction of the concept of convolution for measures and functions.

§22. Products of σ-algebras and measures

Finitely many measurable spaces $(\Omega_j, \mathscr{A}_j)$, $j = 1, \ldots, n \in \mathbb{N}$ are given. We consider the product set

$$\Omega := \mathop{\times}_{j=1}^{n} \Omega_j = \Omega_1 \times \ldots \times \Omega_n$$

and for each j the projection mapping

$$p_j : \Omega \to \Omega_j$$

which assigns to each point $(\omega_1, \ldots, \omega_n) \in \Omega$ its jth coordinate ω_j. The σ-algebra in Ω generated by the mappings $p_1, \ldots, p_n$ is designated

$$\bigotimes_{j=1}^{n} \mathscr{A}_j := \mathscr{A}_1 \otimes \ldots \otimes \mathscr{A}_n := \boldsymbol{\sigma}(p_1, \ldots, p_n)$$

and called the *product of the σ-algebras $\mathscr{A}_1, \ldots, \mathscr{A}_n$.* According to (7.3) we have to do here with the smallest σ-algebra $\mathscr{A}$ in Ω such that each p_j is $\mathscr{A}$-$\mathscr{A}_j$-measurable. The reader may recall that the product of finitely many topological spaces is defined in a very similar way.

An important principle of generation for such products is immediately at hand:

22.1 Theorem. *For each $j = 1, \ldots, n$ let $\mathscr{E}_j$ be a generator of the σ-algebra $\mathscr{A}_j$ in Ω_j which contains a sequence $(E_{jk})_{k \in \mathbb{N}}$ of sets with $E_{jk} \uparrow \Omega_j$. Then the σ-algebra $\mathscr{A}_1 \otimes \ldots \otimes \mathscr{A}_n$ is generated by the system of all sets*

$$E_1 \times \ldots \times E_n$$

with $E_j \in \mathscr{E}_j$ for each $j = 1, \ldots, n$.

Proof. Let $\mathscr{A}$ be any σ-algebra in Ω. What we have to show is that the mappings p_j are all $\mathscr{A}$-$\mathscr{A}_j$-measurable $(j = 1, \ldots, n)$ if and only if $\mathscr{A}$ contains each of the sets $E_1 \times \ldots \times E_n$ described above. According to 7.2 p_j is $\mathscr{A}$-$\mathscr{A}_j$-measurable just exactly if $p_j^{-1}(E_j) \in \mathscr{A}$ for every $E_j \in \mathscr{E}_j$. If this condition is fulfilled for each $j \in \{1, \ldots, n\}$, then the sets

$$E_1 \times \ldots \times E_n = p_1^{-1}(E_1) \cap \ldots \cap p_n^{-1}(E_n)$$

all lie in $\mathscr{A}$. If conversely, $E_1 \times \ldots \times E_n \in \mathscr{A}$ for every possible choice of $E_j \in \mathscr{E}_j$ and $j \in \{1, \ldots, n\}$, then upon fixing $E_j \in \mathscr{E}_j$, the sets

$$F_k := E_{1k} \times \ldots \times E_{j-1,k} \times E_j \times E_{j+1,k} \times \ldots \times E_{nk}, \qquad k \in \mathbb{N},$$

all lie in $\mathscr{A}$. Since the sequence $(F_k)_{k \in \mathbb{N}}$ increases to

$$\Omega_1 \times \ldots \times \Omega_{j-1} \times E_j \times \Omega_{j+1} \times \ldots \times \Omega_n = p_j^{-1}(E_j),$$

this set too lies in $\mathscr{A}$, for each j. The claim is therewith proven. $\square$

Remark. 1. The restriction imposed on the generators $\mathscr{E}_j$ cannot generally be dispensed with. Take, for example, $n := 2$, $\mathscr{A}_1 := \{\emptyset, \Omega_1\}$, $\mathscr{E}_1 := \{\emptyset\}$ and $\mathscr{E}_2 := \mathscr{A}_2$, in which $\mathscr{A}_2$ contains at least four sets.

A particular case of this theorem is the fact that the product $\mathscr{A}_1 \otimes \ldots \otimes \mathscr{A}_n$ is generated by all the sets $A_1 \times \ldots \times A_n$ with each $A_j \in \mathscr{A}_j$. Our further course will be guided by the following example:

Example. For each $j \in \{1, \ldots, n\}$ let $\Omega_j := \mathbb{R}$, $\mathscr{A}_j := \mathscr{B}^1$ and $\mathscr{E}_j := \mathscr{I}^1$. The system of all sets $E_1 \times \ldots \times E_n$ with each $E_j \in \mathscr{I}^1$ is evidently just the system $\mathscr{I}^n$ of all right half-open intervals in $\mathbb{R}^n$. According to 6.1, $\mathscr{I}^n$ generates the σ-algebra $\mathscr{B}^n$ of n-dimensional Borel sets. Taken together with 22.1 – whose hypotheses are clearly satisfied here – this reveals that

$$(22.2) \qquad\qquad \mathscr{B}^n = \mathscr{B}^1 \otimes \ldots \otimes \mathscr{B}^1 \qquad (n \text{ factors on the right}).$$

By 6.2, λ^n is the only measure on $\mathscr{B}^n$ which satisfies

$$\lambda^n(I_1 \times \ldots \times I_n) = \lambda^1(I_1) \cdot \ldots \cdot \lambda^1(I_n)$$

for all $I_1, \ldots, I_n \in \mathscr{I}^1$. This remark and the example preceding it leads to the following question.

Measure spaces $(\Omega_j, \mathscr{A}_j, \mu_j)$ are given, $1 \leq j \leq n$ with $n \geq 2$, and for each $\mathscr{A}_j$ a generator $\mathscr{E}_j$. Under what hypotheses can the existence of a measure π on $\mathscr{A}_1 \otimes \ldots \otimes \mathscr{A}_n$ satisfying

$$(22.3) \qquad \pi(E_1 \times \ldots \times E_n) = \mu_1(E_1) \cdot \ldots \cdot \mu_n(E_n) \qquad \text{for all } E_j \in \mathscr{E}_j, 1 \leq j \leq n$$

be proven?

The accompanying uniqueness question can be settled at once:

22.2 Theorem. *Suppose that for each $j = 1, \ldots, n$ $\mathscr{E}_j$ is an $\cap$-stable generator of $\mathscr{A}_j$ which contains a sequence $(E_{jk})_{k \in \mathbb{N}}$ of sets of finite μ_j-measure satisfying $E_{jk} \uparrow \Omega_j$. Then there is at most one measure π on $\mathscr{A}_1 \otimes \ldots \otimes \mathscr{A}_n$ enjoying property* (22.3).

Proof. Let $\mathscr{E}$ denote the system of all sets $E_1 \times \ldots \times E_n$, where $E_j \in \mathscr{E}_j$ for each j. According to 22.1, $\mathscr{E}$ generates the σ-algebra $\mathscr{A}_1 \otimes \ldots \otimes \mathscr{A}_n$. Since each $\mathscr{E}_j$ is $\cap$-stable, so is $\mathscr{E}$, as the identity

$$\left(\underset{j=1}{\overset{n}{\times}} E_j \right) \cap \left(\underset{j=1}{\overset{n}{\times}} F_j \right) = \underset{j=1}{\overset{n}{\times}} (E_j \cap F_j)$$

makes clear. Moreover $E_k := E_{1k} \times \ldots \times E_{nk}$ $(k \in \mathbb{N})$ defines a sequence in $\mathscr{E}$ that evidently satisfies

$$E_k \uparrow \Omega_1 \times \ldots \times \Omega_n \,.$$

Recalling that $\mu_j(E_{jk}) < +\infty$ for all (relevant) j and k, we see that the uniqueness claim therefore follows from 5.4. (Obviously it would suffice if $\bigcup_{k \in \mathbb{N}} E_{jk} = \Omega_j$ instead of $E_{jk} \uparrow \Omega_j$ were satisfied for each j.) □

Under the hypotheses of 22.2, which obviously entail the σ-finiteness of each measure μ_j, the existence of the desired measure π can also be proven. This proof will be carried out in the next section, first for $n = 2$, then for arbitrary $n \geq 2$.

Remark. 2. In closing it should again be mentioned that a mapping

$$f : \Omega_0 \to \Omega_1 \times \ldots \times \Omega_n$$

of a measurable space $(\Omega_0, \mathscr{A}_0)$ into a product of measurable spaces $(\Omega_j, \mathscr{A}_j)$ is measurable with respect to the σ-algebra $\mathscr{A}_1 \otimes \ldots \otimes \mathscr{A}_n$ *if and only if* each component mapping $f_j := p_j \circ f$ of f is $\mathscr{A}_0$-$\mathscr{A}_j$-measurable – a fact which is immediate from Theorem 7.4.

Exercise.

Finitely many measurable spaces $(\Omega_j, \mathscr{A}_j)$ are given, $j = 1, \ldots, n$. Show that the algebra in $\Omega_1 \times \ldots \times \Omega_n$ generated by all sets $A_1 \times \ldots \times A_n$ with each $A_j \in \mathscr{A}_j$ consists of all finite unions of such product sets.

§23. Product measures and Fubini's theorem

Initially measure spaces $(\Omega_1, \mathscr{A}_1, \mu_1)$, $(\Omega_2, \mathscr{A}_2, \mu_2)$ are given. For every $Q \subset \Omega_1 \times \Omega_2$ the sets

(23.1)
$$Q_{\omega_1} := \{\omega_2 \in \Omega_2 : (\omega_1, \omega_2) \in Q\}$$
$$Q_{\omega_2} := \{\omega_1 \in \Omega_1 : (\omega_1, \omega_2) \in Q\}$$

are called, respectively, the ω_1-section of Q $(\omega_1 \in \Omega_1)$ and the ω_2-section of Q $(\omega_2 \in \Omega_2)$.

This notation is chosen for typographic simplicity and will see us through §23, after which it is not needed. In case $\Omega_1 = \Omega_2$, however, it presents obvious problems, to circumvent which, alternative notations like $_{\omega_1}Q$ or Q^{ω_1} for Q_{ω_1} are also popular in the literature.

About these sets we claim:

23.1 Lemma. *If $Q \in \mathscr{A}_1 \otimes \mathscr{A}_2$, then its ω_1-section lies in $\mathscr{A}_2$ for every $\omega_1 \in \Omega_1$, and its ω_2-section lies in $\mathscr{A}_1$ for every $\omega_2 \in \Omega_2$.*

Proof. For arbitrary subsets $Q, Q_1, Q_2, \ldots$ of $\Omega := \Omega_1 \times \Omega_2$, and points $\omega_1 \in \Omega_1$

$$(\Omega \setminus Q)_{\omega_1} = \Omega_2 \setminus Q_{\omega_1}$$

and

$$\left(\bigcup_{n \in \mathbb{N}} Q_n \right)_{\omega_1} = \bigcup_{n \in \mathbb{N}} (Q_n)_{\omega_1} .$$

Furthermore $\Omega_{\omega_1} = \Omega_2$, and more generally for $A_1 \subset \Omega_1$, $A_2 \subset \Omega_2$ we have

$$(A_1 \times A_2)_{\omega_1} = \begin{cases} A_2 & \text{if } \omega_1 \in A_1 \\ \emptyset & \text{if } \omega_1 \in \Omega_1 \setminus A_1. \end{cases}$$

For each $\omega_1 \in \Omega_1$, therefore, the system of all sets $Q \subset \Omega$ having section $Q_{\omega_1} \in \mathscr{A}_2$ is a σ-algebra in Ω which contains every product set $A_1 \times A_2$ with $A_1 \in \mathscr{A}_1$, $A_2 \in \mathscr{A}_2$. But according to 22.1 $\mathscr{A}_1 \otimes \mathscr{A}_2$ is the smallest σ-algebra which contains all such product sets. This proves the part of the lemma dealing with ω_1-sections. Of course, ω_2-sections are treated the same way. □

Since now $\mu_2(Q_{\omega_1})$ and $\mu_1(Q_{\omega_2})$ make sense for all $Q \in \mathscr{A}_1 \otimes \mathscr{A}_2$, $\omega_1 \in \Omega_1$ and $\omega_2 \in \Omega_2$, we are in a position to take the next step:

23.2 Lemma. *Suppose the measures μ_1 and μ_2 are σ-finite. Then for every $Q \in \mathscr{A}_1 \otimes \mathscr{A}_2$ the functions*

$$\omega_1 \mapsto \mu_2(Q_{\omega_1}) \quad and \quad \omega_2 \mapsto \mu_1(Q_{\omega_2})$$

on Ω_1 and Ω_2, respectively, are $\mathscr{A}_1$-measurable and $\mathscr{A}_2$-measurable, respectively.

Proof. The function $\omega_1 \mapsto \mu_2(Q_{\omega_1})$ will be denoted by s_Q. We will establish the $\mathscr{A}_1$-measurability of s_Q, for each $Q \in \mathscr{A}_1 \otimes \mathscr{A}_2$. The other function can be treated analogously.

First suppose that $\mu_2(\Omega_2) < +\infty$. In this case the set $\mathscr{D}$ of all $D \in \mathscr{A}_1 \otimes \mathscr{A}_2$ whose s_D function is $\mathscr{A}_1$-measurable constitutes a Dynkin system in $\Omega := \Omega_1 \times \Omega_2$. This involves the following easily checked assertions:

$$s_\Omega = \mu_2(\Omega_2);$$

$$s_{\Omega \setminus D} = s_\Omega - s_D \text{ for every } D \in \mathscr{D};$$

$$s_{\cup D_n} = \sum s_{D_n} \text{ for every sequence } (D_n) \text{ of disjoint sets in } \mathscr{D}.$$

Furthermore $\mathscr{D}$ contains $A_1 \times A_2$ for every $A_1 \in \mathscr{A}_1$, $A_2 \in \mathscr{A}_2$, since

$$s_{A_1 \times A_2} = \mu_2(A_2) \cdot 1_{A_1}.$$

The system $\mathscr{E}$ of all such $A_1 \times A_2$ is $\cap$-stable and generates $\mathscr{A}_1 \otimes \mathscr{A}_2$, by 22.1. Therefore 2.4 insures that $\mathscr{A}_1 \otimes \mathscr{A}_2$ is the Dynkin system generated by $\mathscr{E}$. From $\mathscr{E} \subset \mathscr{D} \subset \mathscr{A}_1 \otimes \mathscr{A}_2$ therefore follows that $\mathscr{D} = \mathscr{A}_1 \otimes \mathscr{A}_2$, which is what is being claimed.

If μ_2 is only σ-finite, then there is a sequence (B_n) of sets from $\mathscr{A}_2$, each of finite μ_2-measure, with $B_n \uparrow \Omega_2$. For each n, $A_2 \mapsto \mu_2(A_2 \cap B_n)$ is therefore a finite measure $\mu_{2,n}$ on $\mathscr{A}_2$, to which the already proven result can be applied, showing that $\omega_1 \mapsto \mu_{2,n}(Q_{\omega_1})$ is $\mathscr{A}_1$-measurable for each $Q \in \mathscr{A}_1 \otimes \mathscr{A}_2$. Now

$$\mu_2(Q_{\omega_1}) = \sup_{n \in \mathbb{N}} \mu_{2,n}(Q_{\omega_1})$$

because of the continuity from below of the measure μ_2. From Theorem 9.5 then the mapping $\omega_1 \mapsto \mu_2(Q_{\omega_1})$ is indeed $\mathscr{A}_1$-measurable. $\square$

It is now rather simple to construct the measure π that we seek:

23.3 Theorem. *Let* $(\Omega_j, \mathscr{A}_j, \mu_j)$ *be* σ-*finite measure spaces,* $j = 1, 2$. *Then there is exactly one measure* π *on* $\mathscr{A}_1 \otimes \mathscr{A}_2$ *which satisfies*

$$(23.2) \qquad \pi(A_1 \times A_2) = \mu_1(A_1)\mu_2(A_2) \qquad \text{for all } A_1 \in \mathscr{A}_1, \ A_2 \in \mathscr{A}_2.$$

In addition this measure satisfies

$$(23.3) \quad \pi(Q) = \int \mu_2(Q_{\omega_1})\mu_1(d\omega_1) = \int \mu_1(Q_{\omega_2})\mu_2(d\omega_2) \quad \text{for all } Q \in \mathscr{A}_1 \otimes \mathscr{A}_2$$

and is σ-*finite.*

Proof. As before, for each $Q \in \mathscr{A}_1 \otimes \mathscr{A}_2$ let s_Q denote the $\mathscr{A}_1$-measurable function $\omega_1 \mapsto \mu_2(Q_{\omega_1})$ on Ω_1; it is of course non-negative. Consequently via

$$\pi(Q) := \int s_Q \, d\mu_1$$

a non-negative function π is well defined on $\mathscr{A}_1 \otimes \mathscr{A}_2$. For every sequence $(Q_n)_{n \in \mathbb{N}}$ of pairwise disjoint sets from $\mathscr{A}_1 \otimes \mathscr{A}_2$ the equality $s_{\cup Q_n} = \sum s_{Q_n}$ and 11.5 insure

that

$$\pi\left(\bigcup_{n\in\mathbb{N}} Q_n\right) = \sum_{n=1}^{\infty} \pi(Q_n).$$

Since $s_\emptyset = 0$ we have $\pi(\emptyset) = 0$. This proves that π is indeed a measure on $\mathscr{A}_1 \otimes \mathscr{A}_2$. It has property (23.2) because

$$s_{A_1 \times A_2} = \mu_2(A_2)1_{A_1}, \text{ whence integration yields}$$
$$\pi(A_1 \times A_2) = \mu_1(A_1)\mu_2(A_2).$$

Proceeding analogously, we confirm that

$$\pi'(Q) := \int \mu_1(Q_{\omega_2})\mu_2(d\omega_2)$$

also defines a measure on $\mathscr{A}_1 \otimes \mathscr{A}_2$ having this property. But when Theorem 22.2 is applied to $\mathscr{E}_1 := \mathscr{A}_1$ and $\mathscr{E}_2 := \mathscr{A}_2$ it affirms that there is at most one such measure. Thus $\pi = \pi'$ and (23.3) is confirmed. There is a sequence $(A_{jn})_{n\in\mathbb{N}}$ of sets from $\mathscr{A}_j$, each of finite μ_j-measure, with $A_{jn} \uparrow \Omega_j$, for $j = 1$ and $j = 2$. Using these as the A_1, A_2, respectively, in (23.2) proves the σ-finiteness of π because $\pi(A_{1n} \times A_{2n}) < +\infty$ and $A_{1n} \times A_{2n} \uparrow \Omega_1 \times \Omega_2$. $\square$

23.4 Definition. The measure π on $\mathscr{A}_1 \otimes \mathscr{A}_2$ which is uniquely specified by (23.2) whenever $(\Omega_1, \mathscr{A}_1, \mu_1)$ and $(\Omega_2, \mathscr{A}_2, \mu_2)$ are σ-finite measure spaces is called the *product of the measures* μ_1 and μ_2 and is denoted by

$$\mu_1 \otimes \mu_2.$$

Thus also the question posed in §22 is answered for σ-finite measures μ_1, μ_2. If namely $\mathscr{E}_j$ is a generator of $\mathscr{A}_j$ $(j = 1, 2)$ with the properties formulated in Theorem 22.2, then according to 22.2 and 23.3, $\mu_1 \otimes \mu_2$ is the only measure π on $\mathscr{A}_1 \otimes \mathscr{A}_2$ which satisfies (22.3).

The Example in §22 therefore entails that $\lambda^2 = \lambda^1 \otimes \lambda^1$. Similar considerations lead to the validity of

$$\lambda^{m+n} = \lambda^m \otimes \lambda^n$$

for any $m, n \in \mathbb{N}$, once the appropriate identification of $\mathbb{R}^{m+n}$ with $\mathbb{R}^m \times \mathbb{R}^n$ has been made.

We turn now to integrating with respect to the product measure $\mu_1 \otimes \mu_2$. Our notation for sections can be usefully extended to functions for this purpose. If $f : \Omega_1 \times \Omega_2 \to \Omega_0$ is any mapping, we define its sections f_{ω_1} for each $\omega_1 \in \Omega_1$ and f_{ω_2} for each $\omega_2 \in \Omega_2$ as mappings of Ω_1 and Ω_2, respectively, into Ω_0 by

$$\begin{aligned}
(23.4) \qquad f_{\omega_1}(\omega_2') &:= f(\omega_1, \omega_2') && \text{for all } \omega_2' \in \Omega_2 \\
f_{\omega_2}(\omega_1') &:= f(\omega_1', \omega_2) && \text{for all } \omega_1' \in \Omega_1.
\end{aligned}$$

Notice that if $Q \subset \Omega_1 \times \Omega_2$ and $f := 1_Q$, then these functions satisfy

$$(23.5) \qquad (1_Q)_{\omega_1} = 1_{Q_{\omega_1}}, \quad \text{and} \quad (1_Q)_{\omega_2} = 1_{Q_{\omega_2}}.$$

Note, of course, that these indicator functions have different domains, and, just as with (23.1), further caution is called for with (23.4) in case $\Omega_1 = \Omega_2$. Equations (23.4), and (23.5) lead us to call the mapping f_{ω_j} the ω_j-section of f. It enjoys the expected properties:

23.5 Lemma. *For every measurable space $(\Omega', \mathscr{A}')$ and every measurable mapping*

$$f : (\Omega_1 \times \Omega_2, \mathscr{A}_1 \otimes \mathscr{A}_2) \to (\Omega', \mathscr{A}')$$

f_{ω_1} *is* $\mathscr{A}_2$-$\mathscr{A}'$-*measurable and* f_{ω_2} *is* $\mathscr{A}_1$-$\mathscr{A}'$-*measurable for every* $\omega_1 \in \Omega_1, \omega_2 \in \Omega_2$.

Proof. For every $A' \in \mathscr{A}'$, $\omega_1 \in \Omega_1$

$$f_{\omega_1}^{-1}(A') = \{\omega_2 \in \Omega_2 : (\omega_1, \omega_2) \in f^{-1}(A')\} = (f^{-1}(A'))_{\omega_1}$$

and similarly for every $\omega_2 \in \Omega_2$

$$f_{\omega_2}^{-1}(A') = (f^{-1}(A'))_{\omega_2} \,,$$

so the measurability claims follow from Lemma 23.1. $\square$

Decisive is the following theorem which extends formula (23.3) from indicator functions to non-negative measurable functions. It goes back to L. TONELLI (1885–1946), its corollary to G. FUBINI (1879–1943). Both statements are often combined under the single designation the *theorem of Fubini*.

23.6 Theorem (of Tonelli). *Let $(\Omega_j, \mathscr{A}_j, \mu_j)$ be σ-finite measure spaces $(j = 1, 2)$, and let*

$$f : \Omega_1 \times \Omega_2 \to \overline{\mathbb{R}}_+$$

be $\mathscr{A}_1 \otimes \mathscr{A}_2$-*measurable. Then the functions*

$$\omega_2 \mapsto \int f_{\omega_2} \, d\mu_1 \quad and \quad \omega_1 \mapsto \int f_{\omega_1} \, d\mu_2$$

are $\mathscr{A}_2$-*measurable and* $\mathscr{A}_1$-*measurable, respectively. Moreover,*

$$(23.6) \qquad \int f \, d(\mu_1 \otimes \mu_2) = \int \Big(\int f_{\omega_2} \, d\mu_1 \Big) \mu_2(d\omega_2) = \int \Big(\int f_{\omega_1} \, d\mu_2 \Big) \mu_1(d\omega_1) \,.$$

Proof. Set $\Omega := \Omega_1 \times \Omega_2$, $\mathscr{A} := \mathscr{A}_1 \otimes \mathscr{A}_2$ and $\pi := \mu_1 \otimes \mu_2$. Consider first an $\mathscr{A}$-elementary function f:

$$f := \sum_{j=1}^{n} \alpha_j 1_{Q^j} \qquad (\alpha_j \geq 0, \, Q^j \in \mathscr{A}, \, n \in \mathbb{N}).$$

Then a glance at (23.5) reveals that for each $\omega_2 \in \Omega_2$, $f_{\omega_2} = \sum \alpha_j 1_{Q^j_{\omega_2}}$ and so

$$\int f_{\omega_2} \, d\mu_1 = \sum_{j=1}^{n} \alpha_j \mu_1(Q^j_{\omega_2}) \,,$$

an $\mathscr{A}_2$-measurable function on Ω_2 thanks to 23.2. Its integration is therefore accomplished by (23.3) thus:

$$\int\left(\int f_{\omega_2}\,d\mu_1\right) = \sum_{j=1}^{n} \alpha_j \pi(Q^j) = \int f\,d\pi\,,$$

which confirms the first equation in (23.6), for elementary f.

For an arbitrary $\mathscr{A}$-measurable numerical function $f \geq 0$ let $(u^{(n)})$ be a sequence of $\mathscr{A}$-elementary functions such that $u^{(n)} \uparrow f$. Then, as was noted in the first part of the proof, $(u_{\omega_2}^{(n)})$ is a sequence of $\mathscr{A}_1$-elementary functions, which obviously satisfy $u_{\omega_2}^{(n)} \uparrow f_{\omega_2}$ (for each $\omega_2 \in \Omega_2$). Consequently, the functions

$$\varphi^{(n)}(\omega_2) := \int u_{\omega_2}^{(n)}\,d\mu_1, \qquad\qquad \omega_2 \in \Omega_2,$$

which are $\mathscr{A}_2$-measurable by what has already been proven, increase to the function

$$\omega_2 \mapsto \int f_{\omega_2}\,d\mu_1\,,$$

by 11.3. This function is therefore also $\mathscr{A}_2$-measurable and the monotone convergence theorem 11.4 says that

$$\int\left(\int f_{\omega_2}\,d\mu_1\right)\mu_2(d\omega_2) = \sup_{n\in\mathbb{N}} \int \varphi^{(n)}\,d\mu_2\,.$$

Again, by what has already been proved,

$$\int \varphi^{(n)}\,d\mu_2 = \int u^{(n)}\,d\pi \qquad\qquad \text{for each } n \in \mathbb{N}.$$

By the choice of the sequence $(u^{(n)})$ and definition 11.3

$$\int f\,d\pi = \sup_{n\in\mathbb{N}} \int u^{(n)}\,d\pi\,.$$

Combining the last three equations gives the desired

$$\int\left(\int f_{\omega_2}\,d\mu_1\right)\mu_2(d\omega_2) = \int f\,d\pi\,,$$

and wholly analogous arguments establish the claims about the functions f_{ω_1}. $\quad\square$

Having disposed of non-negative functions, the next step in integration theory is to pass over to integrable functions. For them we get

23.7 Corollary (Theorem of Fubini). *For $j = 1,2$ let $(\Omega_j, \mathscr{A}_j, \mu_j)$ be σ-finite measure spaces, f a $\mu_1 \otimes \mu_2$-integrable numerical function on $\Omega_1 \times \Omega_2$. Then for μ_1-almost every ω_1 the function f_{ω_1} is μ_2-integrable and for μ_2-almost every ω_2 the function f_{ω_2} is μ_1-integrable. The functions*

$$\omega_1 \mapsto \int f_{\omega_1}\,d\mu_2 \quad\text{and}\quad \omega_2 \mapsto \int f_{\omega_2}\,d\mu_1$$

thus defined μ_1-almost everywhere on Ω_1 and μ_2-almost everywhere on Ω_2, respectively, are μ_1-integrable and μ_2-integrable, respectively, and equations (23.6) are valid.

Proof. Evidently for all $\omega_j \in \Omega_j$ ($j = 1, 2$),

$$|f|_{\omega_j} = |f_{\omega_j}|, \qquad (f^+)_{\omega_j} = (f_{\omega_j})^+ \quad \text{and} \quad (f^-)_{\omega_j} = (f_{\omega_j})^-$$

so we will employ parenthesis-free notation. According to (23.6) the product measure $\pi := \mu_1 \otimes \mu_2$ satisfies

$$\int \left(\int |f_{\omega_1}| \, d\mu_2 \right) \mu_1(d\omega_1) = \int \left(\int |f_{\omega_2}| \, d\mu_1 \right) \mu_2(d\omega_2) = \int |f| \, d\pi < +\infty.$$

In particular, the $\mathscr{A}_1$-measurable numerical function $\omega_1 \mapsto \int |f_{\omega_1}| \, d\mu_2$ is μ_1-integrable and so by 13.6 it is μ_1-almost everywhere finite. That is (by 12.1), for μ_1-almost every ω_1 the section f_{ω_1} is μ_2-integrable. Consequently,

$$\omega_1 \mapsto \int f_{\omega_1} \, d\mu_2 = \int f_{\omega_1}^+ \, d\mu_2 - \int f_{\omega_1}^- \, d\mu_2$$

is a μ_1-almost everywhere defined function, which is $\mathscr{A}_1$-measurable because that is assured of each integral on the right by Theorem 23.6. In turn each of these integrals is μ_1-integrable by 23.6. So our μ_1-almost everywhere defined function $\omega_1 \mapsto \int f_{\omega_1} \, d\mu_2$ is μ_1-integrable and

$$\int \left(\int f_{\omega_1} \, d\mu_2 \right) \mu_1(d\omega_1) = \int \left(\int f_{\omega_1}^+ \, d\mu_2 \right) \mu_1(d\omega_1) - \int \left(\int f_{\omega_1}^- \, d\mu_2 \right) \mu_1(d\omega_1)$$

$$= \int f^+ \, d\pi - \int f^- \, d\pi = \int f \, d\pi.$$

Of course, the roles of ω_1 and ω_2 can be interchanged in this argument and we thereby secure the rest of what is being claimed. □

The theorems of Tonelli and Fubini insure, in particular, that under the stated hypotheses the *order of repeated integrations is immaterial*. We can emphasize this by writing the equation (23.6) in the form

(23.6′)
$$\int f \, d(\mu_1 \otimes \mu_2) = \int \int f(\omega_1, \omega_2) \mu_1(d\omega_1) \mu_2(d\omega_2)$$
$$= \int \int f(\omega_1, \omega_2) \mu_2(d\omega_2) \mu_1(d\omega_1).$$

That exceptional sets of measure zero cannot generally be ignored in the conclusions of Fubini's theorem is illustrated by the following example.

Example. 1. Consider L-B measure $\lambda^2 = \lambda^1 \otimes \lambda^1$ on $\mathbb{R}^2$, the set $A := \mathbb{Q} \times \mathbb{R} \in \mathscr{B}^1$, and its indicator function $f := 1_A$. According to 23.3 or 23.6 we have $\lambda^2(A) = \int f \, d\lambda^2 = 0$, so f is λ^2-integrable. Nevertheless, for every $\omega_1 \in \mathbb{Q}$, the section $f_{\omega_1} = 1_\mathbb{R}$ is not λ^1-integrable.

Remark. 1. For certain measures μ_1, μ_2 which are not σ-finite the existence but usually not the uniqueness of a product measure can be proved by other methods. See, e.g., BERBERIAN [1962]. Even if just one of μ_1 or μ_2 fails to be σ-finite, the second equality in (23.3) can fail. Cf. Exercise 1, p. 145 of HALMOS [1974], as well as chapter IV, §16 of HAHN and ROSENTHAL [1948]. Moreover, there exist $f : \Omega_1 \times \Omega_2 \to \mathbb{R}_+$ which are not $\mathscr{A}_1 \otimes \mathscr{A}_2$-measurable yet the "iterated integrals" on the right side of (23.6) make sense (and are finite). For an abundance of illuminating but elementary counterexamples related to this famous theorem, see CHATTERJI [1985–86] and MATTNER [1999].

A useful and at the same time surprising consequence of Tonelli's theorem is that it permits μ-integrals to be expressed by means of λ^1-integrals.

23.8 Theorem. *Let $(\Omega, \mathscr{A}, \mu)$ be a σ-finite measure space and $f : \Omega \to \mathbb{R}_+$ a measurable, non-negative, real function. Further, let $\varphi : \mathbb{R}_+ \to \mathbb{R}_+$ be a continuous isotone function which is continuously differentiable at least on $\mathbb{R}_+^* :=]0, +\infty[$ and satisfies $\varphi(0) = 0$. Then*

$$(23.7) \qquad \int \varphi \circ f \, d\mu = \int_{\mathbb{R}_+^*} \varphi'(t) \mu(\{f \geq t\}) \lambda^1(dt) = \int_0^{+\infty} \varphi'(t) \mu(\{f \geq t\}) \, dt \, .$$

Proof. Consider the L-B measure $\lambda^* := \lambda^1_{\mathbb{R}_+^*}$ on the σ-algebra $\mathscr{B}^* := \mathbb{R}_+^* \cap \mathscr{B}^1$. The function $F : \Omega \times \mathbb{R}_+^* \to \mathbb{R}^2$ defined by

$$F(\omega, t) := (f(\omega), t)$$

is, according to Remark 2 in §22, $\mathscr{A} \otimes \mathscr{B}^*$-measurable, because each of its component functions is. Therefore the F-preimage of the closed half-plane $\{(x, y) \in \mathbb{R}^2 : x \geq y\}$, namely

$$E := \{(\omega, t) \in \Omega \times \mathbb{R}_+^* : f(\omega) \geq t\} \, ,$$

lies in $\mathscr{A} \otimes \mathscr{B}^*$. Theorem 23.6 for the product measure $\mu \otimes \lambda^*$ consequently supplies the equalities

$$(23.8) \qquad \begin{aligned} \int \int \varphi'(t) 1_E(\omega, t) \lambda^*(dt) \mu(d\omega) &= \int \int \varphi'(t) 1_E(\omega, t) \mu(d\omega) \lambda^*(dt) \\ &= \int \varphi'(t) \mu(E_t) \lambda^*(dt) = \int \varphi'(t) \mu(\{f \geq t\}) \lambda^*(dt) \, , \end{aligned}$$

since the t-section of E is just the set of all $\omega \in \Omega$ which satisfy $f(\omega) \geq t$. As φ is isotone, $\varphi'(t) \geq 0$ for all $t > 0$. The continuous function φ' is integrable over $[1/n, a]$ whenever $1/n < a < +\infty$, and since $[1/n, a] \uparrow]0, a]$, and

$$\int_{]0,a]} \varphi'(t) \lambda^*(dt) = \lim_{n \to \infty} \int_{1/n}^a \varphi'(t) \, dt = \varphi(a) - \lim_{n \to \infty} \varphi(1/n) = \varphi(a)$$

$(\varphi(0) = 0$ and φ is continuous on $\mathbb{R}_+)$, we see that φ is also integrable over $]0, a]$ for every $a > 0$. It follows from $f \geq 0$ and the preceding calculation that

$$\int_{]0, f(\omega)]} \varphi'(t) \lambda^*(dt) = \varphi(f(\omega)) \qquad \text{for every } \omega \in \Omega,$$

both expressions being 0 whenever $f(\omega) = 0$. We thus get

$$\int \varphi \circ f \, d\mu = \int \left(\int_{]0, f(\omega)]} \varphi'(t) \lambda^*(dt) \right) \mu(d\omega)$$

$$= \int \int \varphi'(t) 1_{]0, f(\omega)]}(t) \lambda^*(dt) \mu(d\omega)$$

$$= \int \int \varphi'(t) 1_E(\omega, t) \lambda^*(dt) \mu(d\omega),$$

which combined with (23.8) concludes the proof. $\square$

Example. 2. The relevant hypotheses are certainly fulfilled by the functions $\varphi(t) := t^p$ with $p > 0$. Thus for every $\mathscr{A}$-measurable real function $f \geq 0$ on Ω

$$(23.9) \qquad \int f^p \, d\mu = p \cdot \int_0^{+\infty} t^{p-1} \mu(\{f \geq t\}) \, dt.$$

When $p = 1$ we get the especially important formula

$$(23.10) \qquad \int f \, d\mu = \int_{\mathbb{R}_+} \mu(\{f \geq t\}) \lambda^1(dt) = \int_0^{+\infty} \mu(\{f \geq t\}) \, dt.$$

The reader should not overlook the geometric significance of this, which is that the integral $\int f \, d\mu$ is formed "vertically", while the integral on the right-hand side of (23.10) is formed "horizontally".

Now at last we turn back to the general case of §22 and consider finitely many σ-*finite* measure spaces $(\Omega_j, \mathscr{A}_j, \mu_j)$, $j = 1, \ldots, n$ and $n \geq 2$.

The two product sets $(\Omega_1 \times \ldots \times \Omega_{n-1}) \times \Omega_n$ and $\Omega_1 \times \ldots \times \Omega_{n-1} \times \Omega_n$ will be identified via the bijection

$$((\omega_1, \ldots, \omega_{n-1}), \omega_n) \mapsto (\omega_1, \ldots, \omega_{n-1}, \omega_n).$$

The agreed-upon equality of these sets leads at once to the equality of the corresponding products of σ-algebras:

$$(23.11) \qquad (\mathscr{A}_1 \otimes \ldots \otimes \mathscr{A}_{n-1}) \otimes \mathscr{A}_n = \mathscr{A}_1 \otimes \ldots \otimes \mathscr{A}_{n-1} \otimes \mathscr{A}_n.$$

In fact, by 22.1 the sets $A_1 \times \ldots \times A_{n-1}$ with each $A_j \in \mathscr{A}_j$ generate $\mathscr{A}_1 \otimes \ldots \otimes \mathscr{A}_{n-1}$, and by the same theorem the sets

$$(A_1 \times \ldots \times A_{n-1}) \times A_n = A_1 \times \ldots \times A_{n-1} \times A_n$$

then generate $(\mathscr{A}_1 \otimes \ldots \otimes \mathscr{A}_{n-1}) \otimes \mathscr{A}_n$ as well as $\mathscr{A}_1 \otimes \ldots \otimes \mathscr{A}_{n-1} \otimes \mathscr{A}_n$.

In a completely analogous fashion one confirms a general *associativity* in the formation of products of σ-algebras:

$$(23.12) \qquad \left(\bigotimes_{j=1}^{m} \mathscr{A}_j\right) \otimes \left(\bigotimes_{j=m+1}^{n} \mathscr{A}_j\right) = \bigotimes_{j=1}^{n} \mathscr{A}_j \qquad (1 \leq m < n \in \mathbb{N}).$$

The convention (23.11) opens up the possibility of proving the existence of product measures on any finite number $n \geq 2$ of factors via induction on n.

23.9 Theorem. *σ-finite measures $\mu_1, \ldots, \mu_n$ on σ-algebras $\mathscr{A}_1, \ldots, \mathscr{A}_n$ uniquely determine a measure π on $\mathscr{A}_1 \otimes \ldots \otimes \mathscr{A}_n$ such that*

$$(23.13) \qquad \pi(A_1 \times \ldots \times A_n) = \mu_1(A_1) \cdot \ldots \cdot \mu_n(A_n) \quad \text{for all } A_j \in \mathscr{A}_j, \, 1 \leq j \leq n.$$

This measure π is σ-finite.

Corresponding to Definition 23.4, π is called the *product of the measures* $\mu_1, \ldots, \mu_n$ and is denoted by

$$\bigotimes_{j=1}^{n} \mu_j = \mu_1 \otimes \ldots \otimes \mu_n.$$

The question posed in §22 is finally answered in full, by this theorem.

Proof. In 22.2 take for the various generators $\mathscr{E}_j$ the σ-algebra $\mathscr{A}_j$ itself, and learn that there is at most one measure π which satisfies (23.13). The existence question has already been settled for $n = 2$, in 23.3. We make the inductive assumption that $\pi' := \mu_1 \otimes \ldots \otimes \mu_{n-1}$ exists for some $n > 2$ and show how that leads to the existence of $\mu_1 \otimes \ldots \otimes \mu_n$. Evidently the σ-finiteness of $\mu_1, \ldots, \mu_{n-1}$ entails that of π', as in the proof of Theorem 23.3. That theorem therefore supplies us with a measure $\pi := \pi' \otimes \mu_n$ on $(\mathscr{A}_1 \otimes \ldots \otimes \mathscr{A}_{n-1}) \otimes \mathscr{A}_n$ which satisfies

$$\pi(Q' \times A_n) = \pi'(Q')\mu_n(A_n)$$

for all $Q' \in \mathscr{A}_1 \otimes \ldots \otimes \mathscr{A}_{n-1}$ and all $A_n \in \mathscr{A}_n$. Because of (23.11) this measure does what is wanted at level n, completing the induction. Again, σ-finiteness of π is confirmed exactly as in the proof of 23.3. $\square$

This inductive construction of the n-fold product measure builds in the equality

$$(23.14) \qquad (\mu_1 \otimes \ldots \otimes \mu_{n-1}) \otimes \mu_n = \mu_1 \otimes \ldots \otimes \mu_{n-1} \otimes \mu_n.$$

By now familiar considerations show that in fact a general associativity prevails in the formation of product measures:

$$(23.15) \qquad \left(\bigotimes_{j=1}^{m} \mu_j\right) \otimes \left(\bigotimes_{j=m+1}^{n} \mu_j\right) = \bigotimes_{j=1}^{n} \mu_j \qquad (1 \leq m < n \in \mathbb{N}).$$

In particular

$$\lambda^d = \lambda^1 \otimes \ldots \otimes \lambda^1, \qquad\qquad \text{with } d \text{ factors.}$$

In view of (23.15) induction can also be used to extend the theorems of Tonelli and Fubini to multiple factors. We will formulate only the analog of 23.6:

Let $f \geq 0$ be an $\mathscr{A}_1 \otimes \ldots \otimes \mathscr{A}_n$-measurable numerical function on $\Omega_1 \times \ldots \times \Omega_n$. Then for every permutation $j_1, \ldots, j_n$ of $1, \ldots, n$

$$(23.16) \qquad \int f \, d(\mu_1 \otimes \ldots \otimes \mu_n)$$

$$= \int \left(\ldots \left(\int \left(\int f(\omega_1, \ldots, \omega_n) \mu_{j_1}(d\omega_{j_1}) \right) \mu_{j_2}(d\omega_{j_2}) \right) \ldots \right) \mu_{j_n}(d\omega_{j_n}) .$$

Every integral that occurs on the right-hand side is measurable with respect to the product of the appropriate $\mathscr{A}_j$, namely those corresponding to the coordinates in which integration has not yet occurred. This right-hand side is often written in the shorter fashion

$$\int \ldots \int f(\omega_1, \ldots, \omega_n) \mu_{j_1}(d\omega_{j_1}) \ldots \mu_{j_n}(d\omega_{j_n}) .$$

The simple proof of this theorem (involving induction), as well as the formulation and proof of the analog of 23.7, will be left to the reader.

One more piece of notation is convenient:

23.10 Definition. For finitely many σ-finite measure spaces $(\Omega_j, \mathscr{A}_j, \mu_j)$, $1 \leq j \leq n$, the triple $\left(\underset{j=1}{\overset{n}{\times}} \Omega_j, \underset{j=1}{\overset{n}{\bigotimes}} \mathscr{A}_j, \underset{j=1}{\overset{n}{\bigotimes}} \mu_j \right)$ is called the *product of these measure spaces* and is denoted by

$$\bigotimes_{j=1}^{n} (\Omega_j, \mathscr{A}_j, \mu_j) .$$

Remark. 2. Throughout the preceding the index set was finite. But there is also a theory of products of (finite) measures indexed by arbitrary sets, which is particularly important in probability theory; it is treated in detail by BAUER [1996], and somewhat more extensively in HEWITT and STROMBERG [1965]. For p-measures SAEKI [1996] gives a short, elementary proof that uses only 5.1.

In closing we will consider the case where each measure μ_j comes with a *real density* $f_j \geq 0$. According to Theorem 17.11, $\nu_j := f_j \mu_j$ is then a σ-finite measure too.

23.11 Theorem. *Let* $(\Omega_j, \mathscr{A}_j, \mu_j)$ *be* σ*-finite measure spaces and* $f_j \geq 0$ *real-valued* $\mathscr{A}_j$*-measurable functions on* Ω_j. *Set*

$$\nu_j := f_j \mu_j, \qquad\qquad j = 1, \ldots, n.$$

Then the product of these measures is defined and satisfies

$$(23.17) \qquad\qquad \bigotimes_{j=1}^{n} \nu_j = F \cdot \left(\bigotimes_{j=1}^{n} \mu_j \right)$$

with the density function

$$(23.18) \qquad F(\omega_1, \ldots, \omega_n) := \prod_{j=1}^{n} f_j(\omega_j),$$

The function F is the so-called *tensor product* of the densities $f_1, \ldots, f_n$.

Proof. As already noted, 17.11 insures that each measure ν_j is σ-finite, guaranteeing that their product is defined. It suffices to treat the case $n = 2$ and refer the general case to induction. For sets $A_1 \in \mathscr{A}_1$ and $A_2 \in \mathscr{A}_2$

$$\nu_1(A_1)\nu_2(A_2) = \left(\int_{A_1} f_1 \, d\mu_1 \right) \left(\int_{A_2} f_2 \, d\mu_2 \right)$$

$$= \int \int 1_{A_1}(\omega_1) f_1(\omega_1) 1_{A_2}(\omega_2) f_2(\omega_2) \mu_1(d\omega_1) \mu_2(d\omega_2)$$

$$= \int \int 1_{A_1 \times A_2}(\omega_1, \omega_2) F(\omega_1, \omega_2) \mu_1(d\omega_1) \mu_2(d\omega_2).$$

From 23.6 therefore

$$\nu_1(A_1)\nu_2(A_2) = \int_{A_1 \times A_2} F \, d(\mu_1 \otimes \mu_2), \qquad \text{for all } A_1 \in \mathscr{A}_1, A_2 \in \mathscr{A}_2.$$

But then according to 23.3, $\nu_1 \otimes \nu_2$ coincides with the measure $F \cdot (\mu_1 \otimes \mu_2)$. $\square$

Exercises.

1. Consider $\Omega_1 = \Omega_2 := \mathbb{R}$, $\mathscr{A}_1 = \mathscr{A}_2 := \mathscr{B}^1$, $\mu_1 := \lambda^1$ and $\mu_2 := \zeta$, the non-σ-finite counting measure on $\mathscr{B}^1$ (cf. Example 3, §5). Show that equality (23.3) fails to hold for $Q := D$, the diagonal $\{(\omega, \omega) : \omega \in \mathbb{R}\}$ in $\Omega_1 \times \Omega_2$. Why does D lie in $\mathscr{A}_1 \otimes \mathscr{A}_2 = \mathscr{B}^2$?

2. Show that the function

$$(x, y) \mapsto 2e^{2xy} - e^{xy}$$

is not λ^2-integrable over the set $[1, +\infty[\times [0, 1]$.

3. With the aid of Tonelli's theorem find a new proof of Theorem 8.1 along the following lines: If μ is a translation-invariant measure on $\mathscr{B}^d$, $\mu([0, 1[) = 1$, and $f \geq 0$, $g \geq 0$ are Borel measurable numerical functions on $\mathbb{R}^d$, compare the integrals

$$\int \int g(y) f(x + y) \mu(dx) \lambda^d(dy) \quad \text{and} \quad \int \int g(y - x) f(y) \mu(dx) \lambda^d(dy)$$

and, finally, take f to be any indicator function, g the indicator function of $[0, 1[$.

4. Compute

$$I := \int_0^{\infty} e^{-x^2} \, dx,$$

and thereby evaluate anew the important integral $G = 2I$ in (16.1), in the following simple way: $\int_0^{\infty} e^{-t^2} \, dt = \int_0^{\infty} y e^{-x^2 y^2} \, dx$ for every $y > 0$ and therefore

$I^2 = \int_0^\infty (\int_0^\infty f(x,y)\,dx)\,dy$ for the function f on $\mathbb{R}_+ \times \mathbb{R}_+$ defined by $f(x,y) :=$ $ye^{-y^2(1+x^2)}$. Applying Tonelli's theorem leads to $I = \frac{1}{2}\sqrt{\pi}$.

5. Let $|x| := (x_1^2 + \ldots + x_d^2)^{1/2}$ denote the usual euclidean norm of the vector $x := (x_1,\ldots,x_d) \in \mathbb{R}^d$. Show that the function $x \mapsto e^{-|x|^\alpha}$ is λ^d-integrable for every $\alpha > 0$. (Recall Exercise 2 of §16.) In case $\alpha = 2$, show that the λ^d-integral of this function is G^d.

6. $\overline{K}_r(x_0)$ will denote the closed ball in $\mathbb{R}^d$ with center x_0 and radius $r \geq 0$. Set $\alpha_d := \lambda^d(\overline{K}_1(0))$ and prove that

$$\lambda^d(\overline{K}_r(x_0)) = \alpha_d r^d \,.$$

Show also that the numbers α_d can be calculated by

$$\alpha_{2q} = \frac{1}{q!}\pi^q, \quad \text{and} \quad \alpha_{2q-1} = \frac{2^q}{1\cdot 3 \cdot \ldots \cdot (2q-1)}\pi^{q-1} \qquad (q \in \mathbb{N}).$$

[Hint: Use (7.10) and note that every x_d-section of $\overline{K}_r(0)$ is either empty or is a $(d-1)$-dimensional closed ball. Tonelli's theorem then leads to a recursion formula for the α_d. Here, of course, π has its customary geometric meaning.]

How do these relations change if we replace $\overline{K}_r(x_0)$ by the open ball $K_r(x_0)$ in $\mathbb{R}^d$ of radius r and center x_0? [Cf. Exercise 3 in §7.]

7. For every compact interval $[\alpha, \beta] \subset \mathbb{R}_+$ designate by $R(\alpha, \beta)$ the spherical shell

$$\overline{K}_\beta(0) \setminus K_\alpha(0) = \{x \in \mathbb{R}^d : \alpha \leq |x| \leq \beta\}\,.$$

Show that for every continuous real function h on such an interval $[\alpha, \beta] \subset \mathbb{R}_+$

$$\int_{R(\alpha,\beta)} h(|x|)\lambda^d(dx) = d \cdot \alpha_d \int_\alpha^\beta h(t)t^{d-1}\,dt\,,$$

α_d being the number $\lambda^d(\overline{K}_1(0))$ from the preceding exercise. [Hint: The function H defined on $[\alpha, \beta]$ by

$$H(t) := \int_{R(\alpha,t)} h(|x|)\lambda^d(dx), \qquad\qquad t \in [\alpha, \beta],$$

is differentiable with $H'(t) = d \cdot \alpha_d \cdot h(t) \cdot t^{d-1}$ for all such t.]

8. Apply the result of Exercise 7 to the case $d = 2$ and $h(t) := e^{-t^2}$ in order to show, using Exercise 5, once again that $G = \sqrt{\pi}$.

9. Let $(\Omega, \mathscr{A}, \mu)$ be a σ-finite measure space, $f : \Omega \to \mathbb{R}_+$ measurable. Show that the set of all $t \geq 0$ such that $\mu(\{f = t\}) \neq 0$, as well as the set of all $t \geq 0$ such that $\mu(\{f \geq t\}) \neq \mu(\{f > t\})$ is countable. Therefore in the equalities (23.8), (23.9) and (23.10), $\mu(\{f \geq t\})$ can always be replaced by $\mu(\{f > t\})$.

§24. Convolution of finite Borel measures

Consider the d-dimensional Borel measurable space $(\mathbb{R}^d, \mathcal{B}^d)$. Every finite measure μ on $\mathcal{B}^d$ will be called a *finite* or also a *bounded Borel measure*, and the set of all of them will be designated by $\mathcal{M}^b_+(\mathbb{R}^d)$. For every such μ the number

$$(24.1) \qquad \qquad \|\mu\| := \mu(\mathbb{R}^d)$$

is called the *total mass* of μ.

Making critical use of the group structure of $(\mathbb{R}^d, +)$ a so-called convolution product can be assigned to any finitely many measures $\mu_1, \ldots, \mu_n \in \mathcal{M}^b_+(\mathbb{R}^d)$; in contrast to the previously studied product measure, it is again a measure on the original σ-algebra $\mathcal{B}^d$, even an element of $\mathcal{M}^b_+(\mathbb{R}^d)$. What we do below can be carried out in every (abelian) locally compact group. We cannot, however, go into this generalization, but must instead refer interested readers to the excellent monographs of HEWITT and ROSS [1979] and RUDIN [1962]. Initially we consider the product measure $\mu_1 \otimes \ldots \otimes \mu_n$ defined in §23. Since $\mathcal{B}^{nd} = \mathcal{B}^d \otimes \ldots \otimes \mathcal{B}^d$, this measure is an element of $\mathcal{M}^b_+(\mathbb{R}^{nd})$. The mapping $A_n : \mathbb{R}^{nd} \to \mathbb{R}^d$ defined by

$$A_n(x_1, \ldots, x_n) := x_1 + \ldots + x_n$$

is continuous, and so $\mathcal{B}^{nd}$-$\mathcal{B}^d$-measurable. The following definition accordingly makes sense:

24.1 Definition. The image under the mapping A_n of the product measure $\mu_1 \otimes \ldots \otimes \mu_n$ is called the *convolution product* of the measures $\mu_1, \ldots, \mu_n \in \mathcal{M}^b_+(\mathbb{R}^d)$, in symbols

$$(24.2) \qquad \qquad \mu_1 * \ldots * \mu_n := A_n(\mu_1 \otimes \ldots \otimes \mu_n).$$

The theorems on product and image measures combine to yield the most important properties of the *convolution* operation $*$. First of all, $\mu_1 * \ldots * \mu_n$ is again an element of $\mathcal{M}^b_+(\mathbb{R}^d)$ and

$$\mu_1 * \ldots * \mu_n(\mathbb{R}^d) = \mu_1 \otimes \ldots \otimes \mu_n(\mathbb{R}^{nd}) = \|\mu_1\| \cdot \ldots \cdot \|\mu_n\| \,,$$

so that in fact

$$(24.3) \qquad \qquad \|\mu_1 * \ldots * \mu_n\| = \|\mu_1\| \cdot \ldots \cdot \|\mu_n\| \,.$$

In studying the convolution product it suffices to deal with $n = 2$, because

$$(24.4) \qquad \qquad \mu_1 * \ldots * \mu_n * \mu_{n+1} = (\mu_1 * \ldots * \mu_n) * \mu_{n+1}$$

for every $n + 1$ measures from $\mathcal{M}^b_+(\mathbb{R}^d)$. To see this, introduce the continuous mapping $B_{n+1} : \mathbb{R}^{(n+1)d} \to \mathbb{R}^{2d}$ by

$$B_{n+1}(x_1, \ldots, x_n, x_{n+1}) := (x_1 + \ldots + x_n, x_{n+1})$$

and have $A_{n+1} = A_2 \circ B_{n+1}$. Checking that

$$B_{n+1}(\mu_1 \otimes \ldots \otimes \mu_n \otimes \mu_{n+1}) = A_n(\mu_1 \otimes \ldots \otimes \mu_n) \otimes \mu_{n+1},$$

and remembering that the formation of image measures is transitive, we get

$$\mu_1 * \ldots * \mu_n * \mu_{n+1} = A_2(B_{n+1}(\mu_1 \otimes \ldots \otimes \mu_n \otimes \mu_{n+1}))$$
$$= A_2((\mu_1 * \ldots * \mu_n) \otimes \mu_{n+1}),$$

which confirms (24.4). Henceforth therefore $n = 2$.

For any measures $\mu, \nu \in \mathscr{M}_+^b(\mathbb{R}^d)$ and any $\mathscr{B}^d$-measurable numerical function $f \geq 0$ it follows from Theorems 19.1 and 23.6 that

$$\int f \, d(\mu * \nu) = \int f \circ A_2 \, d(\mu \otimes \nu)$$

(24.5)
$$= \int \int f(x+y)\mu(dx)\nu(dy)$$

$$= \int \int f(x+y)\nu(dy)\mu(dx).$$

As this holds for $f := 1_B$, the indicator function of any set $B \in \mathscr{B}^d$, we have

(24.6)
$$\mu * \nu(B) = \int \mu(B-y)\nu(dy) = \int \nu(B-x)\mu(dx)$$

(Recall (7.8) that $B-x = -x+B$.) Consequently $*$ is a *commutative*, and by (24.4) also an *associative* operation in $\mathscr{M}_+^b(\mathbb{R}^d)$.

Due to 19.2 and 23.7, (24.5) are valid as well for every $\mu*\nu$-integrable numerical function f on $\mathbb{R}^d$. Equality (24.6) is frequently taken as the definition of $\mu * \nu$.

Evidently $\mathscr{M}_+^b(\mathbb{R}^d)$ is closed with respect to addition and under multiplication by numbers in $\mathbb{R}_+$. From (24.6) we immediately see the relation of convolution to these two operations: For all $\mu, \nu, \nu_1, \nu_2 \in \mathscr{M}_+^b(\mathbb{R}^d)$, $\alpha \in \mathbb{R}_+$

(24.7)
$$\mu * (\nu_1 + \nu_2) = \mu * \nu_1 + \mu * \nu_2,$$

(24.8)
$$\mu * (\alpha\nu) = (\alpha\mu) * \nu = \alpha(\mu * \nu).$$

The *distributive law* (24.7) even holds in the following generality: For every sequence (ν_n) of measures from $\mathscr{M}_+^b(\mathbb{R}^d)$ satisfying $\sum\limits_{n=1}^{\infty} \|\nu_n\| < +\infty$, the sum $\sum\limits_{n=1}^{\infty} \nu_n$ is also a measure in $\mathscr{M}_+^b(\mathbb{R}^d)$ (cf. Example 4 of §3). Taking account of 11.5, it therefore follows from (24.6) that

(24.9)
$$\mu * \left(\sum_{n=1}^{\infty} \nu_n\right) = \sum_{n=1}^{\infty} \mu * \nu_n$$

for every $\mu \in \mathscr{M}_+^b(\mathbb{R}^d)$.

Let us now compute $\mu * \nu$ in some special cases.

1. We again denote by T_a the translation mapping $x \mapsto x + a$ of $\mathbb{R}^d$ onto itself via $a \in \mathbb{R}^d$, and by ε_a the (Dirac-)measure on $\mathscr{B}^d$ defined by unit mass at the point a. Of course, $\varepsilon_a \in \mathscr{M}_+^b(\mathbb{R}^d)$ and $\|\varepsilon_a\| = 1$. From (24.6) follows that $\varepsilon_a * \mu(B) = \mu(B - a) = \mu(T_a^{-1}(B))$ for all $B \in \mathscr{B}^d$, and so

$$(24.10) \qquad\qquad \varepsilon_a * \mu = T_a(\mu) \qquad\qquad \text{for all } \mu \in \mathscr{M}_+^b(\mathbb{R}^d),\, a \in \mathbb{R}^d.$$

Now T_0 is the identity mapping, so ε_0 is a – and obviously the only – *unit with respect to convolution*. If, namely, ε were also a unit, meaning that $\mu = \varepsilon * \mu$ for every $\mu \in \mathscr{M}_+^b(\mathbb{R}^d)$, then it would follow that $\varepsilon_0 = \varepsilon * \varepsilon_0 = \varepsilon$.

For the special choice $\mu := \varepsilon_b$, (24.10) says that

$$(24.10') \qquad\qquad \varepsilon_a * \varepsilon_b = \varepsilon_{a+b} \qquad\qquad \text{for all } a, b \in \mathbb{R}^d.$$

2. Let $f \geq 0$ be a λ^d-integrable numerical function on $\mathbb{R}^d$ and $\mu := f\lambda^d$. Since $\|\mu\| = \int f\, d\lambda^d < +\infty$, μ also lies in $\mathscr{M}_+^b(\mathbb{R}^d)$. Let us compute $\mu * \nu$ for an arbitrary $\nu \in \mathscr{M}_+^b(\mathbb{R}^d)$. From 17.3 using the translation-invariance of λ^d and the general transformation theorem 19.1, we get

$$\mu * \nu(B) = \int\int 1_B(x + y)f(x)\lambda^d(dx)\nu(dy)$$

$$= \int\int 1_B(x + y)f(x)T_{-y}(\lambda^d)(dx)\nu(dy)$$

$$= \int\int 1_B(x)f(x - y)\lambda^d(dx)\nu(dy)$$

for every $B \in \mathscr{B}^d$. With the help of Tonelli's theorem it further follows that

$$\mu * \nu(B) = \int 1_B(x)q(x)\lambda^d(dx) = \int_B q\, d\lambda^d\,,$$

where q is the non-negative $\mathscr{B}^d$-measurable function $x \mapsto \int f(x - y)\nu(dy)$. This function is also λ^d-integrable, since $\int q\, d\lambda^d = \|\mu * \nu\| < +\infty$. Thus whenever μ has a density with respect to λ^d, so does $\mu * \nu$. We set $f * \nu := q$, that is, we make the definition

$$(24.11) \qquad\qquad f * \nu(x) := \int f(x - y)\nu(dy) \qquad\qquad \text{for } x \in \mathbb{R}^d.$$

The preceding result now assumes the more suggestive form

$$(24.12) \qquad\qquad (f\lambda^d) * \nu = (f * \nu)\lambda^d\,.$$

Naturally $f * \nu$ is called the *convolution* of f and ν.

3. Besides $\mu = f\lambda^d$, let now $\nu = g\lambda^d$ also have a λ^d-integrable density $g \geq 0$. According to 17.3 and the preceding

$$f * (g\lambda^d)(x) = \int f(x - y)g(y)\lambda^d(dy) \qquad\qquad (x \in \mathbb{R}^d)$$

is a density for $\mu * \nu$ with respect to λ^d. We denote this function by $f * g$, that is, we set

(24.13)
$$f * g(x) := \int f(x - y)g(y)\lambda^d(dy) \qquad (x \in \mathbb{R}^d)$$

and get

(24.14)
$$(f\lambda^d) * (g\lambda^d) = (f * g)\lambda^d.$$

Here too $f * g$ is called the *convolution* of f and g. It is defined for every pair of non-negative λ^d-integrable functions and is itself such a function. Nevertheless, it might not be real-valued, even if f and g each are (cf. Remark 1 below). From (24.13) and the translation- and reflection-invariance of λ^d it follows that for every $x \in \mathbb{R}^d$

$$f * g(x) = \int f(x - y)g(y)\lambda^d(dy) = \int f(x + y)g(-y)\lambda^d(dy)$$
$$= \int f(y)g(x - y)\lambda^d(dy) = g * f(x).$$

That is, the $*$ operation between functions is also *commutative*:

(24.15)
$$f * g = g * f.$$

Similar calculations confirm its *associativity*; that is,

(24.16)
$$(f * g) * h = f * (g * h)$$

for all λ^d-integrable, non-negative functions f, g, h.
The *distributive law*

(24.17)
$$f * (g + h) = f * g + f * h$$

and the *homogeneity* property

(24.18)
$$f * (\alpha g) = (\alpha f) * g = \alpha(f * g) \qquad (\alpha \in \mathbb{R}_+)$$

for such functions hold as well and follow immediately from (24.13).

4. For arbitrary functions $f, g \in \mathcal{L}^1(\lambda^d)$ decomposition into their positive and negative parts and appeal to the results secured in 3. show that

$$x \mapsto \int f(x - y)g(y)\lambda^d(dy),$$

while possibly defined only λ^d-almost everywhere (see Remark 1 below), is always λ^d-integrable. One can therefore define $f * g$ by

$$f * g(x) := \int f(x - y)g(y)\lambda^d(dy)$$

but generally only for λ^d-almost all $x \in \mathbb{R}^d$. Once again the expression *convolution* is used for this $f * g$.

Remarks. 1. For real-valued, non-negative functions $f, g \in \mathscr{L}^1(\lambda^d)$ the function $f * g$ need not be finite everywhere. It suffices to consider any real-valued, non-negative, even function f which lies in $\mathscr{L}^1(\lambda^d)$ but not in $\mathscr{L}^2(\lambda^d)$ and to take $g = f$. Then $f * g(0) = +\infty$. In case $d = 1$, such a function is

$$f(x) := \begin{cases} 0 & \text{for } |x| > 1 \text{ or } x = 0 \\ |x|^{-1/2} & \text{for } 0 < |x| \leq 1. \end{cases}$$

2. In passing to $L^1(\lambda^d)$ – cf. Remark 1 in §15 – the difficulties high-lighted above with the definition of $f * g$ disappear. Indeed, let $f \mapsto \tilde{f}$ be the canonical mapping of $\mathscr{L}^1(\lambda^d)$ onto $L^1(\lambda^d)$. One defines $\tilde{f} * \tilde{g}$ for arbitrary $\tilde{f}, \tilde{g} \in L^1(\lambda^d)$ as the image $\tilde{h}$ of a function $h \in \mathscr{L}^1(\lambda^d)$ which coincides λ^d-almost everywhere with $f * g$. This definition is independent of the special choice of representing functions f, g and h from $\mathscr{L}^1(\lambda^d)$. The new operation $*$ renders the vector space $L^1(\lambda^d)$ an *algebra* over $\mathbb{R}$.

Exercises.

1. Show that for any $\mu, \nu \in \mathscr{M}_+^b(\mathbb{R}^d)$ and any linear mapping $T : \mathbb{R}^d \to \mathbb{R}^d$, $T(\mu * \nu) = T(\mu) * T(\nu)$. To this end, first observe that $T \circ A_2 = A_2 \circ (T \otimes T)$, where $T \otimes T$ denotes the mapping $(x, y) \mapsto (T(x), T(y))$ of $\mathbb{R}^d \times \mathbb{R}^d$ into itself.

2. Compute the nth convolution power of the function f defined on $\mathbb{R}$ by $f(x) := e^{-x^2}$, that is, the convolution $f * \ldots * f$ with $n(\in \mathbb{N})$ factors. Is it true that for every $n \in \mathbb{N}$, f has an "n^{th} convolution root"? That is, is f the nth convolution power of some λ^1-integrable function $g \geq 0$?

3. If we set $N_1(f) := \int |f| \, d\lambda^d$ (this is (14.1) for $\mu := \lambda^d$), then

$$N_1(f * g) \leq N_1(f) N_1(g)$$

holds for all $f, g \in \mathscr{L}^1(\lambda^d)$, and for non-negative functions equality prevails.

4. Write out the details of Remark 2 and show that

$$\|\tilde{f} * \tilde{g}\|_1 \leq \|\tilde{f}\|_1 \cdot \|\tilde{g}\|_1$$

holds for all elements $\tilde{f}$ and $\tilde{g}$ of the Banach space $L^1(\lambda^d)$. The latter is therefore a *Banach algebra*.

Chapter IV

Measures on Topological Spaces

In view of many applications in analysis, geometry and probability theory it turns out to be unavoidable to subject the Borel measures on $\mathbb{R}^d$ to more precise analysis. These measures possess a host of remarkable properties involving the topology of $\mathbb{R}^d$. Up to this point topology has only entered the picture in the generation of the Borel sets. We will see that the completeness of the euclidean metric, respectively, the local compactness of $\mathbb{R}^d$ were responsible for the aforementioned properties. But there are more general topological spaces, important in their own right, which share these properties with $\mathbb{R}^d$.

Therefore from the start we will set our exposition in an essentially more general framework: Instead of Borel measures on $\mathbb{R}^d$ we will study Radon measures on Polish and locally compact spaces. In the process new facts, even in the $\mathbb{R}^d$ environment, about the nature of the integral and the measurability concept will emerge. A natural and useful convergence concept will play a role.

In what follows some simple things from general (point-set) topology will be pre-supposed. The textbooks of KELLEY [1955] and WILLARD [1970] are good sources for these, and explicit references to them will be given at the appropriate points in the text.

§25. Borel sets, Borel and Radon measures

Initially E will be an arbitrary *topological space*. The *system of its open subsets* which defines the topology will be denoted $\mathscr{O}$. In the case of $\mathbb{R}^d$ we had determined (cf. 6.4) that the σ-algebra of Borel sets is generated by the open sets. Consonant with this we now make the general

25.1 Definition. The σ-algebra in E generated by $\mathscr{O}$ is denoted by $\mathscr{B}(E)$ and called the *Borel σ-algebra* in E:

(25.1) $$\mathscr{B}(E) := \boldsymbol{\sigma}(\mathscr{O}).$$

The closed sets being the complements of the open ones, $\mathscr{B}(E)$ is also generated by the system of all closed subsets of E. In this respect the analogy with 6.4 extends a bit farther. The intersection of a sequence of open sets is called a G_δ-*set*, and the dual, the union of a sequence of closed sets is called an F_σ-*set*. All such sets are clearly Borel.

From now on E will be a Hausdorff space. Then every compact subset of E is closed, hence Borel. The second example below will show, however, that generally the σ-algebra generated by the class $\mathscr{K}$ of all compact subsets of E is strictly smaller that $\mathscr{B}(E)$. So at this point the analogy with 6.4 falters.

Examples. 1. From 6.4, as has already been mentioned,

$$(25.2) \qquad \mathscr{B}(\mathbb{R}^d) = \mathscr{B}^d ,$$

$E := \mathbb{R}^d$ here carrying its euclidean topology.

2. Let E be a discrete space, meaning that $\mathscr{O} = \mathscr{P}(E)$. Then the system $\mathscr{K}$ of compact sets consists just of the finite subsets of E. Consequently (cf. Examples 2 and 7 in §1) $\sigma(\mathscr{K})$ is the countable and co-countable σ-algebra, comprised of all countable subsets of E and their complements, and so $\sigma(\mathscr{K}) = \mathscr{B}(E)$ if and only if E is countable.

3. Let Q be a subspace of the Hausdorff space E. Then $\mathscr{B}(Q)$ is the trace of $\mathscr{B}(E)$ on Q:

$$\mathscr{B}(Q) = Q \cap \mathscr{B}(E).$$

In fact, by definition the subspace topology of Q consists of the sets $\{Q \cap G : G \in \mathscr{O}\}$, so this system generates $\mathscr{B}(Q)$. Since $Q \cap \mathscr{B}(E)$ is a σ-algebra in Q which contains this system, it follows that $\mathscr{B}(Q) \subset Q \cap \mathscr{B}(E)$. On the other hand, the system $\{A \subset E : Q \cap A \in \mathscr{B}(Q)\}$ is obviously a σ-algebra in E which contains all the open subsets of E, a generating system for $\mathscr{B}(Q)$. Hence $Q \cap \mathscr{B}(E) \subset \mathscr{B}(Q)$.

If Q itself is a Borel set in E, then $\mathscr{B}(Q)$ just consists of all the Borel sets in E which are subsets of Q.

4. The compactified number line $\overline{\mathbb{R}}$ is a topological space which is homeomorphic to the compact interval $[-1, +1]$. For it

$$(25.3) \qquad \mathscr{B}(\overline{\mathbb{R}}) = \overline{\mathscr{B}}^1 .$$

In fact, $\mathbb{R} \cap \mathscr{B}(\overline{\mathbb{R}}) = \mathscr{B}(\mathbb{R}) = \mathscr{B}^1$ by Examples 1 and 3 above. The subsets $\{-\infty\}$ and $\{+\infty\}$ are closed in $\overline{\mathbb{R}}$ and the subset $\mathbb{R}$ is open in $\overline{\mathbb{R}}$, hence all three are Borel sets in $\overline{\mathbb{R}}$. Equality (25.3) therefore follows from the definition of $\overline{\mathscr{B}}^1$, given in §9.

In the sequel we will be studying measures on $\mathscr{B}(E)$ for two important classes of spaces E. In preparation for which we make

25.2 Definition. Let E be a Hausdorff space. A measure μ on the σ-algebra $\mathscr{B}(E)$ is called:
(i)   a *Borel measure* on E if

$$\mu(K) < +\infty \qquad\qquad \text{for every compact } K \subset E;$$

(ii) *locally finite* if every point of E has an open neighborhood of finite μ-measure;
(iii) *inner regular* if for every $B \in \mathscr{B}(E)$

$$(25.4) \qquad \mu(B) = \sup\{\mu(K) : K \text{ compact } \subset B\};$$

(iv) *outer regular* if for every $B \in \mathscr{B}(E)$

(25.5) $$\mu(B) = \inf\{\mu(U) : B \subset U \text{ open}\} \, ;$$

(v) *regular* if it is both inner regular and outer regular.

Note that a Borel measure is more than just a measure defined on $\mathscr{B}(E)$: in addition finiteness on the system of compact sets is demanded. The inner and outer regularity conditions say that the measure is determined on every Borel set by its values on the compact, resp., the open sets. The Borel measures on $E = \mathbb{R}^d$ are already familiar to us from §6.

Every finite measure on $\mathscr{B}(E)$ is obviously a Borel measure; as in §24 where $E = \mathbb{R}^d$, we naturally call it a *finite Borel measure* on E. The notation introduced there for the total mass of a finite Borel measure will be carried over to this more general setting: For every finite Borel measure μ on a Hausdorff space E

(25.6) $$\|\mu\| := \mu(E)$$

is called the *total mass* of μ.

Already at this point we can observe that every locally finite measure μ on $\mathscr{B}(E)$ is a Borel measure, that is, that

(25.7) $$\text{(ii)} \; \Rightarrow \; \text{(i)} \, .$$

Indeed, each point x in the compact set K has an open neighborhood V_x with $\mu(V_x) < +\infty$, and compactness means that finitely many of these, say those corresponding to $x_1, \ldots, x_n$, cover K. Then

$$\mu(K) \le \mu(V_{x_1} \cup \ldots \cup V_{x_n}) \le \sum_{j=1}^{n} \mu(V_{x_j}) < +\infty \, .$$

The converse of (25.7) is, however, not generally valid. Exercise 2 below furnishes an example.

Because of the implication (25.7), instead of locally finite measures defined on $\mathscr{B}(E)$, we will henceforth say simply *locally finite Borel measures*.

For the moment we will be content to illustrate the regularity concept with some examples.

Examples. 5. Let E be an arbitrary Hausdorff space, a a point in E. The measure ε_a on $\mathscr{B}(E)$ defined by unit mass at a:

(25.8) $$\varepsilon_a(A) = 1_A(a) \qquad \text{for } A \in \mathscr{B}(E)$$

is both inner and outer regular on E. Henceforth it will be called the *Dirac measure* on E at a.

6.   As in Example 2, let E be a discrete space, so that $\mathscr{O} = \mathscr{P}(E)$. The compact sets are just the finite ones. The measure defined on $\mathscr{P}(E)$ by

$$\mu(A) := \begin{cases} 0 & \text{if } A \text{ is countable} \\ +\infty & \text{otherwise} \end{cases}$$

is a locally finite Borel measure which is obviously outer regular. It is, however, inner regular if and only if the set E is countable.

7. On $\mathscr{B}^1 = \mathscr{B}(\mathbb{R})$ consider the counting measure. It is not a Borel measure, is however inner regular, but not outer regular. In fact, equality (25.5) fails even for one-point sets B.

8. L-B measure λ^d on $\mathscr{B}^d = \mathscr{B}(\mathbb{R}^d)$ is a (locally finite) Borel measure. In §26 we will see that it – and indeed every Borel measure on $\mathbb{R}^d$ – is regular.

Developments stretching over decades attest to the fact that on a Hausdorff space those Borel measures which are locally finite and inner regular play a distinguished role. Such measures are nowadays named after J. RADON (1887–1956). A work of his from the year 1913 (cf. the bibliography), which has since become classical, set this development in motion.

25.3 Definition. A measure defined on the Borel σ-algebra $\mathscr{B}(E)$ of a Hausdorff space E is called a *Radon measure* on E if it is both locally finite and inner regular.

More precisely the term used is "positive" Radon measure, but in this book we dispense with that adjective because non-negativity is built into our definition of measure, that is, we consider only measures with values in $[0, +\infty]$. Example 5 says that the Dirac measure at any point $a \in E$ is always a Radon measure on E.

We have already noted that Borel measures are not automatically locally finite. Nevertheless for many spaces Radon measures can be defined simply as the inner regular Borel measures. That is the import of

25.4 Lemma. *On a Hausdorff space E in which every point has a countable neighborhood basis, every inner regular Borel measure μ on E is also locally finite and hence a Radon measure.*

Proof. We argue by contradiction: Suppose that μ is not locally finite, which means there is a point $x \in E$ such that $\mu(V) = +\infty$ for every open neighborhood V of x. By hypothesis x has a neighborhood basis consisting of a sequence (V_n) of open sets, and by replacing each V_n with $V_1 \cap \ldots \cap V_n$, we may suppose that $V_n \downarrow \{x\}$. Since $\mu(V_n) = +\infty$ and μ is inner regular, there exists a compact subset $K_n \subset V_n$ such that $\mu(K_n) > n$, and this is true of each $n \in \mathbb{N}$. Now the set

$$K := \{x\} \cup \bigcup_{n \in \mathbb{N}} K_n$$

is compact. For if $\mathscr{U}$ is an open cover of K, then some $U \in \mathscr{U}$ contains x and since (V_n) is a neighborhood basis at x, $V_{n_0} \subset U$ for some $n_0 \in \mathbb{N}$. It follows that $K_n \subset V_n \subset V_{n_0} \subset U$ for all $n \geq n_0$. Since $K_1 \cup \ldots \cup K_{n_0}$ is a compact subset of K, it is covered by finitely many sets in $\mathscr{U}$. These together with U then furnish the desired finite covering of K. On the one hand then $\mu(K) < +\infty$, since μ is a Borel measure, and on the other hand since $K_n \subset K$

$$\mu(K) \geq \mu(K_n) > n \qquad\qquad \text{for all } n \in \mathbb{N}.$$

This is the contradiction sought. $\square$

Exercises.

1. Let $(\Omega, \mathscr{A})$ be a measurable space, $\mathscr{E}$ a generator of $\mathscr{A}$ and Ω' a subset of Ω. Consider the traces $\mathscr{A}'$ and $\mathscr{E}'$ of $\mathscr{A}$ and $\mathscr{E}$, resp., on Ω' and show that $\mathscr{E}'$ is a generator of the σ-algebra $\mathscr{A}'$ in Ω'. Example 3 above is a special case.

2. Equip the set $\mathbb{R}$ with the so-called *right-sided topology* (which is also sometimes named after SORGENFREY [1947]) whose system $\mathscr{O}_r$ of open sets is defined as follows: A subset $U \subset \mathbb{R}$ lies in $\mathscr{O}_r$ if and only if for each $x \in U$ there is an $\varepsilon > 0$ such that $[x, x + \varepsilon[\subset U$. The topological space thus created will be denoted $\mathbb{R}_r$. Establish, one after another, the following claims:

(a) Every right half-open interval $[a, b[$ is both open and closed in $\mathbb{R}_r$. The right-sided topology on $\mathbb{R}$ is strictly finer than the usual topology. In particular, $\mathbb{R}_r$ is a Hausdorff space.

(b) $\mathscr{B}(\mathbb{R}_r) = \mathscr{B}^1$.

(c) Suppose (x_n) is a strictly isotone sequence of real numbers possessing the supremum $b \in \mathbb{R}$. Then the set $\{x_n : n \in \mathbb{N}\} \cup \{b\}$ is closed but not compact in $\mathbb{R}_r$. By contrast, if (y_n) is a strictly antitone sequence of real numbers possessing the infimum $a \in \mathbb{R}$, then $\{a\} \cup \{y_n : n \in \mathbb{N}\}$ is compact in $\mathbb{R}_r$.

(d) Let K be compact in $\mathbb{R}_r$. Then there exists (from the first part of (c)) for every $x \in K$ a $y \in \mathbb{Q}$ with $y < x$ and $[y, x[\cap K = \emptyset$. If for each $x \in K$, $\varrho(x)$ designates such a rational number y, then a mapping $\varrho : K \to \mathbb{Q}$ materializes which is strictly isotone, and hence injective.

(e) Every compact subset of $\mathbb{R}_r$ is countable. (But (c) shows that the converse is not true.)

(f) Consider on $\mathscr{B}(\mathbb{R}_r) = \mathscr{B}^1$ the measure μ which assigns to every countable set the value 0 and to every uncountable set the value $+\infty$ (cf. Example 6). Then μ is a Borel measure on $\mathbb{R}_r$ for which no point of $\mathbb{R}_r$ has a neighborhood of finite measure. In particular, the measure μ is not locally finite and is neither inner regular nor outer regular.

(g) Consider the measure $\nu := f\lambda^1$ with density

$$f(x) := x^{-1} 1_{]0, +\infty[}(x) \qquad\qquad (x \in \mathbb{R})$$

and show that it too is a non-locally-finite Borel measure on $\mathbb{R}_r$.

(h) Investigate the L-B measure λ^1, thought of as a Borel measure on $\mathbb{R}_r$, in respect to its inner and outer regularity.

§26. Radon measures on Polish spaces

For two extensive classes of Hausdorff spaces Borel measures come up very naturally. The first of these classes will be discussed in this section, beginning of course with its

26.1 Definition. A topological space E is called *Polish* when its topology has a countable base and can be defined by a complete metric.

The terminology is due to N. BOURBAKI and commemorates the achievements of Polish topologists in the development of general topology.

A metric is called *complete* when the associated metric space is complete: every Cauchy subsequence in it converges. A *countable base* or *basis* for the topology is a countable system of open sets such that every open set is the union of those from the system which are subsets of it. *For a metrizable space E the existence of such a basis is equivalent to the existence of a countable dense subset.*

Examples. 1. The euclidean spaces $\mathbb{R}^d$ of every dimension $d \geq 1$ are Polish, the ordinary euclidean metric being complete.

2. The product $E' \times E''$ of two Polish spaces is another, when given the product topology. For if d', d'' are complete metrics generating the topologies of E' and E'', resp., then the product topology of $E' \times E''$ is generated by the metric

$$d(x,y) := d'(x',y') + d''(x'',y''), \quad x := (x',x''), \quad y := (y',y''),$$

which moreover is complete. If $\mathscr{G}', \mathscr{G}''$ are countable bases for E', E'', resp., then $\{G' \times G'' : G' \in \mathscr{G}', G'' \in \mathscr{G}''\}$ is a countable basis for $E' \times E''$.

3. *Every closed subspace F of a Polish space E is Polish.* Just restrict to F any complete metric that generates the topology of E.

4. *Every open subspace G of a Polish space E is Polish.*

Proof. We may suppose $G \neq E$. By 1. and 2. $\mathbb{R} \times E$ is Polish. Let d be a complete metric giving the topology of E, and consider the set F of all $(\lambda, x) \in \mathbb{R} \times E$ satisfying $\lambda \cdot d(x, E \setminus G) = 1$. Here, as usual, for $\emptyset \neq A \subset E$, $d(x, A) := \inf\{d(x, a) : a \in A\}$ is the distance from the point $x \in E$ to A. The mapping $x \mapsto d(x, A)$ is continuous on E, in fact, as the reader can easily check, $|d(x, A) - d(y, A)| \leq d(x, y)$ for all $x, y \in E$. Consequently, $(\lambda, x) \mapsto \lambda \cdot d(x, E \setminus G)$ is a continuous real function on $\mathbb{R} \times E$, and F is a closed subset of $\mathbb{R} \times E$, hence itself a Polish space, by 3. Finally, $(\lambda, x) \mapsto x$ maps F homeomorphically onto G. To see surjectivity, we only have to notice that, because $E \setminus G$ is closed, G coincides with the set $\{x \in E : d(x, E \setminus G) > 0\}$.

5. More generally it is true (cf. COHN [1980], Theorem 8.1.4 or WILLARD [1970], Theorem 24.12) that *a subspace A of a Polish space E is Polish if A is a G_δ-set in E, that is, A is the intersection of a sequence of open subsets of E.* Thus, for

example, the set J of all irrational numbers with its topology as a subspace of $\mathbb{R}$ is Polish, since

$$J = \bigcap_{x \in \mathbb{Q}} (\mathbb{R} \setminus \{x\}) .$$

6. *Every compact space E with a countable basis is Polish.* For a famous theorem of P.S. URYSOHN (1889-1924) (cf. KELLEY [1955], p. 125 or WILLARD [1970], Theorem 23.1) guarantees that E is metrizable, and in Remark 3 of §31 we shall even give a proof of this. The compactness of E easily entails that every metric defining its topology is complete.

The key to the further discussion is the following lemma, which is here just a preliminary to the big theorem that follows it, but nevertheless is significant in its own right. In it we encounter our first extensive class of Radon measures.

26.2 Lemma. *Every finite Borel measure μ on a Polish space E is regular.*

Proof. We consider the system $\mathscr{D}$ of all $B \in \mathscr{B}(E)$ which satisfy both

(26.1) $\mu(B) = \sup\{\mu(K) : K \text{ compact } \subset B\}$

and

(26.2) $\mu(B) = \inf\{\mu(U) : B \subset U \text{ open}\} .$

The goal of course is to show that $\mathscr{D} = \mathscr{B}(E)$. We block off the work into five sections. Let d be a complete metric defining the topology of E.

1. $E \in \mathscr{D}$: Only (26.1) needs proof when $B = E$. Let $(x_n)_{n \in \mathbb{N}}$ be a sequence which is dense in E, and for $x \in E$, real $r > 0$ let $K_r(x)$ denote the open ball of center x and d-radius r. For every r then $E = \bigcup_{n \in \mathbb{N}} K_r(x_n)$, because in every ball $K_r(x)$ lies some x_n, so that $x \in K_r(x_n)$. Since μ is continuous from below

$$\mu(E) = \lim_{k \to \infty} \mu\Big(\bigcup_{j=1}^{k} K_r(x_j) \Big) .$$

Therefore, for each $\varepsilon > 0$ and $n \in \mathbb{N}$ there exists $k_n \in \mathbb{N}$ such that

$$\mu\Big(\bigcup_{j=1}^{k_n} K_{1/n}(x_j) \Big) \geq \mu(E) - \varepsilon 2^{-n} .$$

Each set $B_n := \bigcup_{j=1}^{k_n} \overline{K}_{1/n}(x_j)$, hence also their intersection $K := \bigcap_{n \in \mathbb{N}} B_n$ is closed, and we have

$$\mu(E) - \mu(K) = \mu(E \setminus K) = \mu\Big(\bigcup_{n \in \mathbb{N}} (E \setminus B_n) \Big) \leq \sum_{n=1}^{\infty} \mu(E \setminus B_n) \leq \sum_{n=1}^{\infty} \varepsilon 2^{-n} = \varepsilon .$$

This will prove (26.1) if we can confirm that the closed set K is actually compact. For every $n \in \mathbb{N}$

$$K \subset B_n = \overline{K}_{1/n}(x_1) \cup \ldots \cup \overline{K}_{1/n}(x_{k_n})$$

and each set in this union has diameter no greater than $2/n$. This shows that K is pre-compact (=totally bounded) and in a complete metric space that is equivalent to compactness, by very easy arguments (cf. WILLARD [1970], Theorem 39.9 or KELLEY [1955], p. 198).

2. Every closed set C lies in $\mathscr{D}$: Let $\varepsilon > 0$ be given. We already know that there is a compact set K with

$$\mu(E) - \mu(K) < \varepsilon \,.$$

According to 3.5 however

$$\mu(C) - \mu(C \cap K) = \mu(C \cup K) - \mu(K) \leq \mu(E) - \mu(K) < \varepsilon$$

and this proves (26.1) for $B := C$, because $C \cap K$ is compact. As a closed subset of a metric space, C is a G_δ-set, that is, there are open sets $G_n \downarrow C$. To see this we may assume $C \neq \emptyset$, so that $G := E \setminus C$ is an open proper subset of E. Consequently, $x \mapsto d(x, C)$ is a continuous mapping whose zero-set is C, as was shown in treating Example 4. The sets $G_n := \{x \in E : d(x, C) < 1/n\}$ are therefore open and decrease to C. From the finiteness of μ and 3.2(c) we then have that $\mu(G_n) \downarrow \mu(C)$, showing that (26.2) is also satisfied by $B := C$.

3. Whenever B lies in $\mathscr{D}$ so does $\complement B$: First note that for every compact $K \subset B$

$$\mu(\complement K) - \mu(\complement B) = \mu(B) - \mu(K) \,,$$

and so $\complement B$ satisfies (26.2) whenever B satisfies (26.1). Moreover, if G is an open superset of B, then $\complement G$ is a closed subset of $\complement B$ with

$$\mu(\complement B) - \mu(\complement G) = \mu(G) - \mu(B) \,,$$

showing, at least, that $\complement B$ satisfies (26.1) weakened by replacing "compact" there by "closed". But then application of step 2 to these closed sets gives us the full (26.1) for $\complement B$.

4. Whenever pairwise disjoint sets D_n lie in $\mathscr{D}$ $(n \in \mathbb{N})$, their union D also lies in $\mathscr{D}$: First of all

$$\mu(D) = \sum_{n=1}^{\infty} \mu(D_n) \,.$$

Letting $\varepsilon > 0$ be given, we therefore have an $n_\varepsilon \in \mathbb{N}$ such that

$$(26.3) \qquad \mu(D) - \sum_{n=1}^{n_\varepsilon} \mu(D_n) < \varepsilon/2 \,.$$

Every D_j contains a compact K_j such that

$$\mu(D_j) - \mu(K_j) < \frac{\varepsilon}{2 n_\varepsilon} \qquad\qquad (j = 1, \dots, n_\varepsilon)$$

since each $D_j \in \mathscr{D}$. Then $K := K_1 \cup \dots \cup K_{n_\varepsilon}$ is a compact subset of $D_1 \cup \dots \cup D_{n_\varepsilon} \subset D$ which satisfies

$$\mu(D_1 \cup \dots \cup D_{n_\varepsilon}) - \mu(K) \leq \mu\left(\bigcup_{j=1}^{n_\varepsilon} (D_j \setminus K_j)\right) = \sum_{j=1}^{n_\varepsilon} \mu(D_j \setminus K_j) < \varepsilon/2 \,,$$

from which, in view of (26.3),

$$\mu(D) - \mu(K) < \varepsilon.$$

Again, $D_n \in \mathscr{D}$ means there exists open $U_n \supset D_n$ such that

$$\mu(U_n) - \mu(D_n) < \varepsilon/2^n \qquad\qquad \text{for each } n \in \mathbb{N}.$$

Then the open set $U := \bigcup_{n\in\mathbb{N}} U_n$ contains D and satisfies

$$\mu(U) - \mu(D) \le \mu\Big(\bigcup_{n\in\mathbb{N}}(U_n \setminus D_n)\Big) \le \sum_{n=1}^{\infty} \mu(U_n \setminus D_n) < \varepsilon.$$

In summary, we have shown that (26.1) and (26.2) hold for $B := D = \bigcup D_n$.
5. The result of the first four steps is that $\mathscr{D}$ is a Dynkin system which contains the system $\mathscr{F}$ of all closed sets. The claim, namely that $\mathscr{D} = \mathscr{B}(E)$, now follows in the familiar way: Because $\mathscr{F}$ is $\cap$-stable, $\boldsymbol{\delta}(\mathscr{F}) = \boldsymbol{\sigma}(\mathscr{F}) = \mathscr{B}(E)$. From $\mathscr{F} \subset \mathscr{D} \subset \mathscr{B}(E)$ follows $\mathscr{B}(E) = \boldsymbol{\delta}(\mathscr{F}) \subset \mathscr{D} \subset \mathscr{B}(E)$, and thus the equality sought. $\square$

We come now to the principal result of this section. It generalizes the foregoing lemma.

26.3 Theorem. *On a Polish space E every locally finite Borel measure μ is a σ-finite Radon measure.*

Proof. The hypothesis is that every point $x \in E$ has an open neighborhood U_x of finite μ-measure. The family $(U_x)_{x\in E}$ is an open cover of E. Because the topology of E has a countable basis, a theorem of E. LINDELÖF (1879–1946) insures that this cover contains a countable subcover. That is, there is a sequence $(x_n)_{n\in\mathbb{N}}$ in E such that the sequence $(U_{x_n})_{n\in\mathbb{N}}$ already covers E. [It is easy enough to prove Lindelöf's result right here: Let $\mathscr{U}$ be any open cover of E, $\mathscr{A}$ a countable basis for the topology of E, and define $\mathscr{A}'$ to be the system of all $A \in \mathscr{A}$ such that $A \subset U$ for some $U \in \mathscr{U}$ and let $U(A)$ be one such member of $\mathscr{U}$. The subset $\mathscr{A}'$ of $\mathscr{A}$, and therewith the system of all these $U(A)$, is countable. This system covers E. For if $x \in E$, then there is some $U \in \mathscr{U}$ that contains x, and since $\mathscr{A}$ is a basis there is some $A \in \mathscr{A}$ such that $x \in A \subset U$. Thus $A \in \mathscr{A}'$ and $x \in A \subset U(A)$.]
  The system of sets $G_n := U_{x_1} \cup \ldots \cup U_{x_n}$, $n \in \mathbb{N}$, satisfies

$$(26.4) \qquad\qquad \mu(G_n) < +\infty \qquad\qquad \text{for every } n \in \mathbb{N}, \text{ and } G_n \uparrow E.$$

Via

$$\mu_n(A) := \mu(A \cap G_n), \qquad\qquad A \in \mathscr{B}(E)$$

a finite Borel measure μ_n is defined on E for every $n \in \mathbb{N}$. Each such measure is inner regular by the preceding lemma. It follows that for each $A \in \mathscr{B}(E)$

$$\mu(A) = \sup_{n\in\mathbb{N}} \mu(A \cap G_n) = \sup_{n\in\mathbb{N}} \mu_n(A) = \sup_{n\in\mathbb{N}} \sup_{\substack{K\in\mathscr{K} \\ K\subset A}} \mu_n(K).$$

After commuting the two suprema this reads

$$\mu(A) = \sup_{\substack{K \in \mathscr{K} \\ K \subset A}} \sup_{n \in \mathbb{N}} \mu_n(K) = \sup_{\substack{K \in \mathscr{K} \\ K \subset A}} \mu(K),$$

proving the inner regularity of μ. The σ-finiteness of μ is affirmed by (26.4), so the proof is complete. □

The question now suggests itself whether – in analogy with 26.2 – the outer regularity of μ can be proved. This is in fact the case.

26.4 Corollary. *Every Radon measure on a Polish space is outer regular.*

Proof. We have to show that every $B \in \mathscr{B}(E)$ satisfies (25.5). So let $B \in \mathscr{B}(E)$ and $\varepsilon > 0$ be given. Consider the open sets G_n and the finite measures μ_n created in the preceding proof. Lemma 26.2 furnishes open sets $U_n \supset B$ such that

(26.5) $\qquad\qquad \mu((U_n \setminus B) \cap G_n) = \mu_n(U_n \setminus B) \le \varepsilon/2^n \qquad$ for each $n \in \mathbb{N}$.

Let $U := \bigcup_{n \in \mathbb{N}} U_n \cap G_n$, an open set. Since

$$B = B \cap E = B \cap \bigcup_{n \in \mathbb{N}} G_n = \bigcup_{n \in \mathbb{N}} B \cap G_n,$$

it follows from $B \subset U_n$ for every n, that $B \subset U$. Moreover, this representation of B shows that

$$U \setminus B = \bigcup_{n \in \mathbb{N}} (U_n \cap G_n) \setminus \bigcup_{n \in \mathbb{N}} (B \cap G_n) \subset \bigcup_{n \in \mathbb{N}} (U_n \cap G_n) \setminus (B \cap G_n) = \bigcup_{n \in \mathbb{N}} (U_n \setminus B) \cap G_n$$

and consequently

$$\mu(U \setminus B) \le \sum_{n=1}^{\infty} \mu((U_n \setminus B) \cap G_n) \le \sum_{n=1}^{\infty} \varepsilon/2^n = \varepsilon,$$

by (26.5). It follows finally that

$$\mu(U) = \mu(B) + \mu(U \setminus B) \le \mu(B) + \varepsilon,$$

which confirms (25.5). □

The regularity conditions (25.4), (25.5) make sense for outer measures μ^* and together with one other minimal demand on μ^* they assure that all Borel sets are μ^*-measurable. In fact, these conditions on an outer measure come up naturally in the course of proving the famous Riesz representation theorem in §29; cf. also 28.3.

26.5 Lemma. *Let E be a Hausdorff space and μ^* an outer measure on E with the following three properties:*
(i)   *for every set $A \subset E$*

$$\mu^*(A) = \inf\{\mu^*(U) : A \subset U \ open\ \};$$

(ii) *for every open set $U \subset E$*

$$\mu^*(U) = \sup\{\mu^*(K) : K \text{ compact } \subset U\};$$

(iii) *for any two disjoint compact sets $K_1, K_2 \subset E$*

$$\mu^*(K_1 \cup K_2) = \mu^*(K_1) + \mu^*(K_2).$$

Then the restriction of μ^* to $\mathcal{B}(E)$ is a measure.

Proof. We consider the σ-algebra $\mathcal{A}^*$ of all μ^*-measurable sets, that is, according to (5.6) the set of all $A \in \mathcal{P}(E)$ which satisfy

(26.6) $$\mu^*(Q) \geq \mu^*(Q \cap A) + \mu^*(Q \setminus A) \qquad \text{for all } Q \in \mathcal{P}(E).$$

First note that it suffices that this hold for all open sets Q in order that it hold for all Q whatsoever. In other words, what we need to check for an A to be in $\mathcal{A}^*$ is that

(26.6') $$\mu^*(U) \geq \mu^*(U \cap A) + \mu^*(U \setminus A) \qquad \text{for all } U \in \mathcal{O}.$$

Indeed from (26.6') it follows for any $Q \subset E$ that

$$\mu^*(U) \geq \mu^*(Q \cap A) + \mu^*(Q \setminus A)$$

whenever U is an open set containing Q; then (26.6) itself follows by taking the infimum over such U and invoking (i). So now let $A = G$ be an open set; we will use criterion (26.6') to show that G lies in $\mathcal{A}^*$. To this end consider any open $U \subset E$; further, consider any compact $K_1 \subset U \cap G$ and any compact $K_2 \subset U \setminus K_1$. Since then $K_1 \cap K_2 = \emptyset$ and $K_1 \cup K_2 \subset U$, it follows from (iii) that

$$\mu^*(U) \geq \mu^*(K_1 \cup K_2) = \mu^*(K_1) + \mu^*(K_2).$$

The set $U \setminus K_1$ is open, so if we take the supremum over all such K_2 in the preceding inequality and appeal to (ii), we get

$$\mu^*(U) \geq \mu^*(K_1) + \mu^*(U \setminus K_1) \geq \mu^*(K_1) + \mu^*(U \setminus G),$$

the last inequality because $U \setminus K_1 \supset U \setminus G$. This holds for all compact $K_1 \subset U \cap G$, and so after a second appeal to (ii) it yields

$$\mu^*(U) \geq \mu^*(U \cap G) + \mu^*(U \setminus G),$$

holding for all $U \in \mathcal{O}$. That is, (26.6') holds for $A = G$, and consequently $G \in \mathcal{A}^*$. This all proves that $\mathcal{O} \subset \mathcal{A}^*$. But then $\mathcal{B}(E) = \sigma(\mathcal{O}) \subset \mathcal{A}^*$ because the latter is a σ-algebra, by Theorem 5.3. That theorem further affirms that the restriction of μ^* to $\mathcal{A}^*$ is a measure. $\square$

The foregoing Theorem 26.3 and its corollary show in particular that the L-B measure λ^d is a *regular Borel measure* on $\mathbb{R}^d$ in each dimension $d = 1, 2, \ldots$. In fact every Borel measure on $\mathbb{R}^d$ is regular (cf. also Theorem 29.12). Following STROMBERG [1972] we derive from the regularity of λ^d a purely topological result of H. STEINHAUS (1887–1972). It shows, incidentally, that every set of positive L-B measure has the cardinality of $\mathbb{R}$.

26.6 Theorem (of Steinhaus). *Let $A \in \mathscr{B}^d$ be a Borel set in $\mathbb{R}^d$ of positive d-dimensional Lebesgue measure. Then $\mathbf{0}$ is an interior point of the set $A - A$ of differences of elements of A.*

Proof. The inner regularity of λ^d means that A contains a compact subset K with $\lambda^d(K)$ positive. It suffices to prove the claim with K in place of A. Outer regularity furnishes an open set $U \supset K$ with $\lambda^d(U) < 2\lambda^d(K)$. There is an open ball V centered at $\mathbf{0}$ of positive radius such that the sum set satisfies $K + V \subset U$. One only has to choose the radius less than the (positive) distance between the compact set K and the closed set $\complement U$ from which it is disjoint. We will show that $V \subset K - K$, which makes $\mathbf{0}$ an interior point of this difference set. Consider any $v \in V$. The translated set $v + K$ cannot be disjoint from K, for otherwise from $K \cup (v + K) \subset K + V \subset U$ and translation-invariance of λ^d would follow that

$$2\lambda^d(K) = \lambda^d(K) + \lambda^d(v + K) = \lambda^d(K \cup (v + K)) \leq \lambda^d(U),$$

contrary to the choice of U. But $K \cap (v + K) \neq \emptyset$ means that for some $x, y \in K$, $x = v + y$; which says that the given point $v = x - y$ lies in $K - K$. $\square$

In closing we turn to a remarkable consequence of Theorem 26.3 and its Corollary 26.4. It concerns the analogy, pointed out in §7 as measurable mappings were being introduced, between the notions of measurability and continuity. Initially this analogy is merely an analogy. Namely, if $f : E \to E'$ is a mapping of one topological space into another, then f is *Borel measurable* (i.e., $\mathscr{B}(E)$-$\mathscr{B}(E')$-measurable) just if the pre-image $f^{-1}(G')$ of every open set $G' \subset E'$ is a Borel set in E. This follows from Theorem 7.2 and the fact that the Borel σ-algebra $\mathscr{B}(E')$ is generated by the open subsets of E'. By contrast, f is continuous just if $f^{-1}(G')$ is open in E for every open set $G' \subset E'$. What is quite remarkable is that for Polish spaces E a much closer connection between those two concepts exists. This is brought out by the following theorem, discovered in its definitive form by N. LUSIN (1883–1950).

26.7 Theorem (of Lusin). *Let μ be a locally finite Borel measure, thus a Radon measure, on a Polish space E, and E' be a topological space with a countable basis. Then for every mapping $f : E \to E'$ the following are equivalent:*
(a) *f coincides μ-almost everywhere with a Borel measurable mapping of E into E'.*
(b) *There is a decomposition of E into a μ-nullset $N \in \mathscr{B}(E)$ and a sequence $(K_n)_{n \in \mathbb{N}}$ of compact sets, such that the restriction of f to each K_n is continuous.*

If the measure μ is finite, (a) and (b) are further equivalent to:
(c) *For every $\varepsilon > 0$ there is a compact subset $K_\varepsilon \subset E$ such that $\mu(\complement K_\varepsilon) < \varepsilon$ and the restriction of f to K_ε is continuous.*

Proof. Let us first suppose that μ is finite. Let $\mathscr{G}'$ be a countable base for the topology of E' and $(G'_n)_{n \in \mathbb{N}}$ a sequential arrangement of its elements. Notice that $\mathscr{G}'$ is a generator of the Borel σ-algebra because every open subset of E' is a (countable) union of sets from $\mathscr{G}'$.

(a)⇒(c): By hypothesis there is a Borel measurable mapping $g : E \to E'$ and μ-nullset $N \in \mathscr{B}(E)$ with

(26.7) $$f(x) = g(x) \qquad \text{for all } x \in \complement N.$$

For every set G'_n, $g^{-1}(G'_n) \in \mathscr{B}(E)$. Because every Radon measure on E is regular, given $\varepsilon > 0$, there exist compact sets K_n and open sets U_n such that

(26.8) $$K_n \subset g^{-1}(G'_n) \subset U_n \quad \text{and} \quad \mu(U_n \setminus K_n) < 2^{-n}\varepsilon \qquad \text{for each } n \in \mathbb{N}.$$

The set $A := \bigcup_{n \in \mathbb{N}} (U_n \setminus K_n)$ is open, being a union of open sets. For its measure we have the obvious inequality

$$\mu(A) \le \sum_{n=1}^{\infty} \mu(U_n \setminus K_n) < \varepsilon .$$

Using once more the (inner) regularity of μ, we find a compact $K \subset \complement(A \cup N) = \complement A \cap \complement N$ such that

$$\mu(\complement A \cap \complement N \cap \complement K) < \varepsilon - \mu(A) ,$$

thus (since $A \cup N \subset \complement K$ and $A \cup N \cup (\complement A \cap \complement N) = E$) such that

$$\mu(\complement K) = \mu(A \cup N \cup [\complement A \cap \complement N \cap \complement K]) < \mu(A) + \mu(N) + \varepsilon - \mu(A) = \varepsilon .$$

This set K does what is wanted in (c), because by (26.7) f and g coincide in K and because the restriction g_0 of g to $\complement A$ is continuous, as we now confirm. For each set G'_n,

$$g_0^{-1}(G'_n) = g^{-1}(G'_n) \cap \complement A ;$$

from (26.8) and the fact $U_n \setminus K_n \subset A$ follows therefore

$$U_n \cap \complement A = K_n \cap \complement A \subset g_0^{-1}(G'_n) \subset U_n \cap \complement A ,$$

which means that

$$g_0^{-1}(G'_n) = U_n \cap \complement A = K_n \cap \complement A ,$$

showing that the g_0-pre-image of G'_n is open (as well as closed) in $\complement A$. Since $(G'_n)_{n \in \mathbb{N}}$ is a base for the topology of E', this is enough to guarantee the continuity of $g_0 = g \mid \complement A$.

(c)⇒(b): It suffices to find pairwise disjoint compact subsets K_n of E such that $f \mid K_n$ is continuous and

$$\mu\left(\complement \bigcup_{j=1}^{n} K_j\right) < \frac{1}{n}$$

for each $n \in \mathbb{N}$. For then

$$N := \complement \bigcup_{n \in \mathbb{N}} K_n = \bigcap_{n \in \mathbb{N}} \complement K_n$$

is a Borel set disjoint from each K_n and satisfying $\mu(N) < 1/n$ for every $n \in \mathbb{N}$, i.e., $\mu(N) = 0$. The sequence (K_n) is gotten inductively from (c) as follows: To start off, there is a compact $K_1 \subset E$ such that $\mu(\complement K_1) < 1$ and $f \mid K_1$ is continuous.

If $K_1, \ldots, K_n$ have been defined having the desired properties, we will get K_{n+1} from (c) and the inner regularity of μ. By (c) there is a compact $K' \subset E$ such that

$$\mu(\complement K') < (2n+2)^{-1}$$

and $f \mid K'$ is continuous. With $L := K_1 \cup \ldots \cup K_n$ the inner regularity of μ supplies a compact $K_{n+1} \subset K' \setminus L$ such that

$$\mu(K' \setminus L) - \mu(K_{n+1}) = \mu(K' \cap \complement L \cap \complement K_{n+1}) < (2n+2)^{-1}.$$

Because

$$\mu(\complement(L \cup K_{n+1})) = \mu(\complement K' \cap \complement L \cap \complement K_{n+1}) + \mu(K' \cap \complement L \cap \complement K_{n+1})$$
$$\leq \mu(\complement K') + \mu(K' \cap \complement L \cap \complement K_{n+1}) < (n+1)^{-1},$$

with this set K_{n+1} the inductive construction is complete.

(b)$\Rightarrow$(a): If $E = N \cup K_1 \cup K_2 \cup \ldots$ is the given decomposition, one defines a mapping $g : E \to E'$ as follows. In case $N = \emptyset$, let $g := f$. In case $N \neq \emptyset$, choose $y_0 \in f(N)$ arbitrarily and set

$$g(x) := f(x) \text{ for } x \in E \setminus N, \qquad g(x) := y_0 \text{ for } x \in N.$$

What has to be shown is that g is Borel measurable, which is done as follows: For every open $G' \subset E'$

$$g^{-1}(G') = (g^{-1}(G') \cap N) \cup \bigcup_{n \in \mathbb{N}} (g^{-1}(G') \cap K_n) = N_0 \cup \bigcup_{n \in \mathbb{N}} g_n^{-1}(G'),$$

where $N_0 := g^{-1}(G') \cap N$ and $g_n := g \mid K_n$. Now N_0 is either N or $\emptyset$, according as $y_0 \in G'$ or $y_0 \notin G'$. Moreover, g_n coincides with the restriction of f to K_n, so that by hypothesis $g_n^{-1}(G')$ is open in K_n, that is, of the form $K_n \cap U_n$ for some open subset U_n of E. Therefore only Borel sets occur in the above decomposition of $g^{-1}(G')$ and we conclude that $g^{-1}(G')$ is a Borel set. This being true of every open $G' \subset E'$, the Borel measurability of g follows from 7.2.

Now consider an arbitrary locally finite measure μ on $\mathscr{B}(E)$. According to 26.3, μ is σ-finite. Lemma 17.6 therefore furnishes a strictly positive μ-integrable real function h on E. The measure $\nu := h\mu$ is then a finite Borel measure on E which has exactly the same nullsets as μ. The proven equivalence of (a) and (b) for the measure ν therefore entails the validity of this equivalence for the measure μ. Thus the whole theorem is proved. $\square$

Remarks. 1. The equivalence of (a) and (b) in Lusin's theorem may be lost if (a) is strengthened to the $\mathscr{B}(E)$-$\mathscr{B}(E')$-measurability of f. It suffices to take for E the compact set $[0,1] \times [0,1]$ and for μ the L-B measure λ_E^2. As was noted in the second part of Remark 4, §8, E contains a μ-nullset N which contains a non-Borel subset. If M is such a set, its indicator function $f = 1_M$ is not Borel measurable, although f is μ-almost everywhere equal to the Borel measurable function 1_N. On the other hand, if f is $\mathscr{B}(E)$-$\mathscr{B}(E')$-measurable, there is a Polish topology τ on E, stronger than the original but generating exactly the same Borel sets, such that f is τ-continuous. See 3.2.6 of SRIVASTAVA [1998] for the proof, which is not difficult.

2. The Dirichlet jump function (cf. Remark 1 of §16) is continuous at no point of its domain of definition $[0, 1]$, yet it is Borel measurable. This shows that in assertion (c) of Lusin's theorem one cannot hope to be able to replace the continuity of the function $f \mid K$ by the continuity of f at each point of K.

Exercises.

1. Show that every inner regular finite Borel measure on a Hausdorff space is outer regular.

2. Show that in a Polish space E the Dirac measures are the only non-zero Borel measures μ which take only the values 0 and 1. [*Hint*: Show that the system of all compact $K \subset E$ such that $\mu(K) = 1$ is $\cap$-stable and investigate the intersection of all its sets.]

3. Show that $\mathscr{B}(E \times E') = \mathscr{B}(E) \otimes \mathscr{B}(E')$ for any Polish spaces E, E'.

4. Consider K compact $\subset U$ open $\subset \mathbb{R}^d$, and for each $n \in \mathbb{N}$ let V_n denote the open ball of radius $1/n$ and center $\mathbf{0}$. Show that $K + V_n \subset U$ for some n. [*Hint*: If $(K + V_n) \cap \complement U \neq \emptyset$ for every $n \in \mathbb{N}$, find $x_n \in K$, $v_n \in V_n$, $z_n \in \complement U$ such that $x_n + v_n = z_n$, for every $n \in \mathbb{N}$. Some subsequence of (x_n) converges to a point $x_0 \in K$ and because $\complement U$ is closed we even have $x_0 \in K \cap \complement U$, which contradicts the fact that $K \subset U$.]

5. Let μ be a locally finite Borel measure on a Polish space E and $f : E \to E'$ a mapping into a topological space E' with a countable base. Show that assertions (a) and (b) in Lusin's theorem are equivalent to (c'): For every $\varepsilon > 0$ and every compact $K \subset E$ there is a further compact $K_\varepsilon \subset K$ such that $\mu(K \setminus K_\varepsilon) < \varepsilon$ and $f \mid K_\varepsilon$ is continuous.

§27. Properties of locally compact spaces

A topological space is called *locally compact* if it is Hausdorff and if each of its points has at least one compact neighborhood. Examples of such spaces are the euclidean space $\mathbb{R}^d$, every manifold (i.e., every locally euclidean Hausdorff space), every discrete space, and every compact space.

When an arbitrary point is removed from a compact space the remainder is a locally compact space. Actually every locally compact space is of this form. For if $\mathscr{O}$ is the system of all open subsets of the locally compact space E and ω_0 is any (so-called *ideal*) point not in E, then a topology can be defined on $E' := E \cup \{\omega_0\}$ as follows: The system $\mathscr{O}'$ of open sets in E' shall consist of $\mathscr{O}$ together with the sets $E' \setminus K$ for all the compact subsets K of E. This defines a compact topology on E', E is an open subset of E' and the topology that E inherits from $\mathscr{O}'$ is its original topology. E was compact to start with if and only if ω_0 is an isolated point in E'. If E is not compact, then it is dense in E'. These claims are easily confirmed, or the reader can consult KELLEY [1955], p. 150, or WILLARD [1970], 19.2. The space E'

is called, after its creator P.S. ALEXANDROFF (1896–1982), the (Alexandroff) *one-point compactification* of E and ω_0 its *infinitely remote point*.

We will pursue the further theory of locally compact spaces via this compactification. First we study some distinguished continuous functions in this environment. For an arbitrary topological space E we denote by

$$C(E) \quad \text{and} \quad C_b(E)$$

the vector space of all, respectively all bounded, continuous real functions on E.

27.1 Definition. Let $f : E \to \mathbb{R}$ be a real function on a topological space E. The set

(27.1) $$\operatorname{supp}(f) := \overline{\{f \neq 0\}}$$

is called the *support* of f.

The complement of $\operatorname{supp}(f)$ is thus the largest open set at every point of which f takes the value zero. If E is locally compact, we will designate by

$$C_c(E)$$

the set of all $f \in C(E)$ *with compact support* $\operatorname{supp}(f)$. A function $f \in C(E)$ lies in $C_c(E)$ just if there is some compact subset of E in the complement of which f is identically zero.

Clearly

(27.2) $$C_c(E) \subset C_b(E) \subset C(E),$$

since an $f \in C_c(E)$ is bounded on its compact support, hence throughout E.

$C_c(E)$ is a vector subspace of $C_b(E)$. More generally for any $n \in \mathbb{N}$, $\varphi \in C(\mathbb{R}^n)$ with $\varphi(\mathbf{0}) = 0$ and $f_1, \ldots, f_n \in C_c(E)$, the composition $\varphi(f_1, \ldots, f_n)$ lies in $C_c(E)$, and indeed its support is a subset of $\bigcup_{j=1}^{n} \operatorname{supp}(f_j)$. In particular, whenever $u, v \in C_c(E)$ the functions $|u|$, $u \vee v$, $u \wedge v$, and therewith u^+ and u^-, all lie in $C_c(E)$. The needed continuity of $\varphi(x, y) := x \vee y$ on $\mathbb{R}^2$ follows from the identity $x \vee y = \frac{1}{2}(x + y + |x - y|)$.

In the special case of a *compact* space E, all three function spaces in (27.2) coincide.

A fundamental property of the space $C_c(E)$ is the following:

27.2 Theorem (on partitions of unity). *Suppose that the compact subset K of the locally compact space E is covered by the n open sets $U_1, \ldots, U_n$, $n \in \mathbb{N}$. Then there are functions $f_1, \ldots, f_n \in C_c(E)$ with the following properties*

(27.3) $$f_j \geq 0 \qquad\qquad \textit{for } j = 1, \ldots, n;$$

(27.4) $$\operatorname{supp}(f_j) \subset U_j \qquad\qquad \textit{for } j = 1, \ldots, n;$$

(27.5) $$\sum_{j=1}^{n} f_j(x) \leq 1 \qquad\qquad \textit{for all } x \in E;$$

$$(27.6) \qquad\qquad \sum_{j=1}^{n} f_j(x) = 1 \qquad\qquad \text{for all } x \in K.$$

Proof. We work in the one-point compactification $E' := E \cup \{\omega_0\}$ of E. The given open sets together with $U_0 := E' \setminus K$ constitute an open cover of E'. Because compact spaces are normal topological spaces (cf. KELLEY [1955], p. 141 or WILLARD [1970], Theorem 17.10), this covering can be "shrunk" to an open covering $U_1', \ldots, U_n'$ of E' satisfying

$$\overline{U_j'} \subset U_j \qquad\qquad \text{for each } j = 0, \ldots, n,$$

where of course the bar denotes closure in E'. The theorem on partitions of unity in normal spaces (KELLEY [1955], p. 171 or WILLARD [1970], 20 C) provides functions $f_0', \ldots, f_n' \in C(E')$ such that

(i) $\qquad\qquad f_j' \geq 0, \quad \operatorname{supp}(f_j') \subset U_j' \qquad\qquad \text{for } j = 0, \ldots, n;$

(ii) $$\sum_{j=0}^{n} f_j'(x) = 1 \qquad\qquad \text{for all } x \in E'.$$

The restrictions $f_1, \ldots, f_n$ to E of $f_1', \ldots, f_n'$ lie in $C(E)$ and it will be easy to show that they have all the properties wanted. From (i) and (ii) properties (27.3)–(27.5) follow almost immediately. One only has to notice that for each $j = 1, \ldots, n$

$$\operatorname{supp}(f_j) = \operatorname{supp}(f_j') \cap E \subset U_j' \cap E = U_j' \subset U_j$$

since $\overline{U_j'} \subset U_j \subset E$. In particular, $\overline{U_j'}$ being a closed subset of the compact space E', is a compact subset of E. From $\operatorname{supp}(f_j) \subset \overline{U_j'}$ therefore follows the compactness of this support. Thus $f_1, \ldots, f_n$ all lie in $C_c(E)$. The remaining property (27.6) likewise follows from (ii) because $\operatorname{supp}(f_0') \subset U_0 = E \setminus K$ entails that $f_0'(x) = 0$ for all $x \in K$. $\square$

Two consequences of the foregoing will turn out to be especially useful. The first – known as *Urysohn's lemma* – often serves as the starting point for inductive constructions of partitions of unity (see, e.g., RUDIN[1987], p. 39). The second can also be proven directly, as indicated in Exercise 1 below.

27.3 Corollary 1. *In the locally compact space E, U is an open neighborhood of the compact subset K. Then $C_c(E)$ contains a function f which satisfies*

$$(27.7) \qquad\qquad 0 \leq f \leq 1, \quad f(K) = \{1\}, \quad \text{and} \quad \operatorname{supp}(f) \subset U.$$

In particular, $\operatorname{supp}(f)$ is a compact neighborhood of K.

Proof. We have only to apply 27.2 for $n = 1$. Since $K \subset \{f_1 > 0\} \subset \operatorname{supp}(f_1)$, the fact that $\{f_1 > 0\}$ is open means that $\operatorname{supp}(f_1)$ is indeed a neighborhood of K. $\square$

27.4 Corollary 2. *In the locally compact space E the compact subset K is covered by the n open sets $U_1, \ldots, U_n$, $n \in \mathbb{N}$. Then K can be decomposed as $K = K_1 \cup \ldots \cup K_n$ with K_j a compact subset of U_j for each $j = 1, \ldots, n$.*

Proof. Let $f_1, \ldots, f_n \in C_c(E)$ be as provided by 27.2. The compact sets

$$K_j := K \cap \operatorname{supp}(f_j), \qquad\qquad j = 1, \ldots, n$$

do what is wanted; for if $x \in K$, then $1 = f_1(x) + \ldots + f_n(x)$ means that $f_j(x) \neq 0$ for some j, and therefore $x \in K_j$. $\square$

For a locally compact space E there is another function space besides $C_c(E)$ that is of importance. To define it we assign to every bounded real function f on an arbitrary space E its *supremum norm*, also called its *uniform norm*, via

$$\|f\| := \sup_{x \in E} |f(x)| \, .$$

The mapping $(f, g) \to \|f - g\|$ makes $C_b(E)$ – more generally even the vector space of all bounded real functions on E – into a metric space. One speaks of the *metric of uniform convergence* (on E). A sequence (f_n) of bounded real functions on E converges uniformly on E to a bounded function f just means that

$$\lim_{n \to \infty} \|f_n - f\| = 0 \, .$$

27.5 Definition. A continuous real function f on a locally compact space E is said to *vanish at infinity* if it lies in the closure $C_0(E)$ of $C_c(E)$ in $C_b(E)$ with respect to the metric of uniform convergence. Denoting closure in this metric by bar, we thus have

$$C_0(E) := \overline{C_c(E)} \subset C_b(E) \, .$$

The terminology "vanishing at infinity" is both clarified and justified by

27.6 Theorem. *For a real function f on a locally compact space E the following statements are equivalent:*
(a) *$f \in C_0(E)$;*
(b) *$f \in C(E)$ and $\{|f| \geq \varepsilon\}$ is compact for each $\varepsilon > 0$;*
(c) *the function*

$$f'(x) := \begin{cases} f(x), & \text{for all } x \in E \\ 0, & \text{for } x = \omega_0 \end{cases}$$

is continuous on the one-point compactification E' of E.

Proof. (a)$\Rightarrow$(b): Given $\varepsilon > 0$, there is by definition of $f \in C_0(E)$ a $g \in C_c(E)$ with $\|f - g\| \leq \varepsilon/2$. Every $x \in E$ satisfies $|f(x)| - |g(x)| \leq |f(x) - g(x)| \leq \|f - g\|$, so we see that

$$\{|f| \geq \varepsilon\} \subset \{|g| \geq \varepsilon/2\} \subset \operatorname{supp}(g).$$

This shows that $\{|f| \geq \varepsilon\}$ is a relatively compact set. But, due to the continuity of f, it is also closed. Hence it is compact.

(b)$\Rightarrow$(c): Since the subspace topology of E in E' is its original topology and E is an open subset of E', continuity of f' at each point of E is assured by $f \in C(E)$. As to continuity at the ideal point ω_0, given $\varepsilon > 0$, we have $|f'(x) - f'(\omega_0)| = |f'(x)| <$

ε for all x in the set $E' \setminus \{|f| \geq \varepsilon\}$, which by definition of E' is a neighborhood of ω_0, since $\{|f| \geq \varepsilon\}$ is a compact subset of E.

(c)$\Rightarrow$(a): Continuity of f' at ω_0 and the definition of the topology in E' mean that for each $\varepsilon > 0$ there is a compact $K \subset E$ such that $|f(x)| = |f'(x) - f'(\omega_0)| < \varepsilon$ for all $x \in E \setminus K$. 27.3 supplies a $g \in C_c(E)$ with $0 \leq g \leq 1$ and $g(K) = \{1\}$. Then $fg \in C_c(E)$ and satisfies

$$|f(x)g(x) - f(x)| = |f(x)|\,(1 - g(x)) < \varepsilon$$

for all $x \in E$, so $\|fg - f\| \leq \varepsilon$. As $\varepsilon > 0$ is arbitrary, this proves that $f \in \overline{C_c(E)}$. $\square$

Exercises.

1. Without resort to partitions of unity, prove Corollary 27.4 directly. [*Hint* for the case $n = 2$: Separate the disjoint compacta $K \setminus U_1$, $K \setminus U_2$ with disjoint open neighborhoods V_1, V_2 and set $K_1 := K \setminus V_1$, $K_2 := K \setminus V_2$.]

2. Let $E' = E \cup \{\omega_0\}$ be the one-point compactification of a locally compact space E. Describe the Borel sets in E' by means of the Borel sets in E. In particular, see how your description fits into the following general picture: For a measure space $(E, \mathscr{A})$, a point $\omega_0 \notin E$ and the set $E^{\omega_0} := E \cup \{\omega_0\}$, the σ-algebra $\mathscr{A}^{\omega_0}$ in E^{ω_0} generated by $\mathscr{A}$ and $\{\omega_0\}$ consists of all $A' \subset E^{\omega_0}$ such that $A' \cap E \in \mathscr{A}$.

§28. Construction of Radon measures on locally compact spaces

In what follows E will be a locally compact space. We consider a Borel measure μ (defined on $\mathscr{B}(E)$). Here the requirement $\mu(K) < +\infty$ for every compact set K is the same as the local finiteness requirement, because every point of E has a compact neighborhood and the implication (25.7) holds in general. So in the present context the concepts of Borel measure and locally finite measure on $\mathscr{B}(E)$ coincide. *The Radon measures on E are thus* (cf. 25.3) *those Borel measures which are inner regular.*

For a Borel measure μ every $u \in C_c(E)$ turns out to be μ-integrable. For, being continuous, u is Borel measurable. Denoting by K the compact support of u, we have $|u| \leq \|u\|\,1_K$. Since μ is a Borel measure, 1_K is μ-integrable, and the μ-integrability of u follows. Therefore corresponding to the Borel measure is a linear form I_μ on $C_c(E)$ defined by

$$(28.1) \qquad\qquad I_\mu(u) := \int u\,d\mu\,.$$

This is an *isotone* linear form in the sense of (12.3): From $u \leq v$ follows $I_\mu(u) \leq I_\mu(v)$. Because of the linearity of I_μ this is equivalent to

$$0 \leq u \in C_c(E) \quad \Rightarrow \quad I_\mu(u) \geq 0,$$

which is why I_μ is usually called a *positive linear form*.

This brings us to a key question for our further work: Is every positive linear form on $C_c(E)$ an I_μ for some Borel measure μ on E, or are there possibly positive linear forms of a completely different kind? Even for compact intervals $J := [a, b]$ on the number line, answering this question is by no means a trivial task. In this case however, as early as 1909 F. Riesz showed (cf. RIESZ [1911]) that besides the linear forms I_μ arising from Borel measures μ on J, there are no other positive linear forms on $C_c(J) = C(J)$. One of our goals is to show that every locally compact space E shares this property with J. The result in question will, in view of this pioneering work, be called the *Riesz representation theorem*. En route to it we will naturally be led to the construction of Radon measures on E.

Besides the locally compact space E, let now a *positive linear form*

$$I : C_c(E) \to \mathbb{R}$$

be given. What follows will prepare the way for the proof of the Riesz representation theorem.

For every compact $K \subset E$ we set

$$(28.2) \qquad \mu_*(K) := \inf\{I(u) : 1_K \le u \in C_c(E)\}.$$

Such functions u exist thanks to Corollary 27.3. Consequently,

$$(28.3) \qquad 0 \le \mu_*(K) < +\infty.$$

Moreover, the mapping $K \mapsto \mu_*(K)$ is obviously isotone on the system $\mathscr{K}$ of all compact sets. For an arbitrary $A \in \mathscr{P}(E)$ we set

$$(28.4) \qquad \mu_*(A) := \sup\{\mu_*(K) : K \text{ compact } \subset A\}.$$

Because of the above noted isotoneity of μ_* on $\mathscr{K}$, this new definition is consistent with (28.2). Finally, for $A \in \mathscr{P}(E)$ we define

$$(28.5) \qquad \mu^*(A) := \inf\{\mu_*(U) : A \subset U \text{ open}\}.$$

Then μ_* and μ^* are isotone functions on $\mathscr{P}(E)$. Moreover

$$(28.6) \qquad \mu_*(A) \le \mu^*(A) \qquad\qquad \text{for all } A \in \mathscr{P}(E),$$

as follows from the obvious fact that $\mu_*(A) \le \mu_*(U)$ for every open $U \supset A$; and

$$(28.7) \qquad \mu_*(U) = \mu^*(U) \qquad\qquad \text{for all open } U \in \mathscr{P}(E),$$

which follows from (28.5) and the isotoneity of μ_*. Somewhat more effort is required to check that

$$(28.8) \qquad \mu_*(K) = \mu^*(K) \qquad\qquad \text{for all } K \in \mathscr{K}.$$

For every $\varepsilon > 0$ definition (28.2) supplies a $u \in C_c(E)$ with $u \ge 1_K$ and

$$I(u) - \mu_*(K) < \varepsilon.$$

For $0 < \alpha < 1$, $U_\alpha := \{u > \alpha\}$ is an open superset of K and

$$1_{U_\alpha} \leq \frac{1}{\alpha} u \,.$$

If therefore L is a compact subset of U_α, then $1_L \leq \frac{1}{\alpha} u$ and so from (28.2) $\mu_*(L) \leq \frac{1}{\alpha} I(u)$. From definition (28.4) therefore

$$\mu_*(U_\alpha) \leq \frac{1}{\alpha} I(u)$$

and so, since $K \subset U_\alpha$,

$$0 \leq \mu_*(U_\alpha) - \mu_*(K) \leq \frac{1}{\alpha} I(u) - \mu_*(K)$$
$$\leq \frac{1}{\alpha}(\mu_*(K) + \varepsilon) - \mu_*(K)$$
$$= \left(\frac{1}{\alpha} - 1\right)\mu_*(K) + \frac{\varepsilon}{\alpha} \,.$$

As $\alpha \uparrow 1$ this majorant converges to ε, which shows that

$$\inf\{\mu_*(U) : K \subset U \text{ open}\} \leq \mu_*(K) + \varepsilon$$

holds for every $\varepsilon > 0$; that is,

$$\mu^*(K) = \inf\{\mu_*(U) : K \subset U \text{ open}\} \leq \mu_*(K) \,.$$

This confirms (28.8), the reverse inequality being part of (28.6).

Of critical importance is the following result:

28.1 Lemma. μ^* *is an outer measure on* E.

Proof. Obviously $\mu^*(\emptyset) = 0$, so what we have to prove is that

$$(28.9) \qquad \mu^*\left(\bigcup_{n \in \mathbb{N}} Q_n\right) \leq \sum_{n=1}^{\infty} \mu^*(Q_n)$$

holds for every sequence (Q_n) in $\mathscr{P}(E)$. We proceed in three steps.
First step: For any two compact sets K_1, K_2

$$\mu^*(K_1 \cup K_2) \leq \mu^*(K_1) + \mu^*(K_2) \,.$$

Consider any $u_j \in C_c(E)$ with $u_j \geq 1_{K_j}$ for $j = 1, 2$. Then $1_{K_1 \cup K_2} \leq u_1 + u_2$, so (28.2) says that

$$\mu_*(K_1 \cup K_2) \leq I(u_1 + u_2) = I(u_1) + I(u_2) \,.$$

The claimed inequality now follows from (28.2) and (28.8).
Second step: For any finitely many open sets $U_1, \ldots, U_n$

$$\mu^*(U_1 \cup \ldots \cup U_n) \leq \mu^*(U_1) + \ldots + \mu^*(U_n) \,.$$

It suffices to settle the case $n = 2$, as induction then takes care of the rest. If K is a compact subset of $U_1 \cup U_2$, then 27.4 provides compact $K_j \subset U_j$, $j = 1, 2$, such that $K = K_1 \cup K_2$. Then by the result of our first step

$$\mu^*(K) \leq \mu^*(K_1) + \mu^*(K_2) \leq \mu^*(U_1) + \mu^*(U_2).$$

The claimed inequality (with $n = 2$) then follows from (28.8), (28.4) and (28.7). *Third step*: Now we will prove (28.9). In doing so we may obviously assume that $\mu^*(Q_n) < +\infty$ for every $n \in \mathbb{N}$. Given $\varepsilon > 0$, there then exist open $U_n \supset Q_n$ such that

$$\mu^*(U_n) - \mu^*(Q_n) < 2^{-n}\varepsilon \qquad\qquad \text{for every } n \in \mathbb{N}.$$

The open set $U := \bigcup_{n \in \mathbb{N}} U_n$ contains $Q := \bigcup_{n \in \mathbb{N}} Q_n$. If now K is a compact subset of U, then $K \subset U_1 \cup \ldots \cup U_n$ for sufficiently large n. From this it follows that

$$\mu_*(K) = \mu^*(K) \leq \mu^*(U_1 \cup \ldots \cup U_n) \leq \sum_{j=1}^{n} \mu^*(U_j) < \sum_{j=1}^{\infty} \mu^*(Q_j) + \varepsilon,$$

where we used the second step. As this last inequality is satisfied by every compact subset K of U, definition (28.4) and equation (28.7) give

$$\mu_*(U) = \mu^*(U) \leq \sum_{j=1}^{\infty} \mu^*(Q_j) + \varepsilon,$$

and since $Q \subset U$ we will then have as well

$$\mu_*(Q) \leq \sum_{j=1}^{\infty} \mu^*(Q_j) + \varepsilon.$$

Finally, $\varepsilon > 0$ being arbitrary here, (28.9) is proven. $\square$

The next corollary sharpens the inequality proved in the first step above.

28.2 Corollary. *For any two disjoint compact subsets K_1, K_2 of E*

$$\mu^*(K_1 \cup K_2) = \mu^*(K_1) + \mu^*(K_2).$$

Proof. Consider any $u \in C_c(E)$ satisfying

$$u \geq 1_{K_1 \cup K_2} = 1_{K_1} + 1_{K_2}.$$

According to 27.3 there is a $v \in C_c(E)$ with $0 \leq v \leq 1$, $v(K_1) = \{1\}$, and $\mathrm{supp}(v) \subset \complement K_2$, hence with $v(K_2) = \{0\}$. The functions vu and $(1 - v)u$ lie in $C_c(E)$ and satisfy

$$vu \geq 1_{K_1} \quad \text{and} \quad (1 - v)u \geq 1_{K_2}.$$

Therefore

$$\mu_*(K_1) + \mu_*(K_2) \leq I(vu) + I((1 - v)u) = I(u),$$

which, because of (28.2), has the consequence that

$$\mu_*(K_1) + \mu_*(K_2) \le \mu_*(K_1 \cup K_2) \,.$$

In view of (28.8) this inequality is half of the equality being claimed. The other half is simply the subadditivity of the outer measure μ^*. $\square$

The first important consequence of all this is:

28.3 Theorem. *The restriction of μ^* to $\mathscr{B}(E)$ is a Borel measure.*

The proof is immediate from Lemma 26.5 and the facts accumulated to this point. Notice that (28.7) and (28.5) say that hypothesis (i) of 26.5 is fulfilled, while (28.7), (28.8) and (28.4) insure that hypothesis (ii) of 26.5 is fulfilled. $\square$

The Borel measure $\mu^* \mid \mathscr{B}(E)$ has a series of further remarkable properties:

28.4 Theorem. *Every Borel subset $A \subset E$ with $\mu^*(A) < +\infty$ satisfies*

$$\mu_*(A) = \mu^*(A) \,.$$

Proof. Given $\varepsilon > 0$, there is an open $U \supset A$ such that

$$\mu^*(U) - \mu^*(A) < \varepsilon/2 \,,$$

which, due to $\mu^*(A) < +\infty$ and μ^* being a measure on $\mathscr{B}(E)$, can be written as

$$\mu^*(U \setminus A) = \mu^*(U) - \mu^*(A) < \varepsilon/2 \,.$$

From (28.4) we get compact $L \subset U$ such that

$$\mu^*(U \setminus L) = \mu^*(U) - \mu^*(L) < \varepsilon/2 \,.$$

The set

$$Q := (U \setminus A) \cup (U \setminus L)$$

then satisfies $\mu^*(Q) < \varepsilon$. Hence there is an open $G \supset Q$ such that

$$\mu^*(G) < \varepsilon \,.$$

Now $K := L \setminus G$ is a (closed, hence) compact subset of L with the properties

(28.10) $K \subset A$ and $A \setminus K \subset G$.

In fact, on the one hand

$$K = L \setminus G \subset L \setminus Q \subset L \setminus (U \setminus A) = L \cap A \,,$$

since $L \subset U$, and on the other hand

$$A \setminus K = A \setminus (L \setminus G) = (A \cap G) \cup (A \setminus L) \subset G \cup (U \setminus L) = G \,,$$

since $U \setminus L \subset Q \subset G$. From (28.10) we get

$$\mu^*(A) - \mu^*(K) = \mu^*(A \setminus K) \le \mu^*(G) < \varepsilon \,,$$

and so $\mu^*(A) < \mu^*(K) + \varepsilon \leq \mu_*(A) + \varepsilon$. As $\varepsilon > 0$ was arbitrary, this says that $\mu^*(A) \leq \mu_*(A)$, which with (28.6) finishes the proof. $\square$

The finiteness hypothesis in the preceding theorem can be weakened. In doing so we make use of the terminology introduced just before the proof of Theorem 13.6.

28.5 Corollary. *The equality $\mu_*(A) = \mu^*(A)$ also holds for every $A \in \mathscr{B}(E)$ which has σ-finite μ^*-measure.*

Proof. The terminology means that there exist $A_n \in \mathscr{B}(E)$ $(n \in \mathbb{N})$, each of finite μ^*-measure, such that $A_n \uparrow A$. The preceding theorem and the isotoneity yield
$$\mu^*(A_n) = \mu_*(A_n) \leq \mu_*(A),$$
from which and the continuity of μ^* from below on $\mathscr{B}(E)$ follows
$$\mu^*(A) = \sup_n \mu^*(A_n) \leq \mu_*(A).$$
Together with (28.6) this proves the claimed equality. $\square$

Another central result, analogous to 28.3, emerges:

28.6 Theorem. *The restriction of μ_* to $\mathscr{B}(E)$ is also a Borel measure.*

Proof. Since all compact K satisfy $\mu_*(K) = \mu^*(K) < +\infty$, all that has to be proved is that $\mu_* \mid \mathscr{B}(E)$ is a measure, i.e., that μ_* is countably additive on $\mathscr{B}(E)$. To that end, let (A_n) be a sequence of pairwise disjoint sets from $\mathscr{B}(E)$, whose union is A. For every compact $K \subset A$, $K = \bigcup_{n \in \mathbb{N}} (K \cap A_n)$, so from 28.3 and 28.4 we get
$$\mu_*(K) = \mu^*(K) = \sum_{n=1}^{\infty} \mu^*(K \cap A_n) = \sum_{n=1}^{\infty} \mu_*(K \cap A_n) \leq \sum_{n=1}^{\infty} \mu_*(A_n).$$
Taking the supremum over such K on the left, (28.4) gives
$$\mu_*(A) \leq \sum_{n=1}^{\infty} \mu_*(A_n).$$
In proving the reverse inequality we may assume that $\mu_*(A) < +\infty$, and therefore $\mu_*(A_n) < +\infty$ for every $n \in \mathbb{N}$. There is then, given $\varepsilon > 0$, a compact $K_n \subset A_n$ satisfying
$$\mu_*(A_n) - \mu_*(K_n) < 2^{-n}\varepsilon \qquad\qquad \text{for each } n \in \mathbb{N}.$$
Since the sets K_j are pairwise disjoint,
$$\mu_*(A) \geq \mu_*\left(\bigcup_{j=1}^{n} A_j\right) \geq \mu_*\left(\bigcup_{j=1}^{n} K_j\right) = \mu^*\left(\bigcup_{j=1}^{n} K_j\right) = \sum_{j=1}^{n} \mu^*(K_j) = \sum_{j=1}^{n} \mu_*(K_j)$$
$$> \sum_{j=1}^{n} \mu_*(A_j) - \sum_{j=1}^{n} 2^{-j}\varepsilon \qquad\qquad \text{for every } n \in \mathbb{N}.$$

Letting $n \to \infty$ we infer that

$$\mu_*(A) \geq \sum_{j=1}^{\infty} \mu_*(A_j) - \varepsilon \,,$$

holding for every $\varepsilon > 0$. That is, $\mu_*(A) \geq \sum_{j=1}^{\infty} \mu_*(A_j)$, the complementary inequality we needed to finish the proof. □

We now set

(28.11) $\mu_{\mathrm{o}} := \mu_* \mid \mathscr{B}(E)$ and $\mu^{\mathrm{o}} := \mu^* \mid \mathscr{B}(E)$

and, inspired by COURRÈGE [1962], call these the *essential measure* determined by I and the *principal measure* determined by I, respectively. Each is a Borel measure (28.3 and 28.6).

Obviously the *essential measure* μ_{o} is inner regular, hence is a *Radon measure* on E. By contrast the *principal measure* μ^{o} is *outer regular*. It turns out that μ_{o} is the more important of the two.

Thus to the given positive linear form I on $C_c(E)$ we have associated two Borel measures. The further relation of these measures to I and the questions of whether and when they coincide will be clarified in the next section. The closing lemma of this section recasts definition (28.4), when A is open, into a equivalent form. It has a preparatory character.

28.7 Lemma. *Every open set $U \subset E$ satisfies*

(28.12) $\mu_{\mathrm{o}}(U) = \mu^{\mathrm{o}}(U) = \sup\{I(u) : u \in C_c(E), \mathrm{supp}(u) \subset U, 0 \leq u \leq 1\}\,.$

Proof. The first equality is just (28.7). Denote the right side of (28.12) by γ, and consider any compact $K \subset U$. Corollary 27.3 provides a function $u \in C_c(E)$ with $0 \leq u \leq 1$, $u(K) = \{1\}$ and $\mathrm{supp}(u) \subset U$. In particular, $1_K \leq u$ and so by (28.2) $\mu_*(K) \leq I(u) \leq \gamma$, that is, $\mu_*(K) \leq \gamma$ for every such K. It follows that $\mu^{\mathrm{o}}(U) = \mu^*(U) = \mu_*(U) \leq \gamma$, by (28.4). The reverse inequality $\gamma \leq \mu^{\mathrm{o}}(U)$ is derived as follows: Let $u \in C_c(E)$ be a typical function involved in the definition of γ. Set $L := \mathrm{supp}(u)$ and consider a typical $v \in C_c(E)$ involved in the definition (28.2) of $\mu_*(L)$. Evidently then $u \leq v$, so $I(u) \leq I(v)$; that is, $I(u) \leq \mu_*(L) = \mu_{\mathrm{o}}(L) = \mu^{\mathrm{o}}(L) \leq \mu^{\mathrm{o}}(U)$. Taking the supremum over eligible u gives finally the desired complementary inequality $\gamma \leq \mu^{\mathrm{o}}(U)$. □

A sharpening of equality (28.12) will be presented in Exercise 2 of §29. The special case $U = E$ of lemma 28.7 furnishes the following useful description of the total masses of μ_{o} and μ^{o}:

(28.13) $\|\mu_{\mathrm{o}}\| = \|\mu^{\mathrm{o}}\| = \sup\{I(u) : u \in C_c(E), 0 \leq u \leq 1\}\,.$

Exercises.

1. For a locally compact space E and a measure μ defined on $\mathscr{B}(E)$, show that μ is a Borel measure if and only if $C_c(E) \subset \mathscr{L}^1(\mu)$.

2. Let μ be a Radon measure on a locally compact space E and $(G_i)_{i \in I}$ a family of open sets which is upward filtering, that is, for any $i, j \in I$ there is a $k \in I$ such that $G_i \cup G_j \subset G_k$. Show that $G := \bigcup_{i \in I} G_i$ satisfies

$$\mu(G) = \sup\{\mu(G_i) : i \in I\}.$$

3. Using the preceding exercise, show that for any Radon measure μ on a locally compact space E:
(a) There exists a largest open set G with $\mu(G) = 0$. The set $\complement G$ is called the *support of the measure* μ and is denoted $\operatorname{supp}(\mu)$.
(b) A point $x \in E$ lies in $\operatorname{supp}(\mu)$ if and only if every open neighborhood of x has positive μ-measure.
(c) For a non-negative $f \in C(E)$, $\int f \, d\mu = 0$ if and only if $f = 0$ throughout $\operatorname{supp}(\mu)$.
    Determine $\operatorname{supp}(\lambda^d)$ for L-B measure λ^d on $\mathbb{R}^d$, and $\operatorname{supp}(\varepsilon_a)$ for every Dirac measure ε_a on E.

4. Let μ be a Borel measure on a locally compact space E. Show that every set A from the σ-ring $\rho_\sigma(\mathscr{K})$ generated by the system $\mathscr{K}$ of compact subsets of E is a Borel set which satisfies $\mu_\circ(A) = \mu^\circ(A)$. Here a ring $\mathscr{R}$ in a set Ω is called a σ-*ring* if the union of every sequence of sets in $\mathscr{R}$ is itself a set in $\mathscr{R}$. In complete analogy with σ-algebras, every subset of $\mathscr{P}(\Omega)$ is contained in a smallest σ-ring. Sometimes it is only the sets in $\rho_\sigma(\mathscr{K})$ which get called "Borel sets"; this is the case, e.g., in the classic exposition of HALMOS [1974]. Why is it generally the case that $\rho_\sigma(\mathscr{K}) \neq \mathscr{B}(E)$?

§29. Riesz representation theorem

Again let E be a locally compact space. Every Borel measure μ on E defines a positive linear form

$$I_\mu(u) := \int u \, d\mu$$

on $C_c(E)$. The question posed in §28 was: Is it true that for every positive linear form I on $C_c(E)$ there is a Borel measure μ on E such that $I_\mu = I$, that is, such that

$$I(u) = \int u \, d\mu \qquad \text{for all } u \in C_c(E)?$$

Any such Borel measure μ will be called a *representing measure* for I. The answer, leaked earlier, to this question reads:

29.1 Riesz representation theorem. *If E is a locally compact space, every positive linear form I on $C_c(E)$ has at least one representing measure. In fact, both the essential measure μ_o determined by I and the principal measure μ^o determined by I are representing measures for I.*

Proof. μ_o and μ^o are Borel measures. It must be shown that

$$(29.1) \qquad I(u) = \int u\, d\mu_o = \int u\, d\mu^o \qquad \text{for all } u \in C_c(E),$$

and because of linearity and the fact that the positive and negative parts of each $u \in C_c(E)$ also lie in $C_c(E)$, it suffices to show this for non-negative u. So let such a $u \in C_c(E)$ be given and let the real number $b > 0$ be an upper bound for u. For a given $\varepsilon > 0$ choose real numbers $y_0, \ldots, y_n$ with

$$0 = y_0 < y_1 < \ldots < y_n = b$$

and

$$y_j - y_{j-1} < \varepsilon \qquad \text{for each } j = 1, \ldots, n.$$

We set

$$u_j := (u - y_{j-1})^+ \wedge (y_j - y_{j-1}) \qquad (j = 1, \ldots, n)$$

and get non-negative continuous functions, each having its support in $\operatorname{supp}(u)$, which satisfy

$$(29.2) \qquad u = \sum_{j=1}^{n} u_j,$$

as the following deliberations will confirm. If $x \in E$ and $u(x) = 0$, then $u_j(x) = 0$ for each $j = 1, \ldots, n$. If $x \in E$ and $u(x) > 0$, then there is a unique $j \in \{1, \ldots, n\}$ such that $y_{j-1} < u(x) \le y_j$. In that case $u_j(x) = u(x) - y_{j-1}$ and $u_k(x) = y_k - y_{k-1}$ for $k < j$ and $u_k(x) = 0$ for $k > j$. Equality (29.2) follows. Next we set

$$K_0 := \operatorname{supp}(u) \quad \text{and} \quad K_j := \{u \ge y_j\} \qquad \text{for } j = 1, \ldots, n$$

and have

$$(29.3) \qquad (y_j - y_{j-1})1_{K_j} \le u_j \le (y_j - y_{j-1})1_{K_{j-1}}, \qquad \text{for } j = 1, \ldots, n,$$

which becomes clear from considering the three properties

$$(29.4) \qquad 0 \le u_j \le y_j - y_{j-1},$$
$$(29.5) \qquad \complement K_{j-1} \subset \{u_j = 0\},$$
$$(29.6) \qquad K_j \subset \{u_j = y_j - y_{j-1}\},$$

valid for $j = 1, \ldots, n$. Integrating in (29.3) with respect to μ^o gives

$$(29.7') \qquad (y_j - y_{j-1})\mu^o(K_j) \le \int u_j\, d\mu^o \le (y_j - y_{j-1})\mu^o(K_{j-1}),$$

and from (29.3) we will – momentarily – infer the analogous inequalities

$$(29.7'') \qquad (y_j - y_{j-1})\mu^\circ(K_j) \le I(u_j) \le (y_j - y_{j-1})\mu^\circ(K_{j-1}),$$

valid for all $j \in \{1, \ldots, n\}$. The left half of (29.7'') follows from the left half of (29.3) when account is taken of (28.2) and the fact that $\mu_*(K_j) = \mu^*(K_j) = \mu^\circ(K_j)$. From (29.5) we have $\operatorname{supp}(u_j) \subset K_{j-1}$. For every open $U \supset K_{j-1}$, the function $v := (y_j - y_{j-1})^{-1} u_j$ is therefore an element of $C_c(E)$ with $\operatorname{supp}(v) \subset U$ and satisfying, by (29.4), $0 \le v \le 1$. From Lemma 28.7 then $I(v) \le \mu^\circ(U)$ and hence

$$I(u_j) \le (y_j - y_{j-1})\mu^\circ(U).$$

According to (28.7) $\mu^\circ(U) = \mu_*(U)$ and therefore from (28.5) and the arbitrariness of U we have confirmation of the right-hand side of (29.7''). Upon adding up the inequalities in (29.7') and those in (29.7'') and recalling (29.2), we find that both of the numbers $\int u \, d\mu^\circ$ and $I(u)$ lie between

$$\sum_{j=1}^n (y_j - y_{j-1})\mu^\circ(K_j) \quad \text{and} \quad \sum_{j=1}^n (y_j - y_{j-1})\mu^\circ(K_{j-1})$$

and consequently

$$\left| \int u \, d\mu^\circ - I(u) \right| \le \sum_{j=1}^n (y_j - y_{j-1})\mu^\circ(K_{j-1} \setminus K_j),$$

since $K_n \subset K_{n-1} \subset \ldots \subset K_0$. Due to the choice of the y_j it follows that

$$\left| \int u \, d\mu^\circ - I(u) \right| \le \sum_{j=1}^n \varepsilon \mu^\circ(K_{j-1} \setminus K_j) = \varepsilon \mu^\circ(K_0 \setminus K_n) \le \varepsilon \mu^\circ(K_0).$$

The extreme inequality being valid for every $\varepsilon > 0$ and $\mu^\circ(K_0)$ being finite, the desired equality

$$(29.8) \qquad I(u) = \int u \, d\mu^\circ$$

emerges.

The measures of the compact sets K_j, $j = 0, \ldots, n$ do not change, thanks to (28.8), when μ° is replaced by $\mu_\circ$. Another pass through the preceding derivation therefore leads to the conclusion that $\mu_\circ$ is also a representing measure for I. $\square$

These two representing measures can be characterized by extremality properties:

29.2 Lemma. *Every representing measure μ for I satisfies*

$$\mu(K) \le \mu_\circ(K) \quad \text{and} \quad \mu^\circ(U) \le \mu(U)$$

for all compact subsets K and all open subsets U of E.

Proof. Given K and U, consider functions $u, v \in C_c(E)$ with $1_K \leq v$, $0 \leq u \leq 1$, and $\operatorname{supp}(u) \subset U$. Integrating these inequalities,

$$\mu(K) \leq \int v \, d\mu = I(v) \quad \text{and} \quad I(u) = \int u \, d\mu \leq \mu(U).$$

From (28.2) and Lemma 28.7 therefore the claimed inequalities follow. □

After this preparation we can enhance the statement of the Riesz representation theorem by characterizing the measures μ_o and μ^o, thereby putting into relief the role of Radon measures.

29.3 Theorem. *For every positive linear form I on $C_c(E)$ the associated essential measure μ_o is the unique Radon measure among the representing measures of I.*

Proof. Let μ be a representing measure for I which is inner regular, thus a Radon measure. Since μ_o is also inner regular, it follows from the first part of the preceding lemma that

$$\mu(A) \leq \mu_o(A) \qquad\qquad \text{for every } A \in \mathscr{B}(E).$$

In particular then all open $U \subset E$ satisfy $\mu(U) \leq \mu_o(U) \leq \mu^o(U)$ and when this is combined with the second part of 29.2 we have

(29.9) $$\mu(U) = \mu_o(U) \qquad\qquad \text{for every open } U \subset E.$$

If compact $K \subset E$ is given and U is an open, relatively compact neighborhood of K, then $U \setminus K$ is open, so that (29.9) is applicable and

$$\mu(U) - \mu(K) = \mu(U \setminus K) = \mu_o(U \setminus K) = \mu_o(U) - \mu_o(K).$$

Another appeal to (29.9), remembering that $\mu_o(U) < +\infty$, gives the equality

$$\mu(K) = \mu_o(K),$$

valid for every compact $K \subset E$. This fact and the inner regularity of both measures results in their equality. □

29.4 Theorem. *Among all representing measures for a positive linear form I on $C_c(E)$ the principal representing measure μ^o is characterized by each of the following two properties:*
(i) *μ^o is the smallest among all outer regular representing measures.*
(ii) *μ^o is the unique outer regular representing measure μ which is inner regular on open sets, that is, satisfies*

(29.10) $$\mu(U) = \sup\{\mu(K) : K \text{ compact} \subset U\} \qquad \text{for every open } U.$$

Proof. Let μ be an outer regular representing measure. By Lemma 29.2, $\mu^o(U) \leq \mu(U)$ holds for all open sets U. Since, however, μ^o is also outer regular, that inequality passes over to Borel sets generally:

$$\mu^o(B) \leq \mu(B) \qquad\qquad \text{for all } B \in \mathscr{B}(E),$$

which confirms (i). If K is a compact set

$$\mu(K) \leq \mu_\mathrm{o}(K) = \mu^\mathrm{o}(K)$$

by Lemma 29.2 and (28.8), so by what has already been proven equality prevails here. That is, μ and μ^o coincide on the system $\mathscr{K}$ of all compact sets. Now μ^o satisfies the inner regularity condition for open sets in (29.10), as we know from (28.4), (28.7) and (28.8). If μ also satisfies these conditions, then for every open set U

$$\mu(U) = \sup\{\mu(K) : U \supset K \in \mathscr{K}\} = \sup\{\mu^\mathrm{o}(K) : U \supset K \in \mathscr{K}\} = \mu^\mathrm{o}(U),$$

an equality which passes over to all Borel sets via the outer regularity of both measures; i.e., $\mu = \mu^\mathrm{o}$ on $\mathscr{B}(E)$.   □

Remark. 1. Some authors (cf. HEWITT and STROMBERG [1965] and COHN [1980]) employ the adjective "regular" for just those outer regular Borel measures μ that have property (29.10), in contrast to our usage.

The following example shows that in general μ^o is not the only outer regular representing measure.

Example. 1. Let E be an uncountable set and equip it with the discrete topology. For I take the identically 0 form. Then from the last two theorems it follows that $\mu_\mathrm{o} = \mu^\mathrm{o} = 0$. However the measure μ from Example 6 of §25 is an outer regular representing measure which is not identically 0.

Example 1 – there μ_o and μ^o are identical – leads to the important question whether the essential and the principal measures coincide in general, or under appropriate supplemental conditions. Although according to 28.5 $\mu_\mathrm{o}(A) = \mu^\mathrm{o}(A)$ for all $A \in \mathscr{B}(E)$ having σ-finite μ^o-measure, generally $\mu_\mathrm{o} \neq \mu^\mathrm{o}$. An example due to C.H. DOWKER (cf. the reference in EDWARDS [1953], p. 160) will be presented in Exercise 7 below. Nevertheless in many important situations these measures do coincide and we are going to look into this now.

We will encounter two types of supplemental hypotheses which will entail the equality $\mu_\mathrm{o} = \mu^\mathrm{o}$ on $\mathscr{B}(E)$. The first imposes conditions on the space E, but none on the linear form I.

We already know, for example, that for a compact space E the representing measures μ_o and μ^o determined by a given positive linear form I on $C_c(E)$ coincide. This follows immediately from Theorem 28.4. The reasons that underlie this need to be examined more closely.

29.5 Definition. A locally compact space is called *countable at infinity* (also sometimes σ-*compact*) when it can be covered by a sequence of compact subsets.

Examples. 2. The following spaces are countable at infinity:
(i) every *compact* space;
(ii) the *euclidean* spaces $\mathbb{R}^d$, $d \in \mathbb{N}$: The closed balls with any fixed center and integer radii provide a countable covering by compact sets.

(iii) every *locally compact space with a countable basis* $\mathscr{G}$. For $\mathscr{G}_0 := \{\overline{G} : G \in \mathscr{G}, G$ relatively compact$\}$ is a countable system of compact sets which covers E. Indeed, each $x \in E$ possesses by definition a compact neighborhood V, and since $\mathscr{G}$ is a basis, $x \in G \subset V$ for some $G \in \mathscr{G}$. Of course then $\overline{G} \in \mathscr{G}_0$.

3.   A discrete space is countable at infinity just if it is a countable set.

Every subset A of a space E which is countable at infinity is of course covered by a sequence of compact subsets of E, so from 28.5 we immediately get:

29.6 Theorem. *If the locally compact space E is countable at infinity, then the representing measures $\mu_\circ$ and μ° determined by any positive linear form I on $C_c(E)$ coincide.*

A simple consequence is:

29.7 Corollary. *On a locally compact space E which is countable at infinity every Radon measure (inner regular by definition) is also outer regular.*

Proof. Every Radon measure μ on E defines a positive linear form I_μ on $C_c(E)$ of which it is a representing measure. According to 29.3 μ must coincide with the essential measure $\mu_\circ$ determined by I_μ. Since $\mu_\circ = \mu^\circ$ and the latter is outer regular, so must be μ. $\square$

To justify the terminology "countable at infinity" we sharpen the covering condition featuring in Definition 29.5.

29.8 Lemma. *Let E be a locally compact space which is countable at infinity. Then E can be covered by a sequence $(L_n)_{n\in\mathbb{N}}$ of compact subsets each contained in the interior of its successor. Every compact subset of E is therefore a subset of some (hence of all but finitely many) L_n.*

Proof. First of all there is a sequence (K_n) of compact sets K_n such that $K_n \uparrow E$. Using Corollary 27.3 we find $0 \le u_n \in C_c(E)$ with $u_n \uparrow 1_E$. But then the sets

$$L_n := \{u_n \ge 1/n\}, \qquad\qquad n \in \mathbb{N},$$

do what is wanted: Each is closed and, since $L_n \subset \mathrm{supp}(u_n)$, it is compact. Because (u_n) is isotone

$$L_n \subset \{u_{n+1} \ge 1/n\} \subset \{u_{n+1} > 1/(n+1)\} \text{ open } \subset L_{n+1},$$

whence $L_n \subset \overset{\circ}{L}_{n+1}$, where $\overset{\circ}{A}$ denotes the interior of a set A. As a result, $(\overset{\circ}{L}_n)_{n\in\mathbb{N}}$ is an open covering of E, so finitely many of its sets suffice to cover any given compact subset of E. $\square$

A simple interpretation of countability at infinity now emerges: A locally compact space E is countable at infinity if and only if the infinitely remote point ω_0

in the one-point compactification E' has a countable base of neighborhoods. Such a countable neighborhood basis is furnished by the complements $E' \setminus L_n$ of any sequence (L_n) with the properties described in 29.8.

We come now to the second type of supplemental hypotheses. Here E is an arbitrary locally compact space and conditions will be imposed on the positive linear form I on $C_c(E)$.

29.9 Definition. A positive linear form I on $C_c(E)$ is called *bounded* if there is a real number M such that

$$(29.11) \qquad\qquad |I(u)| \leq M \, \|u\| \qquad\qquad \text{for all } u \in C_c(E).$$

Here $\|f\|$ denotes the supremum norm of any bounded real function f on E. The requirement (29.11) means that I is continuous with respect to the metric (of uniform convergence) in $C_c(E)$ derived from this norm.

Remark. 2. If the space E is compact, then every positive linear form I on $C_c(E)$ is bounded, because $C_c(E) = C(E)$ so the constant function 1 lies in $C_c(E)$. Therefore from $-\|u\| \cdot 1 \leq u \leq \|u\| \cdot 1$ and the positivity of I we infer that

$$-\|u\| \cdot I(1) \leq I(u) \leq \|u\| \cdot I(1),$$

so that (29.11) holds with $M := I(1)$.

The next theorem – like its predecessor – covers compact spaces as a special case.

29.10 Theorem. *If I is a bounded positive linear form on a locally compact space E, then its principal representing measure μ° is finite and coincides with the essential measure $\mu_\circ$.*

Proof. According to (28.13)

$$\|\mu^\circ\| = \sup\{I(u) : 0 \leq u \leq 1, u \in C_c(E)\}.$$

Since $0 \leq u \leq 1$ entails $\|u\| \leq 1$, (29.11) says that $0 \leq I(u) \leq M \, \|u\| \leq M$, and so

$$\|\mu^\circ\| \leq M < +\infty.$$

Thus μ° is a finite measure and the rest follows from 28.4. □

Proceeding via I_μ as before (cf. 29.7) yields

29.11 Corollary. *Every finite Radon measure μ on a locally compact space E is also outer regular.*

Indeed, the positive linear form I_μ on $C_c(E)$ defined by μ is bounded, by $M := \|\mu\| < +\infty$:

$$|I_\mu(u)| = \left| \int u \, d\mu \right| \leq \int |u| \, d\mu \leq \|u\| \, M \qquad \text{for every } u \in C_c(E),$$

and we can conclude as in the proof of 29.7. □

Remarks. 3. From the proof of Theorem 29.10 it also follows that the total mass $\|\mu^\circ\|$ of μ° is the smallest real number $M \geq 0$ that can serve in Definition 29.9.

4. It is not to be expected that in every locally compact space E which is countable at infinity every positive linear form on $C_c(E)$ will have exactly one representing measure with no further qualification. Still less is unqualified uniqueness of representing measures for bounded positive linear forms on $C_c(E)$, when E is only a locally compact space, to be expected. There is a counterexample to both in HALMOS [1974], p. 231 – DIEUDONNÉ [1939] is also cited there – in which the space E is even compact: It is the interval $[1, \Omega]$ of all ordinal numbers not greater than the first uncountable ordinal Ω, equipped with the order topology. The positive linear form I_{ε_Ω} on $C([1, \Omega])$ defined by the Dirac measure ε_Ω has a representing measure μ which is neither inner regular nor outer regular. Thus $\int f \, d\varepsilon_\Omega = \int f \, d\mu$ for all $f \in C([1, \Omega])$ although $\mu \neq \varepsilon_\Omega$. Details can be found in PFEFFER [1977], p. 116.

In view of the last remark the following theorem is especially noteworthy, as well as useful:

29.12 Theorem. *If the locally compact space E has a countable base for its topology, then every Borel measure on E is regular, hence in particular a Radon measure.*

Proof. Let μ be a Borel measure, I_μ the associated positive linear form on $C_c(E)$ and μ° the principal representing measure for I_μ. Along with E each of its open subspaces U also has a countable base. From Example 2 therefore U is countable at infinity; there exists a sequence (K_n) of compact sets such that $K_n \uparrow U$. Since the measures μ, μ° are continuous from below, it follows that

$$\mu(U) = \lim_{n \to \infty} \mu(K_n) \quad \text{and} \quad \mu^\circ(U) = \lim_{n \to \infty} \mu^\circ(K_n).$$

But $\mu(K_n) \leq \mu_\circ(K_n) = \mu^\circ(K_n)$ for every $n \in \mathbb{N}$, by Lemma 29.2. So we get $\mu(U) \leq \mu^\circ(U)$, from which and a second appeal to 29.2

$$(29.12) \qquad \qquad \mu(U) = \mu^\circ(U), \qquad \qquad \text{for every open } U \subset E.$$

For an arbitrary Borel set A and open $U \supset A$ we then have $\mu(A) \leq \mu(U) = \mu^\circ(U)$ and so, on account of the outer regularity of μ°,

$$(29.13) \qquad \qquad \mu(A) \leq \mu^\circ(A), \qquad \qquad \text{for every } A \in \mathscr{B}(E).$$

If $A \in \mathscr{B}(E)$ is relatively compact, we can choose an open relatively compact neighborhood U of A and apply the last inequality to $U \setminus A$, getting

$$\mu(U) - \mu(A) = \mu(U \setminus A) \le \mu^{\circ}(U \setminus A) = \mu^{\circ}(U) - \mu^{\circ}(A) \,.$$

Subtracting (29.12) from this gives us the reverse inequality to (29.13). In summary,

$$(29.14) \qquad\qquad \mu(A) = \mu^{\circ}(A), \qquad \text{for every relatively compact } A \in \mathscr{B}(E).$$

Now, E is, as already noted, countable at infinity. So we have a sequence (L_n) of compact sets which increase to E. (29.14) is applicable to $B \cap L_n$ for any Borel set B and any $n \in \mathbb{N}$. We therefore get

$$\mu(B) = \lim_{n \to \infty} \mu(B \cap L_n) = \lim_{n \to \infty} \mu^{\circ}(B \cap L_n) = \mu^{\circ}(B) \,.$$

That is, μ and μ° coincide throughout $\mathscr{B}(E)$. Since the essential measure μ° is a representing measure for I_{μ}, this fact insures (as does Theorem 29.6, for that matter) that $\mu_{\mathrm{o}} = \mu^{\circ}$. From the double equality $\mu = \mu_{\mathrm{o}} = \mu^{\circ}$ follows finally the regularity of μ. $\square$

In this situation the Riesz representation theorem can therefore be expressed thus:

29.13 Corollary. *For a locally compact space E whose topology has a countable base, every positive linear form I on $C_c(E)$ can be represented as*

$$I(u) = \int u \, d\mu, \qquad\qquad u \in C_c(E),$$

by exactly one Borel measure μ on E.

Example. 4. For each $u \in C_c(\mathbb{R})$ choose real numbers $\alpha \le \beta$ such that $\mathrm{supp}(u) \subset [\alpha, \beta]$ and define

$$L(u) := \int_{\alpha}^{\beta} u(x) \, dx \,,$$

the integral being the usual Riemann integral; it is independent of the specific numbers α and β used. Evidently L is a positive linear form on $C_c(\mathbb{R})$. According to 16.4 L-B measure λ^1 represents L, and by 29.13 it is the only representing measure.

Remark. 5. It is also possible to deduce Theorem 29.12 from Theorem 26.3 and its Corollary 26.4 because *every locally compact space E whose topology has a countable basis is Polish.* In fact along with E, its one-point compactification E' also has a countable base, as follows from Lemma 29.8 and the commentary after it. It will be shown in Remark 3 of §31 that E' is consequently metrizable, and completeness of the metric follows easily from compactness (cf. Example 6, §26). Thus E' is Polish and E is an open subset of it. Therefore according to Example 4, §26 E itself is Polish.

Summarizing, we can say that for every locally compact space E, the mapping that associates to each Radon measure μ on E the positive linear form I_μ on $C_c(E)$ is a bijection between the set of Radon measures on E and the set of positive linear forms on $C_c(E)$. That is the reason why in BOURBAKI [1965] the positive linear forms on $C_c(E)$ are themselves designated as (positive) Radon measures.

If the space E is countable at infinity as well, the Radon measures on E are all outer regular. If moreover the topology of E has a countable base, the Radon measures and the Borel measures on E coincide.

We give now an application to integration that is of fundamental importance.

29.14 Theorem. *For any regular Borel measure μ on a locally compact space E and any $p \in [1, +\infty[$, the vector space $C_c(E)$ is dense in $\mathscr{L}^p(\mu)$ with respect to convergence in p^{th} mean.*

Proof. First of all, $C_c(E) \subset \mathscr{L}^p(\mu)$, because $C_c(E) \subset \mathscr{L}^1(\mu)$ by (28.1) and $|u|^p \in C_c(E)$ whenever $u \in C_c(E)$. The denseness claim requires that for each $f \in \mathscr{L}^p(\mu)$ and each number $\varepsilon > 0$, a function $u \in C_c(E)$ be produced with

$$N_p(f - u) := \left(\int |f - u|^p \, d\mu \right)^{1/p} < \varepsilon.$$

We accomplish this by a stepwise simplification of the function f to be approximated. Since along with f, both f^+ and f^- are in $\mathscr{L}^p(\mu)$, and N_p is a semi-norm, we can assume that $f \geq 0$. By 11.3 and 11.6 there is an isotone sequence (f_n) of $\mathscr{B}(E)$-elementary functions such that $f_n \uparrow f$. All these functions also lie in $\mathscr{L}^p(\mu)$, due to $0 \leq f_n \leq f$. Therefore from the dominated convergence theorem

$$\lim_{n \to \infty} N_p(f - f_n) = 0.$$

This makes it clear that only $\mathscr{B}(E)$-elementary functions need be approximated by $C_c(E)$, and because of the semi-norm properties of N_p the matter even comes down to approximating the indicator functions 1_A of Borel sets A having $\mu(A) = [N_p(1_A)]^p < +\infty$. For such an A the outer regularity of μ supplies an open $U \supset A$ such that

$$[\mu(U) - \mu(A)]^{1/p} = N_p(1_{U \setminus A}) = N_p(1_U - 1_A) < \varepsilon/2.$$

In particular, $\mu(U) < +\infty$. Therefore the inner regularity of μ insures that for some compact $K \subset U$

$$\mu(U \setminus K) = \int 1_{U \setminus K} \, d\mu < \left(\frac{\varepsilon}{2} \right)^p,$$

that is,

$$N_p(1_U - 1_K) < \varepsilon/2.$$

Finally, we use 27.3 to select $u \in C_c(E)$ satisfying $1_K \leq u \leq 1_U$, whence

$$0 \leq 1_U - u \leq 1_U - 1_K$$

and so

$$N_p(1_U - u) < \varepsilon/2 \,.$$

For the function $f = 1_A$ to be approximated we now have

$$N_p(f - u) \le N_p(1_A - 1_U) + N_p(1_U - u) < \varepsilon \,,$$

completing the proof. □

The proof actually uses the inner regularity of μ only on open sets. So what is involved here are conditions which according to 29.4(ii) characterize the principal representing measure. We will not pursue this any further but interested readers can in BOURBAKI [1965] and BAUER [1984], where this remark is placed in a more general framework.

Exercises.

1. Let E be an uncountable discrete space. Using the Borel measure from Example 6 in §25, show that every positive linear form on $C_c(E)$ has at least two different representing measures. This sharpens Example 1 of this section.

2. Let E be a locally compact space and I a positive linear form on $C_c(E)$. With the help of the Riesz representation theorem prove the following refinement of equality (28.12): For every open $U \subset E$

$$\mu^\circ(U) = \sup\{I(u) : 0 \le u \le 1_U, u \in C_c(E)\} \,.$$

3. A K_σ-set is a union of countably many compacta. Prove that in a locally compact space in which every open set is a K_σ-set, every Borel measure is regular. [*Hint:* Re-examine the proof of Theorem 29.12.]

4. Show that a locally compact space E is countable at infinity if and only if there exists a strictly positive function in $C_0(E)$.

5. Prove that for an arbitrary Borel measure μ on a locally compact space E the following two assertions are equivalent: (a) μ is finite. (b) $C_b(E) \subset \mathcal{L}^1(\mu)$. Show that if μ is a Radon measure, the assertion $C_0(E) \subset \mathcal{L}^1(\mu)$ is equivalent to each of (a) and (b).

6. Let E be a locally compact space, I a positive linear form on $C_0(E)$. Show that there is exactly one finite Radon measure μ on E such that $I(f) = \int f \, d\mu$ for every $f \in C_0(E)$. [*Hints:* Indirect proof. Or: For every $\varepsilon > 0$ and non-negative $f \in C_0(E)$ there is a $u \in C_c(E)$ with $|f - u| \le \varepsilon\sqrt{f}$.]

7. Let E_1, E_2 be the interval $[0, 1]$ equipped with the discrete topology, respectively, the usual euclidean topology, and consider the product space $E = E_1 \times E_2$. Show that
(a) E is locally compact.
(b) Every product

$$_x E := \{x\} \times [0, 1], \qquad\qquad 0 \le x \le 1,$$

is a compact subspace of E, which is also open in E.

(c)   A set $U \subset E$ is open if and only if $U \cap {}_xE$ is open for each $x \in [0,1]$.

(d)   Every compact subset of E is covered by finitely many of the sets ${}_xE$.

Now consider $u \in C_c(E)$. By (d) u vanishes in the complement of the union of finitely many ${}_xE$ sets, and for each fixed x, $y \mapsto u(x,y)$ is a continuous function on the compact interval $E_2 = [0,1]$. Therefore

$$I(u) := \sum_{0 \leq x \leq 1} \int_0^1 u(x,y)\, dy$$

is a well defined finite sum, evidently a positive linear form on $C_c(E)$. Show that

(e)   The essential and the principal representing measures for I do not coincide. [*Hint:* Show that the set $A := E_1 \times \{0\}$ is closed and that $\mu_\circ(A) = 0$, while $\mu^\circ(A) = +\infty$.]

(f)   In passing from μ° to the Borel measure $1_B \mu^\circ$ for $B \in \mathscr{B}(E)$ outer regularity may be lost. [It suffices to consider $B := E \setminus A$, for the set A in the preceding hint.]

§30. Convergence of Radon measures

For locally compact spaces E we will henceforth use the notation $\mathscr{M}_+(E)$ for the *set of all* (positive) *Radon measures on* E. The Riesz representation theorem furnishes a canonical bijection of $\mathscr{M}_+(E)$ onto the set of all positive linear forms on $C_c(E)$. With $\mu, \nu \in \mathscr{M}_+(E)$ and real numbers $\alpha \geq 0$, $\beta \geq 0$ the measure $\alpha\mu + \beta\nu$ also lies in $\mathscr{M}_+(E)$, as is easily checked. That is, $\mathscr{M}_+(E)$ is what is called a *convex cone*. Besides $\mathscr{M}_+(E)$ we often consider the following subsets

$$\mathscr{M}_+^b(E) := \{\mu \in \mathscr{M}_+(E) : \mu(E) < +\infty\}$$
$$\mathscr{M}_+^1(E) := \{\mu \in \mathscr{M}_+(E) : \mu(E) = 1\},$$

the set of all *finite* (or *bounded*) *Radon measures* and the set of all *Radon p-measures* on E, respectively. Evidently

$$\mathscr{M}_+^1(E) \subset \mathscr{M}_+^b(E) \subset \mathscr{M}_+(E).$$

In $\mathscr{M}_+^1(E)$ are to found all the Dirac measures on E. And $\mathscr{M}_+^b(E)$ is a convex subcone of $\mathscr{M}_+(E)$.

In the special case $E = \mathbb{R}^d$ the set $\mathscr{M}_+^b(\mathbb{R}^d)$ is the set of all finite Borel measures on $\mathbb{R}^d$, already familiar to us from §24. That the definition there is equivalent to the present one is due to Theorem 29.12, according to which every Borel measure on $\mathbb{R}^d$ is a Radon measure.

Depending on whether one thinks of the elements of $\mathscr{M}_+(E)$ as measures on $\mathscr{B}(E)$ or as positive linear forms on $C_c(E)$, two notions of convergence suggest themselves: One can define the convergence of a sequence (μ_n) in $\mathscr{M}_+(E)$ to

$\mu \in \mathscr{M}_+(E)$ by requiring either that

$$\lim_{n \to \infty} \mu_n(A) = \mu(A) \qquad\qquad \text{for all } A \in \mathscr{B}(E)$$

or

$$\lim_{n \to \infty} \int f \, d\mu_n = \int f \, d\mu \qquad\qquad \text{for all } f \in C_c(E).$$

We will forthwith show that the first of these is of limited interest, while the second is of considerable significance.

30.1 Definition. A sequence $(\mu_n)_{n \in \mathbb{N}}$ of Radon measures on E is said to be *vaguely convergent* to a Radon measure μ if

$$(30.1) \qquad\qquad \lim_{n \to \infty} \int f \, d\mu_n = \int f \, d\mu \qquad\qquad \text{for all } f \in C_c(E).$$

A sequence (μ_n) in $\mathscr{M}_+(E)$ is vaguely convergent just when the sequence of real numbers $(\int f \, d\mu_n)$ converges in $\mathbb{R}$ for every $f \in C_c(E)$. For in this case $f \mapsto \lim_n \int f \, d\mu_n$ evidently defines a positive linear form on $C_c(E)$, so by the Riesz representation theorem together with Theorem 29.3 there is a unique Radon measure μ to which (μ_n) vaguely converges. At the same time we see that a sequence in $\mathscr{M}_+(E)$ can have at most one *vague limit*.

Examples. 1. Let (x_n) be a sequence in E, $x \in E$. If (x_n) converges to x, then (ε_{x_n}) converges vaguely to ε_x, for the latter just amounts to $\lim f(x_n) = f(x)$. In general however $\lim \varepsilon_{x_n}(A) = \varepsilon_x(A)$ does not hold for all $A \in \mathscr{B}(E)$; in fact, if all x_n are distinct from x, $A := \{x\}$ is such a set. Conversely, if (ε_{x_n}) vaguely converges to ε_x, then (x_n) converges to x. For if this were not so, there would be a subsequence of (x_n) which remains outside of some neighborhood U of x. 27.3 furnishes an $f \in C_c(E)$ with $f(x) = 1$ and $\text{supp}(f) \subset U$. Evidently the sequence $(\int f \, d\varepsilon_{x_n}) = (f(x_n))$ does not converge to $\int f \, d\varepsilon_x$.

2. Let (α_n) be an arbitrary sequence of non-negative real numbers and (x_n) a sequence in E with the property that $\{n \in \mathbb{N} : x_n \in K\}$ is finite for every compact $K \subset E$. (In other words, E is not compact and $\lim x_n = \omega_0 \in E'$.) Then the sequence of measures $\mu_n := \alpha_n \varepsilon_{x_n}$ $(n \in \mathbb{N})$ is vaguely convergent to the zero measure $\mu := 0$. For $\int f \, d\mu_n = \alpha_n f(x_n) = 0$ for all n except the finitely many for which $x_n \in \text{supp}(f)$, whenever $f \in C_c(E)$.

The fact, illustrated by Example 1, that the vague convergence of (μ_n) to μ does not generally entail the convergence of $(\mu_n(A))$ to $\mu(A)$ for each $A \in \mathscr{B}(E)$, while, as 30.2 will show, the converse is true, seems to indicate that the first mode of convergence mentioned above is too restrictive to be of much use. Actually, vague convergence of (μ_n) to μ follows just from knowing that $(\mu_n(A))$ converges to $\mu(A)$ for certain special sets $A \in \mathscr{B}(E)$. Even more:

30.2 Theorem. *A sequence (μ_n) of Radon measures on a locally compact space E converges vaguely to a Radon measure μ if and only if the following condition is fulfilled:*

$$(30.2) \qquad \limsup_{n\to\infty} \mu_n(K) \leq \mu(K) \quad and \quad \liminf_{n\to\infty} \mu_n(G) \geq \mu(G)$$

for every compact $K \subset E$ and every relatively compact, open $G \subset E$.

Proof. Suppose (μ_n) converges vaguely to μ and that K and G are any compact and open sets, respectively. Consider functions $u, v \in C_c(E)$ with $u \geq 1_K$, $0 \leq v \leq 1$ and $\operatorname{supp}(v) \subset G$. Then for all $n \in \mathbb{N}$

$$\mu_n(K) \leq \int u\,d\mu_n \quad and \quad \int v\,d\mu_n \leq \mu_n(G)\,,$$

whence

$$\limsup_{n\to\infty} \mu_n(K) \leq \int u\,d\mu \quad and \quad \int v\,d\mu \leq \liminf_{n\to\infty} \mu_n(G)\,.$$

From these inequalities (30.2) follows via (28.2) and (28.12). One only has to recall that the Radon measure μ coincides, thanks to Theorem 29.3, with the essential measure μ_o determined by the linear form I_μ.

Now suppose conversely that condition (30.2) is fulfilled and that an $f \in C_c(E)$ has been given. Since our goal is to confirm (30.1), we lose no generality by assuming that $f \geq 0$. For a pre-assigned $\varepsilon > 0$ we choose finitely many numbers

$$0 = y_0 < y_1 < \ldots < y_k$$

with $y_k > \|f\|$ and $y_j - y_{j-1} = \varepsilon$ for each $j = 1, \ldots, k$. Set

$$K := \operatorname{supp}(f) \quad and \quad A_j := \{y_{j-1} \leq f < y_j\} \cap K, \quad j = 1, \ldots, k.$$

Denoting the compact set $\{f \geq y_j\} \cap K$ by K_j for $j = 0, \ldots, k$ (so $K_k = \emptyset$ and $K_0 = K$), we have $K_{j-1} \supset K_j$ and

$$A_j = K_{j-1} \setminus K_j \qquad\qquad (j = 1, \ldots, k).$$

Because of the obvious inequalities

$$\sum_{j=1}^{k} y_{j-1} 1_{A_j} \leq f \leq \sum_{j=1}^{k} y_j 1_{A_j}\,,$$

every Radon measure ν on E satisfies

$$\sum_{j=1}^{k} y_{j-1}\nu(A_j) \leq \int f\,d\nu \leq \sum_{j=1}^{k} y_j\nu(A_j)\,,$$

from which and a simple calculation using the facts $\nu(A_j) = \nu(K_{j-1}) - \nu(K_j)$ and $y_j - y_{j-1} = \varepsilon$, we get

$$\varepsilon \sum_{j=0}^{k} \nu(K_j) - \varepsilon\nu(K) = \varepsilon \sum_{j=1}^{k} \nu(K_j) \leq \int f\, d\nu \leq \varepsilon \sum_{j=0}^{k} \nu(K_j)\,.$$

For $\nu := \mu_n$ the right-hand inequality gives us

$$\int f\, d\mu_n \leq \varepsilon \sum_{j=0}^{k} \mu_n(K_j) \qquad\qquad \text{for all } n \in \mathbb{N},$$

and therefore from the first half of hypothesis (30.2)

$$\limsup_{n\to\infty} \int f\, d\mu_n \leq \varepsilon \sum_{j=0}^{k} \mu(K_j)\,.$$

But this right-hand side can be estimated by using the left end of the earlier chain of inequalities, with $\nu := \mu$. We thereby get

$$\limsup_{n\to\infty} \int f\, d\mu_n \leq \int f\, d\mu + \varepsilon\mu(K)\,,$$

valid for every $\varepsilon > 0$. Consequently,

$$\limsup_{n\to\infty} \int f\, d\mu_n \leq \int f\, d\mu\,.$$

The complementary inequality that we need is

$$\int f\, d\mu \leq \liminf_{n\to\infty} \int f\, d\mu_n$$

and we get it by an analogous procedure, using the second half of hypothesis (30.2). One sets $G_j := \{f > y_j\}$, $j = 0, \ldots, k$, which are open, relatively compact subsets of K with

$$G_{j-1} \setminus G_j = \{y_{j-1} < f \leq y_j\} = \{y_{j-1} < f \leq y_j\} \cap K\,.$$

These sets take over the role of the K_j. $\quad\square$

The second example above (for the case in which, say, all the α_n equal 1) shows that a vaguely convergent sequence of measures from $\mathcal{M}_+^1(E)$ need not converge to a measure in $\mathcal{M}_+^1(E)$: mass can be lost. This illustrates the following general phenomenon:

30.3 Lemma. *If the sequence $(\mu_n)_{n\in\mathbb{N}}$ of Radon measures on the locally compact space E converges vaguely to the measure $\mu \in \mathcal{M}_+(E)$, then the associated total masses satisfy*

(30.3) $$\|\mu\| \leq \liminf_{n\to\infty} \|\mu_n\|\,.$$

Proof. For every $u \in C_c(E)$ with $0 \le u \le 1$

$$\int u \, d\mu_n \le \|u_n\|$$

holds for $n \in \mathbb{N}$, so from (30.1) follows that

$$\int u \, d\mu \le \liminf_{n \to \infty} \|\mu_n\| \, .$$

Take the supremum of these integrals over all such u and you get, according to (28.13), the total mass $\mu(E) = \|\mu\|$ of μ. The inequality persists after this operation $\quad \square$

Vague convergence of sequences in $\mathcal{M}_+(E)$ is convergence in a certain topology on $\mathcal{M}_+(E)$, called, naturally, the *vague topology*. It is defined as the coarsest topology on $\mathcal{M}_+(E)$ with respect to which all the mappings

$$(30.4) \qquad\qquad \mu \mapsto \int f \, d\mu \qquad\qquad (f \in C_c(E))$$

are continuous. A *fundamental system of neighborhoods* of a typical $\mu_0 \in \mathcal{M}_+(E)$ consists of all sets of the form

$$(30.5) \quad V_{f_1,\dots,f_n;\varepsilon}(\mu_0) := \left\{ \mu \in \mathcal{M}_+(E) : \left| \int f_j \, d\mu - \int f_j \, d\mu_0 \right| < \varepsilon, j = 1, \dots, n \right\}$$

in which $n \in \mathbb{N}$, $0 < \varepsilon \in \mathbb{R}$ and $f_1, \dots, f_n \in C_c(E)$ are all arbitrary. The vague topology is *Hausdorff* because the uniqueness aspect of Riesz's theorem says that if μ, ν are different Radon measures, then $I_\mu \ne I_\nu$, which just means that $\int f \, d\mu \ne \int f \, d\nu$ for some $f \in C_c(E)$.

In this context it is now clear too what should be understood by the vague convergence of a mapping $t \mapsto \mu_t$ of a subset A of a topological space T into $\mathcal{M}_+(E)$ when t converges to a point $t_0 \in \overline{A}$. With respect to the vague topology the convergence

$$\lim_{\substack{t \to t_0 \\ t \in A}} \mu_t = \mu$$

for some $\mu \in \mathcal{M}_+(E)$ just means that

$$(30.6) \qquad\qquad \lim_{\substack{t \to t_0 \\ t \in A}} \int f \, d\mu_t = \int f \, d\mu \qquad\qquad \text{for every } f \in C_c(E).$$

Example. 3. Let K be a non-negative λ^d-integrable, real function on $E := \mathbb{R}^d$ with $\int K \, d\lambda^d = 1$ (for example, the indicator function of the unit cube $[\mathbf{0}, \mathbf{1}]$). For every real $r > 0$ set

$$K_r(x) := r^d K(rx) \qquad\qquad (x \in \mathbb{R}^d).$$

Then K_r is also non-negative and λ^d-integrable, and $\int K_r \, d\lambda^d = 1$ as well. To see this we only have to recall (7.10), according to which the homothety $H_r(x) := rx$ on $\mathbb{R}^d$ transforms L-B measure thus: $H_r(\lambda^d) = r^{-d}\lambda^d$. For from that it follows

that

$$\int K_r \, d\lambda^d = r^d \int K \circ H_r \, d\lambda^d = r^d \int K \, d(H_r(\lambda^d)) = \int K \, d\lambda^d = 1 \,.$$

Now $r \mapsto K_r \lambda^d$ is a mapping of $]0, +\infty[$ into $\mathscr{M}_+^1(\mathbb{R}^d)$, and in the sense of the vague topology it satisfies

(30.7) $$\lim_{r \to +\infty} K_r \lambda^d = \varepsilon_0 \,.$$

To confirm this, first notice that for every $f \in C_c(\mathbb{R}^d)$

$$\int f K_r \, d\lambda^d = r^d \int f \cdot (K \circ H_r) \, d\lambda^d = r^d \int (f \circ H_r^{-1}) \cdot K \, dH_r(\lambda^d)$$
$$= \int (f \circ H_r^{-1}) K \, d\lambda^d = \int f(r^{-1}x) K(x) \lambda^d(\,dx) \,.$$

From this and the Lebesgue dominated convergence theorem the claim (30.7) follows upon checking that, on the one hand

$$\lim_{r \to +\infty} f(r^{-1}x) K(x) = f(\mathbf{0}) K(x) \qquad \text{for every } x \in \mathbb{R}^d,$$

and on the other hand for all real $r > 0$ and all $x \in \mathbb{R}^d$

$$\left| f(r^{-1}x) K(x) \right| \leq \|f\| \cdot K(x) \,,$$

so that $\|f\| \cdot K$ is an integrable majorant for all functions. The "approximation of the identity" ε_0 expressed by (30.7) plays an important role in Fourier analysis (cf. the exercises in §23 of Bauer [1996]). For the algebra $L^1(\lambda^d)$ (cf. Remark 2, §24) has no identity element with respect to convolution, but it is not hard to show that $\|\tilde{K}_r * \tilde{f} - \tilde{f}\| \to 0$ as $r \to +\infty$ for each $\tilde{f} \in L^1(\lambda^d)$, and in many situations this is almost as useful as having an identity.

To $\mathscr{M}_+^b(E)$ belong in particular all *discrete Radon measures* on E. These are the measures δ which can be represented in the form

$$\delta = \sum_{j=1}^{k} \alpha_j \varepsilon_{x_j}$$

for some finite number of points $x_1, \ldots, x_k \in E$ and non-negative real numbers $\alpha_1, \ldots, \alpha_k$. Every δ admits many such representations. Every Radon measure can be approximated, in the sense of the vague topology, by such δ, as we next show.

30.4 Theorem. *For every locally compact space E the set of discrete Radon measures on E is dense in $\mathscr{M}_+(E)$ in the vague topology.*

Proof. Let a measure $\mu_0 \in \mathscr{M}_+(E)$ and a vague neighborhood V of μ_0 be given. As noted after (30.5), we can suppose V is $V_{f_1, \ldots, f_n; 1}(\mu_0)$ for some non-zero $f_1, \ldots, f_n \in C_c(E)$. We have to find a discrete measure δ in V. To that end, consider the compact set

$$K := \bigcup_{j=1}^{n} \operatorname{supp}(f_j)$$

and $\eta > 0$ such that $\eta\mu_0(K) < 1$. Every $y \in K$ has an open neighborhood U_y in E such that $|f_j(y') - f_j(y'')| \leq \eta$ for all $y', y'' \in U_y$ and all $j \in \{1, \ldots, n\}$. Finitely many U_y, say $U_{y_1}, \ldots, U_{y_k}$ suffice to cover K. Set

$$A_1 := K \cap U_{y_1}, \quad A_2 := (K \cap U_{y_2}) \setminus A_1, \ldots, A_k := (K \cap U_{y_k}) \setminus (A_1 \cup \ldots \cup A_{k-1}).$$

These are pairwise disjoint, relatively compact Borel sets whose union is K, and for all $j \in \{1, \ldots, n\}$, $i \in \{1, \ldots, k\}$ and $y', y'' \in A_i$ the inequality $|f_j(y') - f_j(y'')| \leq \eta$ holds. Since only these properties of the A_i are used in the sequel, we can discard those that are empty (not all are because $\emptyset \neq K = A_1 \cup \ldots \cup A_k$), and re-index the others. That is, we can suppose all the A_i are non-empty and then select a point $x_i \in A_i$ for each i. The discrete measure

$$\delta := \sum_{i=1}^{k} \mu_0(A_i)\varepsilon_{x_i}$$

(notice that $\mu_0(A_i)$ is finite because A_i is relatively compact) will be shown to lie in V and that will complete the proof:

$$\left| \int f_j \, d\mu_0 - \int f_j \, d\delta \right| = \left| \sum_{i=1}^{k} \int_{A_i} f_j \, d\mu_0 - \sum_{i=1}^{k} \mu_0(A_i) f_j(x_i) \right|$$

$$= \left| \sum_{i=1}^{k} \int_{A_i} (f_j - f_j(x_i)) \, d\mu_0 \right|$$

$$\leq \sum_{i=1}^{k} \int_{A_i} |f_j - f_j(x_i)| \, d\mu_0 \leq \sum_{i=1}^{k} \eta\mu_0(A_i) = \eta\mu_0(K),$$

using the fact that $|f_j(x) - f_j(x_i)| \leq \eta$ for all $x \in A_i$, all $i \in \{1, \ldots, k\}$. This holds for each $j \in \{1, \ldots, n\}$, and $\eta\mu_0(K) < 1$ by choice of η. Therefore $\delta \in V_{f_1, \ldots, f_n; 1}(\mu_0) = V$, as was to be shown. $\square$

30.5 Corollary. *The discrete p-measures on E are dense in $\mathcal{M}_+^1(E)$ in the vague topology.*

Proof. We take over the notation of the preceding proof. Now μ_0 is a measure in $\mathcal{M}_+^1(E)$, but the discrete measure $\delta = \sum \mu_0(A_i)\varepsilon_{x_i}$ may not be a p-measure, so more work is required. Set $\alpha_i := \mu_0(A_i)$, $i = 1, \ldots, k$. If $K = E$ (in which case E had to be a compact space), then $\alpha_1 + \ldots + \alpha_k = \mu_0(K) = 1$ and δ actually is a p-measure. In general what we have is

$$\alpha_1 + \ldots + \alpha_k = \mu_0(K) \leq \mu_0(E) = 1$$

and if $K \neq E$ we can choose another point, $x_{k+1} \in E \setminus K$, and set

$$\alpha_{k+1} := 1 - (\alpha_1 + \ldots + \alpha_k),$$

which is non-negative. Then

$$\delta' := \sum_{i=1}^{k+1} \alpha_i \varepsilon_{x_i}$$

is a discrete p-measure with $\int f_j \, d\delta = \int f_j \, d\delta'$ for each $j = 1, \ldots, n$, since x_{k+1} lies outside the supports of all these functions. Consequently, $\delta \in V = V_{f_1, \ldots, f_n; 1}(\mu_0)$ yields that also $\delta' \in V$. $\square$

Next we will investigate whether the equality (30.1) and the continuity assertion (30.4) remain valid for classes of continuous functions more general than $C_c(E)$. Recall in this connection that for a measure $\mu \in \mathcal{M}_+^b(E)$, every $f \in C_b(E)$ is μ-integrable: it is $\mathcal{B}(E)$-measurable and its modulus is majorized by a real constant, hence μ-integrable, function. We will formulate the relevant results for sequences only; their extensions to mappings $t \mapsto \mu_t$ are routine.

30.6 Theorem. *If a sequence $(\mu_n)_{n \in \mathbb{N}}$ in $\mathcal{M}_+(E)$ is vaguely convergent to $\mu \in \mathcal{M}_+(E)$ and if the sequence $(\|\mu_n\|)_{n \in \mathbb{N}}$ of total masses is bounded, then along with all the μ_n the measure μ is also finite, and for every $f \in C_0(E)$*

$$\lim_{n \to \infty} \int f \, d\mu_n = \int f \, d\mu.$$

Proof. If we set $\alpha := \sup\{\|\mu_n\| : n \in \mathbb{N}\}$, which is finite, then $\|\mu\| \leq \alpha$, by (30.3), so μ is a finite measure. Definition 27.5 says that for each $\varepsilon > 0$ there is a $g = g_\varepsilon \in C_c(E)$ such that $\|f - g\| \leq \varepsilon$. Therefore

$$\left| \int f \, d\mu_n - \int g \, d\mu_n \right| \leq \|f - g\| \, \|\mu_n\| \leq \alpha\varepsilon \qquad \text{for each } n \in \mathbb{N}$$

and

$$\left| \int f \, d\mu - \int g \, d\mu \right| \leq \alpha\varepsilon,$$

so that via the triangle inequality

$$\left| \int f \, d\mu_n - \int f \, d\mu \right| \leq 2\alpha\varepsilon + \left| \int g \, d\mu_n - \int g \, d\mu \right| \qquad \text{for all } n \in \mathbb{N}.$$

Since the hypothesis of vague convergence means that $\int g \, d\mu_n \to \int g \, d\mu$, we get

$$\limsup_{n \to \infty} \left| \int f \, d\mu_n - \int f \, d\mu \right| \leq 2\alpha\varepsilon,$$

valid for every $\varepsilon > 0$. That is, the limit exists and is 0. $\square$

Remarks. 1. If one considers measures μ_n and $\mu \in \mathcal{M}_+^b(E)$ without the hypothesis $\sup \|\mu_n\| < +\infty$, the above conclusion can fail. The special case of Example 2 in which $E := \mathbb{R}$, $x_n := n$ and $\alpha_n := n$ for all $n \in \mathbb{N}$ illustrates this. For the function f defined by

$$f(x) := \min\{1, |x|^{-1}\} \qquad \text{for } x \neq 0, \;\; f(0) := 1$$

lies in $C_0(\mathbb{R})$. But $\int f \, d\mu_n = 1$ for every $n \in \mathbb{N}$, while $\int f \, d\mu = 0$, because here the vague limit μ is the 0-measure.

2. Example 2, again with $E := \mathbb{R}$ and $x_n := n$ for all n, considered earlier, but this time with the constant sequence $\alpha_n := 1$, shows that indeed $\lim \int f \, d\varepsilon_{x_n} = \int f \, d\mu$ for the measure $\mu := 0$ and all $f \in C_0(\mathbb{R})$, but this equality is already false for the constant function $f := 1_E$ in $C_b(\mathbb{R})$.

The passage from $C_0(E)$ to $C_b(E)$ therefore calls for a special investigation, which we stress by introducing a new definition:

30.7 Definition. Let $\mu, \mu_1, \mu_2, \ldots$ be measures in $\mathcal{M}^b_+(E)$. The sequence $(\mu_n)_{n \in \mathbb{N}}$ is said to be *weakly convergent* to μ if

$$(30.8) \qquad\qquad \lim_{n \to \infty} \int f \, d\mu_n = \int f \, d\mu \qquad\qquad \text{for all } f \in C_b(E).$$

30.8 Theorem. *Suppose the sequence $(\mu_n)_{n \in \mathbb{N}}$ in $\mathcal{M}^b_+(E)$ converges vaguely to the measure $\mu \in \mathcal{M}^b_+(E)$. Then the following statements are equivalent:*
(i)   *The sequence (μ_n) converges weakly to μ.*
(ii) $\lim\limits_{n \to \infty} \|\mu_n\| = \|\mu\|.$
(iii) *For every $\varepsilon > 0$ there exists a compact subset $K = K_\varepsilon$ of E such that*

$$\mu_n(E \setminus K) < \varepsilon \qquad\qquad \text{for all } n \in \mathbb{N}.$$

Proof. (i)$\Rightarrow$(ii) is obvious because $1 \in C_b(E)$.

(ii)$\Rightarrow$(iii): Let $\varepsilon > 0$ be given. The inner regularity and finiteness of μ yield that there is a compact subset L of E such that $\mu(E \setminus L) < \varepsilon$. According to 27.3, L has a compact neighborhood K_0, so there is an open set G with $L \subset G \subset K_0$. By (30.2)

$$\liminf_{n \to \infty} \mu_n(G) \geq \mu(G) \geq \mu(L) > \|\mu\| - \varepsilon,$$

so if we choose $\alpha \in \,] \|\mu\| - \varepsilon, \mu(L) [$ there will be an $n_0 \in \mathbb{N}$ such that $\mu_n(G) > \alpha$ for all $n \geq n_0$. Moreover, in view of (ii) this n_0 may be supposed large enough that $\|\mu_n\| < \alpha + \varepsilon$ for all $n \geq n_0$. Consequently, $\mu_n(K_0) \geq \mu_n(G) > \alpha > \|\mu_n\| - \varepsilon$, so that $\mu_n(E \setminus K_0) < \varepsilon$, for all $n \geq n_0$. For each $n \in \{1, \ldots, n_0\}$ inner regularity and finiteness of μ_n give us a compact $K_n \subset E$ such that $\mu_n(E \setminus K_n) < \varepsilon$. The compact set $K := K_0 \cup K_1 \cup \ldots \cup K_{n_0}$ then satisfies (iii).

(iii)$\Rightarrow$(i): Given $\varepsilon > 0$, let $K = K_\varepsilon$ be as described. Again from (30.2) we have $\mu(E \setminus K) \leq \liminf \mu_n(E \setminus K) \leq \varepsilon$. There is a function $u \in C_c(E)$ with $0 \leq u \leq 1$ and $u(K) = \{1\}$. It satisfies $0 \leq 1 - u \leq 1_{\complement K}$ and so for each $f \in C_b(E)$

$$\left| \int (1-u) f \, d\mu_n \right| \leq \|f\| \int (1-u) \, d\mu_n \leq \|f\| \, \mu_n(\complement K) \leq \|f\| \, \varepsilon \qquad \text{for all } n \in \mathbb{N}$$

and by the same argument

$$\left| \int (1 - u) f \, d\mu \right| \leq \|f\| \, \varepsilon \,.$$

As in the preceding proof, the triangle inequality then gives

$$\left| \int f \, d\mu_n - \int f \, d\mu \right| \leq 2 \|f\| \, \varepsilon + \left| \int u f \, d\mu_n - \int u f \, d\mu \right| \qquad \text{for all } n \in \mathbb{N}.$$

Since $uf \in C_c(E)$, the hypothesis of vague convergence insures that $(\int u f \, d\mu_n)$ converges to $\int u f \, d\mu$, so the preceding inequality yields

$$\limsup_{n \to \infty} \left| \int f \, d\mu_n - \int f \, d\mu \right| \leq 2 \|f\| \, \varepsilon \,,$$

valid for every $\varepsilon > 0$. That is, this limit exists and equals 0, for every $f \in C_b(E)$. Which proves (i). $\square$

30.9 Corollary. *A sequence $(\mu_n)_{n \in \mathbb{N}}$ in $\mathcal{M}_+^1(E)$ is vaguely convergent to $\mu \in \mathcal{M}_+^1(E)$ if and only if it is weakly convergent to μ.*

Remark. 3. A sequence (μ_n) in $\mathcal{M}_+^b(E)$ which satisfies condition (iii) is called *tight*, whether or not any convergence is going on. If a tight sequence from $\mathcal{M}_+^1(E)$ vaguely converges to a measure $\mu \in \mathcal{M}_+(E)$, then first of all, $\|\mu\| \leq 1$ by (30.3), so that $\mu \in \mathcal{M}_+^b(E)$. The preceding theorem then guarantees the weak convergence of (μ_n) to μ and therewith $\mu \in \mathcal{M}_+^1(E)$. In particular, with vaguely convergent tight sequences in $\mathcal{M}_+^1(E)$ no mass is lost (cf. the remark preliminary to Lemma 30.3). Consequences like these constitute the real significance of the tightness concept.

At this point it is worth returning once more to Theorem 30.2. If the measures μ, μ_n there are all finite and of the same total mass, e.g., if they are all p-measures, then the two components of the compound condition (30.2) become equivalent. The result is the following *portmanteau-theorem*:

30.10 Theorem. *Let $\mu, \mu_1, \mu_2, \ldots$ be measures in $\mathcal{M}_+^1(E)$. Then the following three assertions are equivalent:*
(i)   *The sequence $(\mu_n)_{n \in \mathbb{N}}$ converges vaguely (and therefore also weakly) to μ.*
(ii) *For every closed $F \subset E$*

(30.9)
$$\limsup_{n \to \infty} \mu_n(F) \leq \mu(F) \,.$$

(iii) *For every open $G \subset E$*

(30.9')
$$\liminf_{n \to \infty} \mu_n(G) \geq \mu(G) \,.$$

Proof. The first paragraph of the proof of 30.2 actually established that (i)$\Rightarrow$(iii), under the less restrictive hypotheses prevailing there. Since that theorem further shows that the conjunction of (ii) and (iii) implies (i), it only remains to establish the equivalence of (ii) and (iii). That follows from the trivial observation that

$$\nu(\complement A) = \nu(E) - \nu(A) = 1 - \nu(A)$$

holds for all $A \in \mathscr{B}(E)$ and all $\nu \in \mathscr{M}_+^1(E)$. $\square$

Example 1 in this section shows that the weak convergence of a sequence (μ_n) in $\mathscr{M}_+^b(E)$ to a $\mu \in \mathscr{M}_+^b(E)$ does not imply the convergence of $(\int f \, d\mu_n)$ to $\int f \, d\mu$ for every bounded Borel measurable function f. Nevertheless the continuity of the functions f which define weak convergence can be relaxed somewhat. To this end, we consider bounded, real-valued, Borel measurable functions f on E which are *μ-almost everywhere continuous* for a $\mu \in \mathscr{M}_+^b(E)$: After excision of a μ-nullset $N \in \mathscr{B}(E)$, f is continuous at each point of $E \setminus N$. Important examples of such are the indicator functions of boundaryless Borel sets. The latter are defined as follows:

30.11 Definition. A Borel subset Q of a locally compact space E is called bound-aryless with respect to a measure $\mu \in \mathscr{M}_+^b(E)$, *$\mu$-boundaryless* (or *$\mu$-quadrable*) for short, if the boundary $Q^* := \overline{Q} \setminus \mathring{Q}$ of Q is μ-null:

$$(30.10) \qquad\qquad \mu(Q^*) = 0 \, .$$

Examples. 4. Every interval of the number line $\mathbb{R}$ is λ^1-boundaryless.

5. A set $Q \in \mathscr{B}(E)$ is boundaryless with respect to a Dirac measure ε_a if and only if $a \in E \setminus Q^*$. Look back at Example 1 with this observation and the following theorem in mind.

30.12 Theorem. *Suppose the sequence $(\mu_n)_{n \in \mathbb{N}}$ in $\mathscr{M}_+^b(E)$ converges weakly to $\mu \in \mathscr{M}_+^b(E)$. Then*

$$(30.11) \qquad\qquad \lim_{n \to \infty} \int f \, d\mu_n = \int f \, d\mu$$

holds for every bounded Borel measurable function f that is μ-almost everywhere continuous on E. In particular,

$$(30.12) \qquad\qquad \lim_{n \to \infty} \mu_n(Q) = \mu(Q)$$

holds for every μ-boundaryless set $Q \in \mathscr{B}(E)$.

Proof. By hypothesis there is a Borel set $E_0 \subset E$ with $\mu(E \setminus E_0) = 0$ such that f is continuous at the each point of E_0. Let $\varepsilon > 0$ be given. Since μ is a Radon measure, there is a compact $K \subset E_0$ with

$$\mu(E_0 \setminus K) < \varepsilon \, .$$

Every $x \in K$ has an open neighborhood U_x on which the oscillation of f is at most ε, meaning that

$$|f(y_1) - f(y_2)| \le \varepsilon \qquad\qquad \text{for all } y_1, y_2 \in U_x.$$

Choose a compact neighborhood V_x of x with $V_x \subset U_x$ and then use the compactness of K to find finitely many points $x_1, \ldots, x_n \in K$ such that $V_{x_1}, \ldots, V_{x_n}$ cover K. If we now set

$$\alpha := \inf f(E), \quad \beta := \sup f(E), \quad \alpha_j := \inf f(U_{x_j}), \quad \beta_j := \sup f(U_{x_j})$$

for $j = 1, \ldots, n$, then for each such j there exist functions $g_j, h_j \in C_b(E)$ satisfying

$$g_j(x) = \begin{cases} \alpha_j & \text{if } x \in V_{x_j} \\ \alpha & \text{if } x \in \complement U_{x_j} \end{cases} \quad \text{and} \quad h_j(x) = \begin{cases} \beta_j & \text{if } x \in V_{x_j} \\ \beta & \text{if } x \in \complement U_{x_j} \end{cases}$$

as well as

$$\alpha \le g_j \le \alpha_j \le \beta_j \le h_j \le \beta.$$

This follows at the once from 27.3 and the application of an appropriate affine transformation in the range space $\mathbb{R}$. From these properties and definitions it follows in particular that $g_j \le f \le h_j$ for all j. Therefore if we set

$$g := g_1 \vee \ldots \vee g_n \quad \text{and} \quad h := h_1 \wedge \ldots \wedge h_n,$$

then both these functions lie in $C_b(E)$ and they satisfy $\alpha \le g \le f \le h \le \beta$. Moreover,

$$0 \le h(x) - g(x) \le \varepsilon \qquad \text{for all } x \in K.$$

For each $x \in K$ lies in some $V_{x_j} \subset U_{x_j}$ and because of the way U_{x_j} was chosen with respect to the oscillation of f, it follows that $h(x) - g(x) \le h_j(x) - g_j(x) = \beta_j - \alpha_j \le \varepsilon$. We are now in a position to finish the proof, as follows:

$$\int (h - g)\, d\mu = \int_K (h - g)\, d\mu + \int_{E \setminus K} (h - g)\, d\mu$$

$$\le \varepsilon \mu(K) + (\beta - \alpha)\mu(E \setminus K) \le \varepsilon(\mu(E) + \beta - \alpha);$$

and, because $g \le f \le h$ and $g, h \in C_b(E)$, the weak convergence hypothesis gives

$$\int g\, d\mu = \lim_{n \to \infty} \int g\, d\mu_n \le \liminf_{n \to \infty} \int f\, d\mu_n \le \limsup_{n \to \infty} \int f\, d\mu_n$$

$$\le \lim_{n \to \infty} \int h\, d\mu_n = \int h\, d\mu.$$

Of course we also have $\int g\, d\mu \le \int f\, d\mu \le \int h\, d\mu$. Putting all this together shows that any pair of the numbers $\int f\, d\mu$, $\liminf \int f\, d\mu_n$ and $\limsup \int f\, d\mu_n$ differ by at most $\varepsilon(\mu(E) + \beta - \alpha)$. Since $\varepsilon > 0$ is arbitrary, (30.11) holds. $\square$

Let us now look at an application of this theorem which relates the vague convergence of p-measures on the number line to their Theorem 6.6 description in terms of distribution functions. This is the way that weak (and hence vague) convergence made its original historical appearance.

30.13 Theorem. Let $\mu, \mu_1, \mu_2, \ldots$ be measures in $\mathcal{M}_+^1(\mathbb{R})$, that is, probability measures on $\mathscr{B}^1$, and $F, F_1, F_2 \ldots$ their distribution functions. If the sequence $(\mu_n)_{n \in \mathbb{N}}$

converges weakly to μ, then

(30.13) $$\lim_{n\to\infty} F_n(x) = F(x)$$

holds for every $x \in \mathbb{R}$ at which F is continuous. If F is continuous throughout $\mathbb{R}$, then this convergence is uniform on $\mathbb{R}$.

Proof. According to Theorem 30.12, $\lim \mu_n(Q) = \mu(Q)$ for every μ-boundaryless set $Q \in \mathscr{B}^1$ and thus, after (6.11), $\lim F_n(x) = F(x)$ for every $x \in \mathbb{R}$ such that the interval $Q_x :=\,]-\infty, x[$ is μ-boundaryless. We have

$$]-\infty, x] = \overline{Q}_x = \bigcap_{k\in\mathbb{N}} Q_{x+1/k}$$

and therefore

$$\mu(\overline{Q}_x) = \lim_{k\to\infty} \mu(Q_{x+1/k}) = \lim_{k\to\infty} F(x + 1/k)\,.$$

Consequently, Q_x is μ-boundaryless just if the (isotone) function F is right-continuous at x, that is (since distribution functions are everywhere left-continuous), just if x is a point of continuity of F. This proves the first assertion.

Let us now hypothesize that F is continuous on the whole line, and let $\varepsilon > 0$ be given. First of all, (6.13) supplies numbers $a < b$ such that $F(a) < \varepsilon$ and $1 - F(b) < \varepsilon$. The uniform continuity of F on the compact interval $[a, b]$ insures that points $a = x_0 < x_1 < \ldots < x_k = b$ exist such that

$$F(x_j) - F(x_{j-1}) < \varepsilon \qquad\qquad \text{for } j = 1, \ldots, k.$$

From what has already been proven we know that there exists $n_\varepsilon \in \mathbb{N}$ such that

$$|F_n(x_j) - F(x_j)| < \varepsilon \qquad \text{for each } j \in \{0, \ldots, k\} \text{ and all } n \geq n_\varepsilon.$$

But then, as we will show, the inequality $|F_n(x) - F(x)| < 2\varepsilon$ prevails for every $x \in \mathbb{R}$ and all $n \geq n_\varepsilon$, which proves the uniform convergence of (F_n) to F. For if $x < x_0$, then

$$0 \leq F(x) \leq F(x_0) < \varepsilon \quad \text{and} \quad 0 \leq F_n(x) \leq F_n(x_0) < F(x_0) + \varepsilon < 2\varepsilon,$$

that is, $|F_n(x) - F(x)| < 2\varepsilon$. And a similar argument works if $x \geq x_k$. The remaining x fall into $[x_{j-1}, x_j[$ for an appropriate $j \in \{1, \ldots, k\}$, so

$$F(x_{j-1}) \leq F(x) \leq F(x_j) < F(x_{j-1}) + \varepsilon$$

and

$$F(x_{j-1}) - \varepsilon < F_n(x_{j-1}) \leq F_n(x) \leq F_n(x_j) < F(x_j) + \varepsilon < F(x_{j-1}) + 2\varepsilon\,,$$

confirming that in this case too $|F_n(x) - F(x)| < 2\varepsilon$. $\quad\square$

Remarks. 4. At a point $x \in \mathbb{R}$ of discontinuity of F limit relation (30.13) generally fails, as the example $\mu_n := \varepsilon_{-1/n}$, $n \in \mathbb{N}$, confirms.

5. Condition (30.12) for every μ-boundaryless set $Q \in \mathscr{B}(E)$ is also sufficient for the weak convergence of the sequence (μ_n) to μ (cf. Exercise 6 below). The same is true of condition (30.13) (cf. Exercise 7).

The concept of *weak convergence* (with the same definition) is also meaningful if E is a *Polish* space (or even just a metric space) if the measures involved in Definition 30.7 are all finite Borel measure on E. Only the uniqueness of limits calls for discussion:

30.14 Lemma. *Finite Borel measures μ and ν on a metric space E are equal if $\int f\, d\mu = \int f\, d\nu$ for all $f \in C_b(E)$.*

Proof. Let d be a metric giving the topology of E and consider closed subsets $F \subset E$. Suppose we can always find a sequence (f_n) in $C_b(E)$ with $f_n \downarrow 1_F$. Then it would follow from the hypothesis and from Lebesgue's dominated convergence theorem that $\mu(F) = \nu(F)$. The system of closed subsets F of E is an $\cap$-stable generator of the Borel σ-algebra $\mathscr{B}(E)$ and it contains the whole space E. The equality $\mu = \nu$ would thus follow from the uniqueness theorem 5.4.

It remains therefore to prove the existence of such sequences (f_n) and we can suppose $F \neq \emptyset$. For this purpose we use the (uniformly) continuous antitone function $h : \mathbb{R} \to \mathbb{R}$ which is constantly 1 on $]-\infty, 0]$, constantly 0 on $[1, +\infty[$ and defined by $h(t) := 1 - t$ on $[0, 1]$, together with the function $x \mapsto d(x, F) := \inf\{d(x, y) : y \in F\}$. The latter is a (uniformly) continuous function on E, as we showed in the proof of Example 4, §26. Moreover, its zero-set is exactly F, because F is closed. Apparently then the sequence of (uniformly) continuous functions

$$f_n(x) := h(n \cdot d(x, F)), \qquad\qquad x \in E,\, n \in \mathbb{N}$$

does what is wanted. □

Remarks. 6. The concept of μ-boundaryless sets is also meaningful for finite Borel measures μ on Polish spaces. One easily convinces himself that Theorem 30.12 remains valid in this new situation. In the proof one merely has to secure the existence of the needed functions g_j and h_j somewhat differently: To this end one engages Urysohn's lemma (WILLARD [1970], p. 102 or KELLEY [1955], p. 115).

7. Weak convergence in the set of finite Radon measures on a Polish or a locally compact space E derives from a topology in the same way that vague convergence does. It is called, naturally, the *weak topology* and it is defined by letting $C_b(E)$ take over the role of $C_c(E)$ in (30.4).

Weak convergence in (non-locally compact) Polish spaces plays only a marginal role in this book, but is thoroughly investigated in BILLINGSLEY [1968] and PARTHASARATHY [1967].

Exercises.

1. Let E be a locally compact space, $(\mu_n)_{n\in\mathbb{N}}$ a sequence in $\mathcal{M}_+^b(E)$ which is vaguely convergent to $\mu \in \mathcal{M}_+^1(E)$. If $\|\mu_n\| \leq 1$ for every $n \in \mathbb{N}$, then $\lim\limits_{n\to\infty}\|\mu_n\|$ exists and equals 1.

2. Let $(a_n)_{n\in\mathbb{N}}$ be a convergent sequence of real numbers, with $\lim\limits_{n\to\infty} a_n = a \in \mathbb{R}$. Further, let $(\sigma_n)_{n\in\mathbb{N}}$ be a sequence of non-negative real numbers such that $\sigma_1 > 0$ and the series $\sum \sigma_n$ is divergent. Then

$$\lim_{n\to\infty} \frac{\sigma_1 a_1 + \ldots + \sigma_n a_n}{\sigma_1 + \ldots + \sigma_n} = a\,,$$

the case in which all $\sigma_n = 1$ being the best known instance. Here is an outline for a measure-theoretic proof: The equations

$$\mu_n := \frac{\sigma_1 \varepsilon_1 + \ldots + \sigma_n \varepsilon_n}{\sigma_1 + \ldots + \sigma_n}\,, \qquad\qquad n \in \mathbb{N},$$

define a sequence of measures in $\mathcal{M}_+^1(\mathbb{N})$ which vaguely converges to 0. Therefore according to 30.6, $\lim \int f \, d\mu_n = 0$ holds for every $f \in C_0(\mathbb{N})$. The relevant f is the one defined by $f(n) := a_n - a$.

3. Let E be a locally compact space and T a subset of $C_c(E)$ with the following properties: Each compact $K \subset E$ has a relatively compact neighborhood U such that every $f \in C_c(E)$ with $\mathrm{supp}(f) \subset K$ is uniformly approximable on E by functions $t \in T$ whose supports lie in U; and further, there exists a $t \in T$ with $0 \leq t \leq 1$ and $t(K) = \{1\}$. Show that:
(a) A sequence (μ_n) in $\mathcal{M}_+(E)$ is vaguely convergent if and only if the sequence $(\int t \, d\mu_n)$ is convergent in $\mathbb{R}$ for every $t \in T$.
(b) For $E := \mathbb{R}$, the set of all continuously differentiable real-valued functions with compact support is a T with the above properties.

4. With the help of Exercise 3 show that for the functions $f_n(x) := 1 - \sin(nx)$ on $\mathbb{R}$, the sequence $(f_n \lambda^1)_{n\in\mathbb{N}}$ converges vaguely to λ^1, and deduce from this the Riemann–Lebesgue lemma:

$$\lim_{n\to\infty} \int f(x) \sin(nx) \, dx = 0 \qquad\qquad \text{for every } f \in \mathscr{L}^1(\lambda^1).$$

5. Let μ be a finite Radon measure on a locally compact space E. Prove that:
(a) The system $\mathcal{Q}_\mu$ of all μ-boundaryless sets is an algebra in E.
(b) For every $f \in C_b(E)$ there is a countable set $A_f \subset \mathbb{R}$ such that $\{f > \alpha\} \in \mathcal{Q}_\mu$ for every $\alpha \in \mathbb{R} \setminus A_f$. [*Hint*: For every finite set $\{\alpha_1, \ldots, \alpha_n\}$ of real numbers

$$\sum_{j=1}^{n} \mu(\{f = \alpha_j\}) \leq \mu(E) < +\infty\,.]$$

6. $\mu, \mu_1, \mu_2, \ldots$ are finite Radon measures on the locally compact space E. Show that condition (30.12) is also sufficient for weak convergence; that is, from $\lim \mu_n(Q) = \mu(Q)$ for every μ-boundaryless set $Q \subset E$ follows the weak convergence of (μ_n) to μ. This is also true if E is a Polish space. [*Hints*: Imitate the proof

of Theorem 11.6 and show with the help of Exercise 5 that every $0 \leq f \in C_b(E)$ is the uniform limit on E of an isotone sequence (u_n) in the vector space spanned by the indicator functions of the sets in $\mathcal{Q}_\mu$.]

7. As an application of Exercise 6 show that in the context of Theorem 30.13 condition (30.13) there is also sufficient for the weak convergence of (μ_n) to μ.

8. Let $(\alpha_n)_{n \in \mathbb{N}}$ be a sequence of real numbers in $]0, 1[$. From $[0, 1]$ delete the open interval I_{11} centered at $1/2$ having length α_1. There remain two disjoint closed intervals J_{11}, J_{12}. From J_{1j} delete the open interval I_{2j} of length $\alpha_2 \lambda^1(J_{1j})$ whose midpoint is that of J_{1j} $(j = 1, 2)$. Then there remain four pairwise disjoint closed intervals $J_{21}, J_{22}, J_{23}, J_{24}$. From J_{2j} delete the open interval I_{3j} of length $\alpha_3 \lambda^1(J_{2j})$ whose midpoint is that of J_{2j} $(j = 1, 2, 3, 4)$. Then there remain $8 = 2^3$ pairwise disjoint closed intervals $J_{3j}, j = 1, \ldots, 8$. Continuing in this way one gets for each $n \in \mathbb{N}$ pairwise disjoint closed intervals $J_{nj}, j = 1, \ldots, 2^n$. The set

$$C := \bigcap_{n \in \mathbb{N}} (J_{n1} \cup \ldots \cup J_{n2^n})$$

is called a *generalized Cantor discontinuum*, and if all $\alpha_n = 1/3$ it is simply called the *Cantor discontinuum*. Prove that:

(a) C is compact and non-void, but C has void interior.

(b) $\lambda^1(C) = \lim\limits_{n \to \infty} \prod_{j=1}^{n}(1 - \alpha_j)$.

(c) $\lambda^1(C) = 0 \Leftrightarrow \sum\limits_{n=1}^{\infty} \alpha_n = +\infty$

[*Hint*: Recall the inequalities $1 + \alpha < (1 - \alpha)^{-1}$ and $1 - \alpha < e^{-\alpha}$ for $0 < \alpha < 1$.]

(d) In case $\sum\limits_{n=1}^{\infty} \alpha_n < +\infty$, $U :=]0, 1[\setminus C$ is an open subset of $\mathbb{R}$ whose boundary $U^* := \overline{U} \setminus U$ is not a λ^1-nullset.

9. Construct an open subset of $]0, 1[\times]0, 1[$ whose boundary has positive λ^2-measure.

10. Let E be a metric space, with metric d, and let $\mu, \mu_1, \mu_2, \ldots$ be p-measures on $\mathscr{B}(E)$. Show that each of the following is necessary and sufficient for the weak convergence of the sequence (μ_n) to μ:

(a) $\lim \int f \, d\mu_n = \int f \, d\mu$ for all bounded functions f which are *uniformly continuous* on E.

(b) $\limsup \mu_n(F) \leq \mu(F)$ for all closed $F \subset E$.

(c) $\liminf \mu_n(G) \geq \mu(G)$ for all open $G \subset E$.

[*Hints* for (a)⇒(b): Re-examine the proof of 30.14. There it was shown how, for a closed non-empty $F \subset E$, to construct uniformly continuous functions f_n satisfying $f_n \downarrow 1_F$.]

§31. Vague compactness and metrizability questions

We again consider a locally compact space E along with its space $\mathscr{M}_+ = \mathscr{M}_+(E)$ of Radon measures, equipped with the vague topology. Our interest here is in the subsets of $\mathscr{M}_+$ which are compact or relatively compact in this topology. They are naturally called *vaguely compact*, resp., *vaguely relatively compact*.

A necessary condition for the vague relative compactness of a set $H \subset \mathscr{M}_+$ can be inferred at once from the very definition of the vague topology. According to it, for each $f \in C_c(E)$ the real function $\mu \mapsto \int f\, d\mu$ is continuous on $\mathscr{M}_+$. Therefore the image of any relatively compact H under each such mapping must be a relatively compact subset of $\mathbb{R}$, that is, a bounded set. This observation leads to the following definition:

31.1 Definition. A set $H \subset \mathscr{M}_+(E)$ is called *vaguely bounded* (sometimes simply *bounded*) if

$$(31.1) \qquad\qquad \sup_{\mu \in H} \left| \int f\, d\mu \right| < +\infty \qquad\qquad \text{for every } f \in C_c(E).$$

Thus vague boundedness of a set $H \subset \mathscr{M}_+$ is a necessary condition for its vague relative compactness. We want to show that it is also sufficient:

31.2 Theorem. *A set $H \subset \mathscr{M}_+(E)$ is vaguely relatively compact if and only if it is vaguely bounded.*

Proof. In view of the preceding, all that has to be shown in that vague relative compactness follows from the vague boundedness of H. To this end, let α_f denote the real number in (31.1), for each $f \in C_c(E)$, and J_f the compact interval $[-\alpha_f, \alpha_f]$ in $\mathbb{R}$. Also denote the (vague) closure of H in $\mathscr{M}_+$ by $\overline{H}$. First observe that

$$\int f\, d\mu \in J_f$$

for all $f \in C_c(E)$ and all $\mu \in \overline{H}$. In fact, if $f \in C_c(E)$ and $\varepsilon > 0$ are given

$$V_{f;\varepsilon}(\mu) := \left\{ \nu \in \mathscr{M}_+ : \left| \int f\, d\nu - \int f\, d\mu \right| < \varepsilon \right\}$$

is a vague neighborhood of μ, so if $\mu \in \overline{H}$ then $H \cap V_{f;\varepsilon}(\mu) \neq \emptyset$. For any ν in this intersection, $\int f\, d\nu \in J_f$ and therefore

$$\left| \int f\, d\mu \right| \leq \left| \int f\, d\nu \right| + \left| \int f\, d\mu - \int f\, d\nu \right| < \alpha_f + \varepsilon.$$

As the extreme inequality holds for every $\varepsilon > 0$, we see that $\left| \int f\, d\mu \right| \leq \alpha_f$, that is, $\int f\, d\mu \in J_f$.

Now consider the product space

$$P := \mathbb{R}^{C_c} = \underset{f \in C_c}{\times} \mathbb{R}_f$$

in which for each $f \in C_c = C_c(E)$ a copy $\mathbb{R}_f := \mathbb{R}$ of the number line appears as a factor. The product

$$J := \underset{f \in C_c}{\times} J_f$$

is a subspace of P which, as a product of compact spaces, is compact, by the famous Tychonoff theorem (KELLEY [1955], p. 139 or WRIGHT [1994]). To each Radon measure $\mu \in \mathcal{M}_+$ we assign the mapping $f \mapsto \int f \, d\mu$ of $C_c(E)$ into $\mathbb{R}$. This is a point in P. In this way a mapping

$$\Phi : \mathcal{M}_+ \to P$$

is defined which is injective by the Riesz representation theorem. On the basis of what was shown in the opening campaign

$$\Phi(\overline{H}) \subset J.$$

Our goal will be realized if we can show that
  (a) Φ maps $\mathcal{M}_+$ homeomorphically onto $\Phi(\mathcal{M}_+)$, and
  (b) $\Phi(\mathcal{M}_+)$ is closed in P.
  For then $\Phi(\overline{H})$, as a closed subset of $\Phi(\mathcal{M}_+)$, is also closed in P. From $\Phi(\overline{H})$ lying in the compact set J it therefore follows that $\Phi(\overline{H})$ is compact, hence too its homeomorphic image $\overline{H}$.

As to (a): Continuity of a mapping Φ into a product means continuity of every "component" of Φ, that is, of each mapping $\mu \mapsto \int f \, d\mu$ ($f \in C_c(E)$). But this is true right from the definition of the vague topology. Continuity of the mapping Ψ inverse to Φ means continuity of each mapping

$$\Phi(\mu) \mapsto \int f \, d(\Psi(\Phi(\mu))) = \int f \, d\mu$$

of $\Phi(\mathcal{M}_+)$ into $\mathbb{R}$ ($f \in C_c(E)$). But this mapping is just the restriction to $\Phi(\mathcal{M}_+)$ of the projection of $P = \mathbb{R}^{C_c}$ onto its coordinate specified by f.

As to (b): Let $I \in P$ be a point in the closure of $\Phi(\mathcal{M}_+)$ in P. Then I is a positive linear form on $C_c(E)$. To see its additivity, for example, let $f, g \in C_c(E)$ and $\varepsilon > 0$ be given. The set of all $I' \in P$ which satisfy

$$|I'(u) - I(u)| < \varepsilon \qquad\qquad \text{for } u \in \{f, g, f + g\}$$

is a neighborhood of I in P, and therefore contains a point $I' = \Phi(\mu)$ from $\Phi(\mathcal{M}_+)$. I' is thus the positive linear form

$$u \mapsto I'(u) = \int u \, d\mu$$

on $C_c(E)$. That means that we have

$$|I(f + g) - I(f) - I(g)| \leq |I(f + g) - I'(f + g)| + |I'(f + g) - I(f) - I(g)|$$
$$= |I(f + g) - I'(f + g)| + |I'(f) - I(f) + I'(g) - I(g)|$$
$$< \varepsilon + |I'(f) - I(f)| + |I'(g) - I(g)| < 3\varepsilon,$$

and because $\varepsilon > 0$ is arbitrary, the extreme inequality means that its left-hand side must be 0. In a completely analogous way one proves that $I(\alpha f) = \alpha I(f)$ for every $\alpha \in \mathbb{R}$, $f \in C_c(E)$, and $I(g) \geq 0$ for every non-negative $g \in C_c(E)$. With the linearity of I confirmed, the Riesz representation theorem supplies a Radon measure $\nu \in \mathcal{M}_+$ such that $\Phi(\nu) = I$. That is, I lies in $\Phi(\mathcal{M}_+)$, confirming that the latter is closed in P. $\square$

31.3 Corollary. *For every real number $\alpha \geq 0$ the set*

$$\mathcal{E}_\alpha := \{\mu \in \mathcal{M}_+^b(E) : \|\mu\| \leq \alpha\}$$

is vaguely compact.

Proof. For every $f \in C_c(E)$ and $\mu \in \mathcal{E}_\alpha$, $\left|\int f \, d\mu\right| \leq \int |f| \, d\mu \leq \alpha \|f\|$. Consequently, $\mathcal{E}_\alpha$ is vaguely bounded, hence vaguely relatively compact. What therefore remains to be confirmed is the closedness of $\mathcal{E}_\alpha$ in $\mathcal{M}_+$. According to (28.13) $\mathcal{E}_\alpha$ is just the set of all $\mu \in \mathcal{M}_+$ such that $\int u \, d\mu \leq \alpha$ holds for all $[0, 1]$-valued $u \in C_c(E)$. Because the mapping $\mu \mapsto \int u \, d\mu$ of $\mathcal{M}_+$ into $\mathbb{R}$ is continuous, the set $\{\mu \in \mathcal{M}_+ : \int u \, d\mu \leq \alpha\}$ is closed, for each $u \in C_c(E)$, and by the preceding observation $\mathcal{E}_\alpha$ is an intersection of such sets, those for which $u(E) \subset [0, 1]$. Thus $\mathcal{E}_\alpha$ is indeed (vaguely) closed. $\square$

Remark. 1. The set of all measures $\mu \in \mathcal{M}_+(E)$ with $\|\mu\|$ equal to a fixed positive number α is vaguely closed if E is compact (because in that case $1_E \in C_c(E)$). Example 2 of §30, with all the α_n there equal to α, illustrates this.

For a variety of applications it is important to know when, in terms of E, the vague topology of $\mathcal{M}_+(E)$ is metrizable. One reason is that sequences suffice for dealing with metric topologies, but generally not for non-metric ones. The following remark will prove useful in answering this question.

Remark. 2. For every locally compact space E the, obviously injective, mapping

(31.2) $$\varphi : E \to \mathcal{M}_+(E)$$

defined by $\varphi(x) := \varepsilon_x$ is a homeomorphism of E with $\varphi(E) = \{\varepsilon_x : x \in E\}$. For every point $x \in E$ the (open) sets

$$M_{f_1, \ldots, f_n; \eta}(x) := \{y \in E : |f_j(x) - f_j(y)| < \eta, j = 1, \ldots, n\}$$

form a neighborhood basis at x as the f_j run through all finite subsets of $C_c(E)$ and η through all positive real numbers. In fact, if U is a neighborhood of some

$x \in E$, 27.3 furnishes a $u \in C_c(E)$ with $0 \le u \le 1$, $u(x) = 1$ and $\mathrm{supp}(u) \subset U$, which implies that $M_{u;1/2}(x) \subset U$. Using the notation (30.5) it is obvious that

$$\varphi(M_{f_1,\dots,f_n;\eta}(x)) = \varphi(E) \cap V_{f_1,\dots,f_n;\eta}(\varepsilon_x)$$

for all relevant functions, $\eta \in \mathbb{R}_+$ and $x \in E$. Together with the injectivity this clearly shows that φ is a homeomorphism.

As a result of the foregoing, the metrizability of the locally compact space E is clearly a necessary condition for the metrizability of the vague topology on $\mathscr{M}_+(E)$. For the former the existence of a countable basis in E is sufficient, as was noted in Remark 5 of §29. It is useful to formulate this in terms of $C_c(E)$:

31.4 Lemma. *For any locally compact space E the following assertions are equivalent:*
(a) *E has a countable basis.*
(b) *There is a countable subset of $C_c(E)$ which is dense with respect to uniform convergence.*

Proof. (a)$\Rightarrow$(b): Let $\mathscr{G}$ be a countable base for (the topology of) E, $\mathscr{R}$ the set of all open intervals in $\mathbb{R}$ with rational endpoints. For every natural number n let us say that an n-tuple $(G_1,\dots,G_n) \in \mathscr{G}^n$ and an n-tuple $(I_1,\dots,I_n) \in \mathscr{R}^n$ are *compatible with each other* if a function $f \in C_c(E)$ exists such that $f(G_j) \subset I_j$ for each $j = 1,\dots,n$ and $\mathrm{supp}(f) \subset G_1 \cup \dots \cup G_n$. Any such f will be called a *compatibility function* for the pair of n-tuples. Obviously, the set

$$\bigcup_{n\in\mathbb{N}} (\mathscr{G}^n \times \mathscr{R}^n)$$

is countable; there are therefore only countably many such pairs of n-tuples ($n \in \mathbb{N}$) that are compatible with each other. We choose a compatibility function for each such pair and designate by F the set of functions chosen. It suffices to prove that F is a countable dense subset of $C_c(E)$. To prove its denseness, let $u \in C_c(E)$ and $\varepsilon > 0$ be given. Denote the support of u by K. Every $x \in K$ lies in an open neighborhood from $\mathscr{G}$ each point y of which satisfies $|u(x) - u(y)| < \varepsilon$. The compact set K is covered by finitely many such neighborhoods, say by $G_1,\dots,G_n$. The diameter of each image set $u(G_j)$ is at most 2ε. Consequently there are intervals $I_j \in \mathscr{R}$ of length less that 3ε such that $u(G_j) \subset I_j$, for $j = 1,\dots,n$. Thus u is a compatibility function for the pair of n-tuples $(G_1,\dots,G_n)$, $(I_1,\dots,I_n)$. Hence there must also be such a compatibility function f in the representative set F. Every $x \in G_j$ therefore satisfies $|u(x) - f(x)| \le \lambda^1(I_j) < 3\varepsilon$; that is, $|u(x) - f(x)| < 3\varepsilon$ for all $x \in G_1 \cup \dots \cup G_n$. But this latter inequality prevails as well for all $x \in E \setminus (G_1 \cup \dots \cup G_n)$ for the simple reason that both f and u vanish identically in this complement. In summary, $\|u - f\| \le 3\varepsilon$. This proves that F is dense in $C_c(E)$.

(b)$\Rightarrow$(a): Let D be a dense subset of $C_c(E)$. We will show that the system $\mathscr{G}$ of all sets $\{u > 1/2\}$ with $u \in D$ is a base for the topology of E. For every open $U \subset E$ and every point $x \in U$ Corollary 27.3 furnishes an $f \in C_c(E)$ with $f(x) = 1$

and $\operatorname{supp}(f) \subset U$. Since D is dense, there is a $u \in D$ with $\|u - f\| < 1/2$. Then

$$x \in \{u > 1/2\} \subset \{f > 0\} \subset \operatorname{supp}(f) \subset U.$$

If D is countable, so is $\mathscr{G}$. $\square$

Remark. 3. It is easy to show directly that (b) implies the metrizability of E. To this end, let D be a countable dense subset of $C_c(E)$. Now (cf. Corollary 27.3) $C_c(E)$ separates the points of E, so D must also; that is, for any two distinct points $x, y \in E$ there is a $u \in D$ with $u(x) \neq u(y)$. The functions in $D \setminus \{0\}$ may be organized into a sequence $u_1, u_2, \dots$ and we may then define

$$(31.3) \qquad d(x, y) := \sum_{n=1}^{\infty} \frac{|u_n(x) - u_n(y)|}{2^n \|u_n\|}, \qquad x, y \in E.$$

Point-separation by D means that $d(x, y) > 0$ whenever $x \neq y$. All the other properties of a metric on E are obvious for d. This function d on $E \times E$ is a uniform limit of continuous functions and is consequently continuous. Therefore the topology generated by d, which we will call the d-topology, is coarser than the original topology of E. For any given point $x \in E$ and neighborhood U of x in the original topology of E there is, as was shown in the "(b)$\Rightarrow$(a)" part of the preceding proof, a $u \in D$ with

$$x \in V := \{u > 1/2\} \subset U.$$

This function u is however a u_n, so that by (31.3) u is d-continuous and V is d-open. Therefore the d-topology is finer than the original topology of E. Consequently the two topologies in fact coincide.

Now we can provide the final answer to the question posed after Remark 1.

31.5 Theorem. *The following assertions about a locally compact space E are equivalent:*
(a) *$\mathscr{M}_+(E)$ is a Polish space in its vague topology.*
(b) *The vague topology of $\mathscr{M}_+(E)$ is metrizable and has a countable base.*
(c) *The topology of E has a countable base.*
(d) *E is a Polish space.*

Proof. (a)$\Rightarrow$(b): This follows from Definition 26.1 of a Polish space.
(b)$\Rightarrow$(c): In Remark 2 we learned that $x \mapsto \varepsilon_x$ is a homeomorphic mapping of E onto the subspace $\{\varepsilon_x : x \in E\}$ of all Dirac measures in $\mathscr{M}_+(E)$. Since the property of having a countable basis clearly passes to subspaces, (c) follows.
(c)$\Rightarrow$(d): This was shown in Remark 5 of §29.
(d)$\Rightarrow$(a): Lemma 31.4 provides a countable $D_0 \subset C_c(E)$ which is dense in $C_c(E)$ with respect to uniform convergence. Furthermore, according to Example 2 of §29, E is countable at infinity, so that by 29.8 there is a sequence $(L_n)_{n \in \mathbb{N}}$ of compact sets such that $L_n \uparrow E$ and every compact subset K of E satisfies $K \subset L_n$ for all but finitely many n. For each $n \in \mathbb{N}$ choose an $e_n \in C_c(E)$ satisfying $0 \leq e_n \leq 1$,

$e_n(L_n) = \{1\}$. The subset

$$D := D_0 \cup \{u \cdot e_n : u \in D_0, n \in \mathbb{N}\} \cup \{e_n : n \in \mathbb{N}\}$$

of $C_c(E)$ is still only countable and, of course, is dense in $C_c(E)$. Let $d_1, d_2, \ldots$ be an enumeration of its elements:

$$D = \{d_n : n \in \mathbb{N}\}.$$

Using this enumeration we define a mapping

$$\varrho : \mathcal{M}_+ \times \mathcal{M}_+ \to \mathbb{R}_+$$

by

(31.4) $$\varrho(\mu, \nu) := \sum_{n=1}^{\infty} 2^{-n} \min\{1, |\int d_n \, d\mu - \int d_n \, d\nu|\}, \qquad \mu, \nu \in \mathcal{M}_+.$$

All the properties of a metric save perhaps one are obvious for ϱ. What needs checking is that $\mu = \nu$ follows from $\varrho(\mu, \nu) = 0$. In view of the uniqueness part of the Riesz representation theorem this amounts to showing that from

$$\int d_n \, d\mu = \int d_n \, d\nu \qquad\qquad \text{for all } n \in \mathbb{N}$$

follows the equality

$$\int f \, d\mu = \int f \, d\nu \qquad\qquad \text{for every } f \in C_c(E).$$

So let us show this. Given $f \in C_c(E)$ there is $k \in \mathbb{N}$ such that

$$\mathrm{supp}(f) \subset L_k \subset \{e_k = 1\}.$$

Further, given $\varepsilon > 0$ there is $u \in D_0$ with $\|f - u\| \le \varepsilon$, whence, since $f = f e_k$,

(31.5) $$|f - u e_k| \le \varepsilon e_k.$$

Integration yields

(31.6) $$\left| \int f \, d\mu - \int u e_k \, d\mu \right| \le \varepsilon \int e_k \, d\mu,$$

(31.6′) $$\left| \int f \, d\nu - \int u e_k \, d\nu \right| \le \varepsilon \int e_k \, d\nu.$$

As the functions e_k and $u e_k$ are in D, the assumption that $\varrho(\mu, \nu) = 0$ entails that their μ- and the ν-integrals coincide, and it follows that

$$\left| \int f \, d\mu - \int f \, d\nu \right| \le 2\varepsilon \int e_k \, d\mu,$$

holding for every $\varepsilon > 0$. That is, the desired equality $\int f \, d\mu = \int f \, d\nu$ must hold.

The next step is to show that the topology determined by ϱ is none other than the vague topology. We will, to that end, make use of the fact that the sets $V_{f_1, \ldots, f_n; \varepsilon}(\nu)$ defined in (30.5) are a neighborhood base at $\nu \in \mathcal{M}_+$ in the vague topology, when all possible finite subsets $\{f_1, \ldots, f_n\}$ of $C_c(E)$ and all numbers $\varepsilon > 0$ are considered. We will denote by $U_\varepsilon(\nu)$ the open ball of center ν and radius ε with respect to the metric ϱ.

1. Given $\varepsilon > 0$ there exists $m \in \mathbb{N}$ such that

(31.7) $$V_{d_1,\ldots,d_m;\varepsilon/2}(\nu) \subset U_\varepsilon(\nu) \qquad \text{for every } \nu \in \mathscr{M}_+.$$

Indeed, one may take any $m \in \mathbb{N}$ such that

$$\sum_{n=m+1}^{\infty} 2^{-n} < \varepsilon/2$$

and every $\mu \in V_{d_1,\ldots,d_m;\varepsilon/2}(\nu)$ will then satisfy

$$\varrho(\mu,\nu) < \sum_{n=1}^{m} 2^{-n}\frac{\varepsilon}{2} + \frac{\varepsilon}{2} < \varepsilon$$

and consequently lie in $U_\varepsilon(\nu)$.

2. For finitely many $f_1, \ldots, f_n \in C_c(E)$, for every number $\varepsilon > 0$ and every $\nu \in \mathscr{M}_+$, there is a number $\eta > 0$ such that

(31.8) $$U_\eta(\nu) \subset V_{f_1,\ldots,f_n;\varepsilon}(\nu).$$

First of all, choose $k \in \mathbb{N}$ so that

$$\bigcup_{j=1}^{n} \operatorname{supp}(f_j) \subset L_k \subset \{e_k = 1\}.$$

We can find a number δ, dependent on ν, so that

$$0 < \delta < 1 \quad \text{and} \quad \delta^2 + (1 + 2 \int e_k \, d\nu)\delta < \varepsilon.$$

For each j there is a function $u_j \in D_0$ with $\|f_j - u_j\| < \delta$, hence with

$$|f_j - u_j e_k| \le \delta e_k \qquad (j = 1, \ldots, n).$$

Integration with respect to ν and any $\mu \in \mathscr{M}_+$ gives

(31.9) $$\left| \int f_j \, d\mu - \int u_j e_k \, d\mu \right| \le \delta \int e_k \, d\mu,$$

(31.9') $$\left| \int f_j \, d\nu - \int u_j e_k \, d\nu \right| \le \delta \int e_k \, d\nu$$

for $j = 1, \ldots, n$. Choose m so large that all the functions $e_k, u_1 e_k, \ldots u_n e_k$ show up among the first m functions $d_1, \ldots, d_m$ in the enumeration of D, to which they all belong. Finally, set

$$\eta := \delta 2^{-m}$$

and consider any $\mu \in U_\eta(\nu)$. It satisfies

$$2^{-i} \min\{1, |\int d_i \, d\mu - \int d_i \, d\nu|\} \le \varrho(u,\nu) < \eta \le \delta 2^{-i},$$

whence, since $\delta < 1$

$$\left| \int d_i \, d\mu - \int d_i \, d\nu \right| < \delta \qquad\qquad \text{for } i = 1, \ldots, m.$$

Because of the way m was chosen

(31.10) $$\left| \int u_j e_k \, d\mu - \int u_j e_k \, d\nu \right| < \delta \qquad\qquad \text{for } j = 1, \ldots, n$$

and

(31.10′) $$\left| \int e_k \, d\mu - \int e_k \, d\nu \right| < \delta.$$

From (31.9) and (31.9′), as well as from (31.10) it follows, via the triangle inequality that

$$\left| \int f_j \, d\mu - \int f_j \, d\nu \right| < \left(1 + \int e_k \, d\mu + \int e_k \, d\nu \right) \delta;$$

while from (31.10′)

$$\int e_k \, d\mu < \delta + \int e_k \, d\nu,$$

so the preceding implies that

$$\left| \int f_j \, d\mu - \int f_j \, d\nu \right| < \delta^2 + \left(1 + 2 \int e_k \, d\nu \right) \delta < \varepsilon.$$

As this holds for every $j \in \{1, \ldots, n\}$, it asserts that $\mu \in V_{f_1, \ldots, f_n; \varepsilon}(\nu)$ and confirms (31.8). Together (31.7) and (31.8) assert the equality of the vague and the ϱ-topologies.

The next step will be to prove the completeness of the metric ϱ, and we can do that via slight modifications in the foregoing arguments. Let $(\mu_n)_{n \in \mathbb{N}}$ be a ϱ-Cauchy sequence in $\mathcal{M}_+$. Instead of the functions $f_1, \ldots, f_n$ and the number $\varepsilon > 0$ in 2. above, let an $f \in C_c(E)$ and a number $\delta \in \,]0, 1[$ be given. We aim first to prove that the numerical sequence $(\int f \, d\mu_n)_{n \in \mathbb{N}}$ converges in $\mathbb{R}$. Choose $k \in \mathbb{N}$ with $\mathrm{supp}(f) \subset \{e_k = 1\}$ and $u \in D_0$ with $\|f - u\| < \delta$. Then choose $m \in \mathbb{N}$ large enough that the two functions e_k and ue_k are among $d_1, \ldots, d_m$ and set $\eta := \delta 2^{-m}$. Since (μ_n) is a ϱ-Cauchy sequence, there is a natural number N, dependent on η, thus on f and δ, such that

$$\varrho(\mu_r, \mu_s) < \eta \qquad\qquad \text{for all } r, s \geq N.$$

Just as in the earlier deduction scheme, we get that for such r, s

$$\left| \int d_i \, d\mu_r - \int d_i \, d\mu_s \right| < \delta \qquad\qquad \text{for all } i \in \{1, \ldots, m\},$$

which contains in particular the inequalities

(31.11) $$\left| \int u e_k \, d\mu_r - \int u e_k \, d\mu_s \right| < \delta \quad \text{and} \quad \left| \int e_k \, d\mu_r - \int e_k \, d\mu_s \right| < \delta.$$

Of course we also have the f-analogs of (31.9) and (31.9′), so that reasoning similar to that used earlier delivers the inequality

$$\left| \int f \, d\mu_r - \int f \, d\mu_s \right| < \delta^2 + (1 + 2 \int e_k \, d\mu_s) \delta \qquad \text{for all } r, s \geq N.$$

The second of the (valid for all $r, s \geq N$) inequalities in (31.11) shows that the numerical sequence $(\int e_k \, d\mu_n)_{n \in \mathbb{N}}$ is bounded, say by $M \in \mathbb{R}_+$:

$$\int e_k \, d\mu_n \leq M < +\infty \qquad \text{for all } n \in \mathbb{N}.$$

The earlier inequality therefore yields

$$\left| \int f \, d\mu_r - \int f \, d\mu_s \right| < \delta^2 + (1 + 2M) \delta \qquad \text{for all } r, s \geq N.$$

Notice that M depends only on k, hence only on f. Furthermore N depends only on δ and f. Therefore this last inequality affirms that $(\int f \, d\mu_n)_{n \in \mathbb{N}}$ is a Cauchy sequence in $\mathbb{R}$. According to the remark following Definition 30.1 the sequence (μ_n) is therefore vaguely convergent to some $\mu_0 \in \mathscr{M}_+$. Since the vague topology coincides with the ϱ-topology, as we have already confirmed, this means that the ϱ-Cauchy sequence (μ_n) converges to μ_0 in the ϱ-metric.

We finally need to prove that, like the topology of E, the vague topology of $\mathscr{M}_+$ has a countable base. Since the vague topology is generated by the metric ϱ, it is enough to find a countable set $\mathscr{D}_0$ which is dense in $\mathscr{M}_+$; because it is obvious that the set of all open balls with respect to the metric ϱ centered at points of $\mathscr{D}_0$ and having rational radii is then a countable base for the ϱ-topology of $\mathscr{M}_+$. Our candidate for $\mathscr{D}_0$ is the set of all discrete measures

$$\delta := \sum_{i=1}^{k} \alpha_i \varepsilon_{x_i}$$

with positive rational α_i and points x_i drawn from a countable set E_0 which is dense in E. We get such a set E_0 simply by taking a point from each set in a countable base for the topology of E. Evidently, this $\mathscr{D}_0$ is countable. We have to show that for every $\mu \in \mathscr{M}_+$, every real $\varepsilon > 0$, and every finite set $F := \{f_1, \ldots, f_n\} \subset C_c(E)$, the basic vague neighborhood $V_{f_1, \ldots, f_n; \varepsilon}(\mu)$ contains a measure from $\mathscr{D}_0$. At least, according to 30.4, this neighborhood contains a

$$\overline{\delta} := \sum_{i=1}^{k} \overline{\alpha}_i \varepsilon_{\overline{x}_i}$$

with positive real $\overline{\alpha}_i$ and $\overline{x}_i \in E$. Thus

(31.12) $$\left| \int f \, d\mu - \int f \, d\overline{\delta} \right| = \left| \int f \, d\mu - \sum_{i=1}^{k} \overline{\alpha}_i f(\overline{x}_i) \right| < \varepsilon \qquad \text{for all } f \in F.$$

Now for such f and δ as above

$$\left| \int f\, d\mu - \int f\, d\delta \right| \leq \left| \int f\, d\mu - \int f\, d\overline{\delta} \right| + \left| \int f\, d\overline{\delta} - \int f\, d\delta \right|$$

$$(31.13) \qquad \leq \left| \int f\, d\mu - \int f\, d\overline{\delta} \right| + \sum_{i=1}^{k} \overline{\alpha}_i \left| f(\overline{x}_i) - f(x_i) \right| + \sum_{i=1}^{k} |\overline{\alpha}_i - \alpha_i|\, \|f\| \, .$$

Inequality (31.12) says that the number

$$\varepsilon - \max_{f \in F} \left| \int f\, d\mu - \int f\, d\overline{\delta} \right|$$

is positive. If we choose α_i from $\mathbb{Q}_+$ sufficiently close to $\overline{\alpha}_i$ and x_i from the dense set E_0 sufficiently close to $\overline{x}_i$ $(i = 1, \ldots, k)$, then because of the continuity of the (finitely many) functions f, we can obviously see to it that the two sums in (31.13) together are less than this, so that the right side of inequality (31.13) is less than ε, for each $f \in F$. But that means that $\delta \in \mathscr{D}_0 \cap V_{f_1,\ldots,f_n;\varepsilon}(\mu)$. $\quad\square$

Remarks. 4. The reader should recall the rather elementary fact that for a metric space compactness and sequential compactness are equivalent (see (6.37) in HEWITT and STROMBERG [1965]). In view of this, a very useful consequence of Theorems 31.2 and 31.5 for a locally compact space E with a countable base is that every vaguely bounded sequence in $\mathscr{M}_+(E)$ contains a vaguely convergent subsequence.

In particular, for such E every sequence (μ_n) in $\mathscr{M}_+^1(E)$, that is, every sequence of p-measures, contains a vaguely convergent subsequence. Moreover, in case all convergent subsequences have the same limit μ, the original sequence (μ_n) itself converges vaguely to μ: Otherwise there would be an $f \in C_c(E)$ for which $(\int f\, d\mu_n)$ does not converge to $\int f\, d\mu$, and so an $\varepsilon > 0$ and integers $1 \leq n_1 < n_2 < \ldots$ such that $\left| \int f\, d\mu_{n_j} - \int f\, d\mu \right| \geq \varepsilon$ for all $j \in \mathbb{N}$. The sequence $(\mu_{n_j})_{j \in \mathbb{N}}$ would have a vaguely convergent subsequence and its vague limit could not be μ. If we further hypothesize of (μ_n) that it is *tight*, then with the aid of Remark 3 in §30 we can conclude that $\mu \in \mathscr{M}_+^1(E)$ as well, and that (μ_n) even converges weakly to μ.

5. The foregoing deliberations show (for locally compact E with a countable base) that tight sequences in $\mathscr{M}_+^1(E)$ always contain weakly convergent subsequences. Explicitly formulated this says: A set $H \subset \mathscr{M}_+^1(E)$ is relatively compact (= relatively sequentially compact) in the weak topology if it is tight, meaning that for every $\varepsilon > 0$ a compact $K_\varepsilon \subset E$ exists such that $\mu(E \setminus K_\varepsilon) < \varepsilon$ for every $\mu \in H$. A theorem of YU.V. PROHOROV asserts that the *tightness of H is even equivalent to its weak relative compactness*. More is true: This equivalence prevails as well whenever E is any Polish space. For details the reader can consult BILLINGSLEY [1968].

The ideas employed in the proofs of Theorems 31.4 and 31.5, slightly modified, lead to a further interesting result. It concerns the space

$$C := C(\mathbb{R}_+, E)$$

of all continuous mappings f of $\mathbb{R}_+ := [0, +\infty[$ into a Polish space E, for example, $\mathbb{R}^d$. We endow C with the topology of *uniform convergence on compact subsets* of $\mathbb{R}_+$.

31.6 Theorem. *Along with E, the space $C(\mathbb{R}_+, E)$ is also Polish.*

Proof. Consider any complete metric ϱ which generates the topology of E. Another such metric is given by $(x, y) \mapsto \min\{1, \varrho(x, y)\}$, and using it if need be, we can simply assume that $\varrho \le 1$. This lets us define d_n in C for each $n \in \mathbb{N}$ by

$$d_n(f, g) := \sup\{\varrho(f(x), g(x)) : x \in [0, n]\}, \qquad f, g \in C;$$

and

$$(31.14) \qquad d(f, g) := \sum_{n=1}^{\infty} 2^{-n} d_n(f, g), \qquad f, g \in C.$$

Just as earlier (cf. (31.3) and (31.4)), one easily confirms that d is a metric on C (with all its values in $[0, 1]$) which satisfies

$$(31.15) \qquad 2^{-n} d_n(f, g) \le d(f, g) \le d_n(f, g) + 2^{-n} \qquad \text{for all } n \in \mathbb{N},$$

the right-most inequality following from the fact that $d_i \le d_{i+1}$ for all $i \in \mathbb{N}$, resulting in

$$d(f, g) \le \sum_{i=1}^{n} 2^{-i} d_i(f, g) + \sum_{i=n+1}^{\infty} 2^{-i}.$$

It follows from (31.15) via by-now-familiar reasoning that the d-topology coincides with the original topology of C, and moreover that d is a complete metric.

So it only remains to prove that the topology of C has a countable base. As we showed in the very last phase of the proof of Theorem 31.5, the Polish space E contains a countable dense subset E_0. The system $\mathscr{G}$ of all open balls with respect to the metric ϱ with centers in E_0 and with positive rational radii is then a countable base for E. Together with it we consider a countable base $\mathscr{O}$ for $\mathbb{R}_+$. Thus n-tuples $(O_1, \ldots, O_n) \in \mathscr{O}^n$ and $(G_1, \ldots, G_n) \in \mathscr{G}^n$ are called compatible if there is a function $f \in C$ such that $f(O_j) \subset G_j$ for each $j = 1, \ldots, n$. And, as before, any such f will be called a compatibility function. Because

$$\bigcup_{n \in \mathbb{N}} (\mathscr{O}^n \times \mathscr{G}^n)$$

is countable, there is a countable set $F \subset C$ which contains a compatibility function for each pair of compatible n-tuples, for each $n \in \mathbb{N}$. The open d-balls having centers in F and rational radii are a countable set, and it is easy to see that they constitute a base for the d-topology of C once we confirm that F is dense in C.

So that is now our goal. Consider then an arbitrary $f_0 \in C$ and $N \in \mathbb{N}$. Set $\varepsilon := 2^{-N-2}$. Since f is continuous, every $x \in [0, N]$ lies in a set $O \in \mathscr{O}$ such that

$$\varrho(f_0(y), f_0(x)) < \varepsilon/2 \qquad \text{for all } y \in O.$$

Finitely many such sets O suffice to cover $[0, N]$, say $O_1, \ldots, O_n$. By the triangle inequality

$$\varrho(f_0(y), f_0(x)) < \varepsilon \qquad \text{for all } x, y \in O_j, \, j \in \{1, \ldots, n\}.$$

Choose a point x_j from each O_j. Then

$$\varrho(f_0(x), f_0(x_j)) < \varepsilon \qquad \text{for all } x \in O_j, \, j \in \{1, \ldots, n\}.$$

The open ϱ-ball of radius ε centered at $f_0(x_j)$ meets the dense set E_0, say in the point z_j. As ε is rational, the open ϱ-ball of center z_j and radius 2ε is a set $G_j \in \mathscr{G}$. Then every $x \in O_j$ satisfies

$$\varrho(f_0(x), z_j) \le \varrho(f_0(x), f_0(x_j)) + \varrho(f_0(x_j), z_j) < 2\varepsilon ,$$

which means that $f_0(O_j) \subset G_j$, all this for each $j \in \{1, \ldots, n\}$. This shows that f_0 is a compatibility function for $(O_1, \ldots, O_n)$ and $(G_1, \ldots, G_n)$. Consequently, this pair of n-tuples has a compatibility function $f \in F$, that is, $f \in F$ satisfies

$$f(O_j) \subset G_j \qquad \text{for } j = 1, \ldots, n.$$

It follows that $f(x), f_0(x)$ both lie in G_j whenever $x \in O_j$ and so

$$\varrho(f(x), f_0(x)) < 4\varepsilon .$$

As the O_j cover $[0, N]$, this inequality holds for every $x \in [0, N]$. It affirms that $d_N(f, f_0) < 4\varepsilon$, and so thanks to (31.15) and the definition of ε, $d(f, f_0) < 4\varepsilon + 2^{-N} = 2^{-N+1}$. As $N \in \mathbb{N}$ is arbitrary, this shows that F is d-dense in C, which, as noted earlier, completes the proof. $\square$

The significance of Theorem 31.5 lies partly in the fact that for a locally compact space E whose topology has a countable base the space $\mathscr{M}_+(E)$ of all (positive) Radon measures – which according to 29.12 is the set of all Borel measures on E – being also a Polish space, is itself an environment in which measure theory can be pursued. And this happens in convex analysis, in integral geometry, and in stochastic geometry, a meeting point between geometry and probability theory. The path-space $C(\mathbb{R}_+, E)$ of all continuous paths or curves $t \mapsto f(t)$, $t \in \mathbb{R}_+$, in a Polish space E (Theorem 31.6) plays a fundamental role in the theory of stochastic processes. For example, the Polish space $C(\mathbb{R}_+, \mathbb{R}^d)$ carries the famous Wiener measure; it is the steering mechanism of the Brownian motion in $\mathbb{R}^d$ (cf. Bauer [1996]).

Exercises.

1. Let E be a locally compact space, $\nu \in \mathscr{M}_+(E)$. Show that the set of all $\mu \in \mathscr{M}_+(E)$ which satisfy $0 \le \int u \, d\mu \le \int u \, d\nu$ for every non-negative $u \in C_c(E)$ is vaguely compact.

2. Let E be a locally compact space with a countable base. Prove that there is a countable subset of $C_c(E)$ that has the properties of the set T in Exercise 3, §30. [*Hint*: Try the set D that featured in the proof of Theorem 31.5.]

3. (Selection theorem of E. HELLY (1884–1943)). Prove the original form of Corollary 31.3: To every sequence $(F_n)_{n \in \mathbb{N}}$ of distribution functions on $\mathbb{R}$ corresponds a measure-generating function $F : \mathbb{R} \to \mathbb{R}$ and a subsequence $(F_{n_k})_{k \in \mathbb{N}}$ of the original sequence such that $\lim_{k \to \infty} F_{n_k}(x) = F(x)$ for every continuity point x of F. Why is F generally not a distribution function? How does one recover 31.3 (for the case $E := \mathbb{R}$) from Helly's theorem?

4. For a Polish space E consider the topology (introduced in Remark 7 of §30) of weak convergence on the set of finite Borel measures (the finite Radon measures – cf. 26.2) on E. By adapting the ideas in the proof of Theorem 31.5, show that this topology is metrizable.

5. For what more general spaces taking over the role of $\mathbb{R}_+$ in the definition of $C(\mathbb{R}_+, E)$ does Theorem 31.6 remain valid?

Bibliography

U. ANONYME [1889]: "Sur l'intégrale $\int_0^\infty e^{-x^2}\,dx$", *Bull. Sci. Math.* (2)13, 84.

G. AUMANN [1969]: *Reelle Funktionen.* Grundlehren Math. Wiss. 68 (2nd edition), Springer-Verlag, Berlin–Heidelberg–New York.

S. BANACH [1923]: "Sur le problème de la mesure", *Fund. Math.* 4, 7–33.

R.G. BARTLE and J.T. JOICHI [1961]: "The preservation of convergence of measurable functions", *Proc. Amer. Math. Soc.* 12, 122–126.

H. BAUER [1984]: *Maße auf topologischen Räumen,* Kurs der Fernuniversität-Gesamthochschule-Hagen.

— [1996]: *Probability Theory,* de Gruyter Stud. Math. 23. Walter de Gruyter, Berlin–New York.

S.K. BERBERIAN [1962]: "The product of two measures", *Amer. Math. Monthly* 69, 961–968.

P. BILLINGSLEY [1968]: *Convergence of Probability Measures.* John Wiley & Sons, Inc., New York–London–Sydney–Toronto.

G. BIRKHOFF and S. MACLANE [1965]: *A Survey of Modern Algebra* (3rd edition). The Macmillan Co., New York.

N. BOURBAKI [1965]: *Intégration, Chap. 1–4.* Hermann, Paris.

A. BROUGHTON and B.W. HUFF [1977]: "A comment on unions of σ-fields", *Amer. Math. Monthly* 84, 553–554.

S.D. CHATTERJI [1985–86]: "Elementary counter-examples in the theory of double integrals", *Atti Sem. Mat. Fis. Univ. Modena* 34, 363–384.

G. CHOQUET [1969]: *Lectures on Analysis.* Vol. I. W.A. Benjamin, New York–Amsterdam.

J.P.R. CHRISTENSEN [1974]: *Topology and Borel Structure.* Mathematical Studies 10. North-Holland Publ. Co., Amsterdam–London.

D.L. COHN [1980]: *Measure Theory.* Birkhäuser Verlag, Basel–Boston–Stuttgart.

P. COURRÈGE [1962]: *Théorie de la mesure.* Les cours de Sorbonne. Centre de Documentation Universitaire, Paris 5ᵉ.

C. DELLACHERIE et P.-A. MEYER [1975]: *Probabilités et potentiel,* Chap. I à IV. Hermann, Paris.

P. DIEROLF and V. SCHMIDT [1998]: "A proof of the change of variable formula for d-dimensional integrals", *Amer. Math. Monthly* 105, 654–656.

J. DIEUDONNÉ [1939]: "Un exemple d'espace normal non susceptible d'une structure uniforme d'espace complet", *C. R. Acad. Sci. Paris Sér. I Math.* 209, 145–147.

— [1978]: *Abrégé d'Histoire des Mathématiques, 1700-1900,* tome II. Hermann, Paris.

E.B. DYNKIN [1965]: *Markov Processes,* I, II. Grundlehren Math. Wiss. 121, 122. Springer-Verlag, Berlin–Heidelberg–New York.

R.E. EDWARDS [1953]: "A theory of Radon measures on locally compact spaces", *Acta Math.* 89, 133-164.

B.W. GNEDENKO [1988]: *The Theory of Probability* (translated from Russian by G. Yankovsky) 6th printing. Mir Publishers, Moscow.

C. GOFFMAN and G. PEDRICK [1975]: "A proof of the homeomorphism of Lebesgue-Stieltjes measure with Lebesgue measure", *Proc. Amer. Math. Soc.* 52, 196–198.

H. HAHN and A. ROSENTHAL [1948]: *Set Functions.* The University of New Mexico Press, Albuquerque.

P.R. HALMOS [1974]: *Naive Set Theory.* Undergrad. Texts Math., Springer-Verlag, New York–Heidelberg.

— [1974]: *Measure Theory.* Grad. Texts in Math. 18 , Springer-Verlag, New York–Heidelberg–Berlin.

F. HAUSDORFF [1914]: *Grundzüge der Mengenlehre.* Verlag von Veit und Comp., Leipzig; reprinted (1949), Chelsea Publishing Comp., New York.

T. HAWKINS [1970]: *Lebesgue's Theory of Integration.* University of Wisconsin Press, Madison–Milwaukee–London.

J. HENLE and S. WAGON [1983]: "A translation-invariant measure", *Amer. Math. Monthly* 90, 62–63.

E. HEWITT and K.A. ROSS [1979]: *Abstract Harmonic Analysis I.* Grundlehren Math. Wiss. 115 (2nd edition). Springer-Verlag, Berlin–Heidelberg–New York.

E. HEWITT and K. STROMBERG [1965]: *Real and Abstract Analysis.* Grad. Texts in Math. 25. Springer-Verlag, New York-Heidelberg-Berlin.

J.L. KELLEY [1955]: *General Topology,* Grad. Texts in Math. 27. D. Van Nostrand Co., Inc. Princeton; reprinted (1975), Springer-Verlag, New York–Heidelberg–Berlin.

L. MATTNER [1999]: "Product measurability, parameter integrals, and a Fubini counterexample", *Enseign. Math.* (2) 45, 271–279.

P.-A. MEYER [1966]: *Probability and Potentials.* Blaisdell Publ. Comp., Waltham, Massachusetts–Toronto–London.

L. NACHBIN [1965]: *The Haar Integral.* The University Series in Higher Mathematics. (Translated from Portugese by L. Bechtolsheim.) D. Van Nostrand Co., Inc. Princeton; reprinted (1976), R.E. Krieger Publ. Comp., Huntington, New York.

J. VON NEUMANN [1929]: "Zur allgemeinen Theorie des Maßes", *Fund. Math.* 13, 73–116+333.

W.P. NOVINGER [1972]: "Mean convergence in L^p-spaces", *Proc. Amer. Math. Soc.* 34, 627–628.

D.A. OVERDIJK, F.H. SIMONS and J.G.F. THIEMANN [1979]: "A comment on unions of rings", *Indag. Math.* 41, 439–441.

J.C. OXTOBY and S. ULAM [1941]: "Measure-preserving homeomorphisms and metrical transitivity", *Ann. of Math.* (2) 42, 874–920.

K.R. PARTHASARATHY [1967]: *Probability Measures on Metric Spaces,* Academic Press, New York–London.

W.F. PFEFFER [1977]: *Integrals and Measures.* Marcel Dekker. New York–Basel.

J. RADON [1913]: "Theorie und Anwendungen der absolut additiven Mengenfunktionen", *Sitzungsber. Kaiserl. Akad. Wiss. Wien, Math.-Naturwiss. Kl.* 122, 1295–1438.

H. RICHTER [1966]: *Wahrscheinlichkeitstheorie.* Grundlehren Math. Wiss. 86 (2nd edition). Springer-Verlag, Berlin–Heidelberg–New York.

F. RIESZ [1911]: "Sur certaines systèmes singuliers d'équations intégrales", *Ann. Sci. École Norm. Sup.* (3) 28, 33-62.

J.B. ROBERTSON [1967]: "Uniqueness of measures", *Amer. Math. Monthly* 74, 50–53.

W. RUDIN [1962]: *Fourier Analysis on Groups.* Interscience Tracts in Pure Appl. Math. 12. John Wiley & Sons, New York–London.

— [1987]: *Real and Complex Analysis* (3rd edition). McGraw-Hill Book Comp., New York–Hamburg–Tokyo–Toronto.

S. SAEKI [1996]: "A proof of the existence of infinite product probability measures", *Amer. Math. Monthly* 103, 682–683.

W. SIERPINSKI [1928]: "Un théorème général sur les familles d'ensembles", *Fund. Math.* 1, 206–210.

R.M. SOLOVAY [1970]: "A model of set-theory in which every set of reals is Lebesgue measurable", *Ann. of Math.* (2) 92, 1–56.

R.H. SORGENFREY [1947]: "On the topological product of paracompact spaces", *Bull. Amer. Math. Soc.* 53, 631–632.

S.M. SRIVASTAVA [1998]: *A Course on Borel Sets.* Grad. Texts in Math. 180. Springer-Verlag, New York–Berlin.

K. STROMBERG [1972]: "An elementary proof of Steinhaus's theorem", *Proc. Amer. Math. Soc.* 36, 308.

— [1979]: "The Banach-Tarski paradox", *Amer. Math. Monthly* 86, 151–161.

— [1981]: *An Introduction to Classical Real Analysis.* Wadsworth International, Belmont, California.

H.G. TUCKER [1967]: *A Graduate Course in Probability.* Academic Press, New York–San Francisco–London.

J. VAN YZEREN [1979]: "Moivre's and Fresnel's integrals by simple integration", *Amer. Math. Monthly* 86, 691–693.

D.E. VARBERG [1971]: "Change of variables in multiple integrals", *Amer. Math. Monthly* 18, 42–45.

S. WAGON [1985]: *The Banach-Tarski Paradox*. Encyclopedia Math. Appl. 24. Cambridge University Press, Cambridge.

S. WILLARD [1970]: *General Topology*. Addison-Wesley Publishing Co., Reading, Massachusetts.

J. YAM TING WOO [1971]: "An elementary proof of the Lebesgue decomposition theorem", *Amer. Math. Monthly* 78, 783.

D.G. WRIGHT [1994]: "Tychonoff's theorem", *Proc. Amer. Math. Soc.* 120, 985–987.

Symbol Index

The numbers beside the symbols refer to the pages where the symbol in question is defined.

Name Index

Subject Index

de Gruyter Studies in Mathematics

Volume 23

Heinz Bauer

Probability Theory

Translated from the German by Robert Burckel

1996. 24 x 17 cm. XV, 523 pages. 6 figures. Cloth.
DM 157,– / € 80,27 / öS 1.146,–* / sFr 135,– / US$ 79.95
• ISBN 3-11-013935-9

Graduate-level introduction to probability theory and the theory of stochastic processes. Basic requirement is a course in real analysis, including measure and integration theory.

The German original edition is a standard textbook for a period of nearly 30 years. This English translation is based on the 4th German edition published in 1991.

Contents:

Price is subject to change
*suggested retail price

WALTER DE GRUYTER GMBH & CO. KG
Genthiner Straße 13 · 10785 Berlin
Telefon +49-(0)30-2 60 05-0
Fax +49-(0)30-2 60 05-251
www.deGruyter.de

de Gruyter
Berlin · New York